P9-CRN-752

ESSENTIAL

Math

Basic Math For Everyday Use

Second Edition

Edward Williams
Formerly Assistant Principal—Supervision, Mathematics Department
Washington Irving High School, New York

Dr. Robert A. Atkins
Associate Professor of Mathematics
St. John's University, Queens, New York

© Copyright 1994 by Barron's Educational Series, Inc.

Prior edition © Copyright under the title *Survival Mathematics Basic Math to Help You Cope*, 1983 by Barron's Educational Series, Inc.

All rights reserved.
No part of this book may be reproduced in any form, by photostat, micro-film, xerography, or any other means, or incorporated into any information retrieval system, electronic or mechanical, without any written permission of the copyright owner.

All inquiries should be addressed to:
Barron's Educational Series, Inc.
250 Wireless Boulevard
Hauppauge, New York 11788

Library of Congress Catalog Card No. 94-14129
International Standard Book No. 0-8120-1337-9

Library of Congress Cataloging-in-Publication Data
Williams, Edward, 1926-1988
   Essential math : basic math for everyday use / Edward Williams and
Robert A. Atkins.
      p. cm.
   Rev. ed of : Survival mathematics. 1983.
   Includes index.
   ISBN 0-8120-1337-9
   1. Mathematics.   I. Atkins, Robert A.   II. Williams, Edward,
1926–1988   Survival mathematics.   III. Title.
QA39.2.W55833   1994
513—dc20   94-14129
                                                        CIP
                                                        AC

PRINTED IN THE UNITED STATES OF AMERICA
567   100   9876543

*To our fellow mathematics supervisor, Jacob Cohen,
for his friendship, help, and support.*

Contents

THE WHEELS-AND-DEALS AUTOMOBILE SALESROOM

43

You will *learn* how to: add, subtract, multiply, and divide whole numbers; add fractions; compute the perimeters and areas of rectangles using whole numbers, fractions, decimals, and units of measure; convert measures; write ratios and solve proportions relating to similar triangles; use the Pythagorean theorem; read, interpret, and compute costs from printed advertisements and tables of auto insurance.

You will *review* how to: add and multiply decimals; round off numbers.

THE EXTRA-FINE & FUSSY MACHINE SHOP

95

You will *learn* how to: subtract and divide fractions; change improper fractions to mixed numbers; write ratios and simplify the fractions; change a fraction to higher terms; find the least common denominator; read and interpret horizontal and vertical bar graphs; convert measures; measure with a ruler; learn different types of angles and their measures; find the sum of the angles of a triangle; use scale drawings to write ratios and solve proportions.

You will *review* how to: add and subtract whole numbers; add and multiply fractions; add and multiply decimals

INTERMISSION: FUN TIME 155

You will *learn* how to: recognize a prime and a composite number; compute probability; locate points on a grid.

THE MORE-FOR-LESS DEPARTMENT STORE 165

You will *learn* how to: change fractions to decimals; compare fractions; compute commissions, discounts, and interest; convert measures and compute equivalent measures; read and interpret bar graphs, line graphs, and pictographs; compute averages by three different methods.

You will *review* how to: multiply and divide whole numbers and fractions; add, subtract, multiply, and divide decimals; round off numbers; find a percent of a number; change percents to decimals.

THE GREASE, GAS, AND GADGETS SERVICE STATION AND AUTO SUPPLY SHOP

231

You will *learn* how to: solve simple equations of the type $ax = b$; use metric measures of temperature, length, volume or capacity, and weight; order, add, subtract, multiply, and divide directed numbers; read, interpret, and construct circle graphs.

You will *review* how to: multiply and divide whole numbers; multiply fractions; change improper fractions to mixed numbers; change mixed numbers to improper fractions; change fractions to decimals; change percents to fractions and decimals; add, subtract, multiply, and divide decimals; round off numbers; convert measures.

THE MONEY BAGS SAVINGS BANK 297

You will *learn* how to: use the interest formula; compute using postal rates.

You will *review* how to: add, subtract, multiply, and divide decimals; change fractions to decimals; multiply fractions; add, subtract, multiply, and divide whole numbers.

Preface

The purpose of this book is to help high school students and adults improve their skills in mathematics. You know that mathematics is essential to all occupations. Basic mathematics—adding, subtracting, multiplying, and dividing—applies to aspects of employment such as buying and selling, wages, banking, taxes, and measurement. As you deal with more advanced problems in your career area, your home, or your contacts with stores or other business firms, you need more advanced skills.

You selected this book because you wanted help. The six career areas will help you remember the material you thought you forgot, and they will enable you to see how essential mathematics is in your everyday living.

If you are a high school student, the material in this book will also help you develop the concepts and skills needed to pass the statewide mathematics minimum competency examination.

The advertisements reproduced in this book were taken from the following sources:

AMES Department Stores, Inc.: advertisements on pp. 166, 169, 170, 172–180, 175, 177, 181, 183–185, 187, 190, 191, 198, 214, 278, 280, 281

SBLI Fund, New York, NY 10001: advertisement on p. 315

TViews a Division of CIS, New Milford, CT: advertisements on pp. 176, 191, 242, 243, 246

United Stationers Supply Co.: advertisements on pp. 57–66, 68, 71–79, 97, 129, 130, 132, 136

We acknowledge with appreciation the cooperation of these organizations.

Edward Williams
Robert A. Atkins

New York City

Introduction

Essential Math reviews and explains basic mathematics as it pertains to career areas. For those of you who want to improve your chances for a better job or to be better informed when you buy something or do business with someone, mathematical skills are important. With this book, you can move through six career areas at your own speed. You can use this text on your own or in a class with a teacher to help.

WHAT THIS BOOK CAN DO

Essential Math will:

1. Review and explain the basic ideas and skills required in the six career areas.

2. Give many illustrative examples and model solutions for you to read and understand.

3. Provide many practice exercises to reinforce ideas and skills.

4. Prepare you to meet the requirements needed to pass statewide competency examinations in mathematics.

HOW THE BOOK IS ORGANIZED

The emphasis is on applied mathematics and its necessity for survival in this mathematically oriented world. The six career areas are only symbols of the overwhelming need for mathematics in your everyday life.

The mathematics covered includes the four basic operations—adding, subtracting, multiplying, and dividing—as applied to whole numbers, fractions, and decimals. Percents, graphing, proportions, indirect measurement, signed numbers, and introductory algebra are some other topics included. The ideas and skills are related to problems that may be confronted on the job or at home.

Since the calculator has become such an important tool in our modern world, you are urged to obtain one and learn how to use it. Each chapter has exercises specifically designed to be done on a calculator.

Each career area covers many topics in that field of employment. Each topic and each method are carefully explained. Practice exercises are provided to help you learn the principle involved; answers are given at the end of the career unit.

After studying each career, you can check how well you have learned the material by turning to "Exam Time" and taking the test. You may find some problems that you cannot do. To help you, there is a rating scale at the end of the unit that will inform you how well you did on the career exam. If you did well, you can feel confident about moving to the next career. If you did not do well, you should reread the material, rework all the practice exercises, and retake the exam.

The standard for doing well varies for each test. Answers are given so that you can check whether you did the problem correctly. Obviously, there is no value in looking at the answer unless you have worked the problem. You want to know how to do the problem and to learn the technique so that you can use it on similar problems.

For a little relaxation, there is an interlude entitled "Fun Time" in the middle of the book. The games and puzzles included there will give you additional knowledge on other topics, as well as practice in the basic mathematical skills.

Doing arithmetic on a calculator is quite easy because the operations on a calculator can be done the same way you do them on paper. For example, when you add 4.98 and 3.79, you could write the example this way:

$$4.98 + 3.79 =$$

and then add to get the answer (8.77). On the calculator, the problem is done the same way. Press the ④ ⊙ ⑨ ⑧ keys (you should see 4.98 on the display), then ⊕ , then ③ ⊙ ⑦ ⑨ (which you will now see on the display). To get the answer, simply press the ⊜ key. The answer (8.77) will appear on the display. Subtraction, multiplication, and division are done in the same way: enter the first number, press the key for the desired operation (⊕ , ⊖ , ⊕ , ⊗), enter the second number, and then press ⊜ to get the answer.

A few calculators may operate in a slightly different manner. Read the manual that came with your calculator and become familiar with its operation.

YE OLDE EAT SHOPPE

The restaurant workers who are in contact with the public are the waiters, waitresses, and cashiers. They play an important role in any restaurant operation. Each one of them must be able to compute accurately and quickly. The waiter and waitress must know the cost of the items ordered, total the check, and figure the sales tax. The cashier looks over the check, accepts money, and may return change to the customer.

Behind the scenes are the chef and the baker. The chef prepares the food and computes increases and decreases in the quantities of ingredients when changing recipes. The baker knows how to figure the areas and volumes of differently shaped cake and pie pans.

THE CHECK

Margie and Juan work at Ye Olde Eat Shoppe. All of the items listed in the menu are a la carte. "A la carte" means that each item chosen is charged for separately. A copy of the menu is shown on page 2.

YE OLDE EAT SHOPPE

TO START WITH

JUICES:	Orange	Regular	1.00
	Grapefruit		
	Tomato	Large	1.50

SOUP DU JOUR 2.40

HALF GRAPEFRUIT 1.25

TASTY SANDWICHES

AMERICAN CHEESE & LETTUCE	2.60
BACON, LETTUCE, & TOMATO	3.75
CHICKEN SALAD	4.40
CHOPPED CHICKEN LIVER..............	3.95
CREAM CHEESE	2.00
with Olives or Jelly	2.25
BOLOGNA....................................	3.60
HAM.......................................	3.95
with Imported Swiss Cheese.........	4.60
EGG SALAD.................................	2.75
LETTUCE & TOMATO	2.40
LIVERWURST	3.60
PEANUT BUTTER............................	1.75
with Jelly............................	2.10
SALAMI	3.60
SHRIMP SALAD SANDWICH	5.35
SWISS CHEESE (Switzerland)..........	3.15
GRILLED SWISS CHEESE.................	3.75
TUNA FISH SALAD	4.40
FRIED EGG SANDWICH [2 Eggs].......	2.65
with Bacon or Ham.....................	3.55
GRILLED AMERICAN CHEESE..........	3.55
with Bacon 4.55 with Tomato 3.90	

HOT SANDWICHES

ROAST BEEF	5.65
CORNED BEEF..............................	5.65
BAKED VIRGINIA HAM	5.65
SLICED STEAK	6.40
TURKEY....................................	5.65

BROILED STEAKBURGERS

PURE BEEF STEAKBURGER	
on Toasted Bun	3.15
STEAKBURGER on Toasted Bun with	
French Fries, Lettuce, and	
Tomato......................................	5.25
CHEESEBURGER on Toasted Bun.......	3.55
CHEESEBURGER on Toasted Bun	
with French Fries, Lettuce, and	
Tomato......................................	5.40

SIDE DISHES

FRENCH FRIES...............................	1.65
POTATO SALAD..............................	1.65
COLE SLAW...................................	1.65

REFRESHING BEVERAGES

ICED COFFEE, Whipped Cream	1.00
TEA or COFFEE...............................	.70
ICED TEA......................................	1.00
JUMBO IND. MILK	1.00
CHOCOLATE MILK	1.25
HOT CHOCOLATE, Whipped Cream	1.00

DELICIOUS PASTRIES

PURE BUTTER DANISH: Plain, Prune,	
Cheese, Custard, or Cinnamon.......	1.25
PIES: Apple, Blueberry, Cherry,	
Chocolate, or Pineapple	1.85
FRENCH CHEESE CAKE: Blueberry,	
Cherry, or Chocolate	1.95
A la Mode	2.75
NAPOLEON...................................	1.85
CHOCOLATE ECLAIR	1.75
CHOCOLATE SEVEN-LAYER CAKE.....	2.95
BOSTON CREAM CAKE......................	2.95
FRENCH NUT CAKE	2.95
OVEN FRESH FRUIT CAKE: Blueberry,	
Peach, or Plum (in Season)	1.55

Tony and Sally Marino ate lunch at Ye Olde Eat Shoppe. Margie was their waitress. Her order check read as follows:

CHECK NO. 43590	SERVER NO. 4	NO. PERSONS	AMOUNT

YE OLDE EAT SHOPPE

TABLE NO.	NO. PERSONS	CHECK NO. 43590	SERVER NO. 4
1	Soup du Jour		2	40
1	Half Grapefruit		1	25
1	Ham + Cheese		4	60
1	Tuna Fish Salad		4	40
2	Tea		1	40
1	Blueberry Pie		1	85
1	Seven-Layer Cake		2	95
			$18	85
	TAX			

PRACTICE EXERCISE 1

Use the blank order checks on page 4 to write up each order. Figure the amount of the check, using the prices on the menu.

1. Angela and Sonia ate lunch and ordered 2 large orange juices; 1 bacon, lettuce, and tomato sandwich; 1 steakburger with french fries, lettuce, and tomato; 1 milk; and 1 iced tea.

2. Juan served a table of three who ordered 2 soups du jour; 1 large tomato juice; 2 cheeseburgers with lettuce, tomato, and french fries; 1 grilled American cheese with bacon; 1 cole slaw; and 3 iced coffees.

3. Margie served a party of four people at one of her tables. Her order check, shown on page 5, was much larger than usual. Figure the food bill.

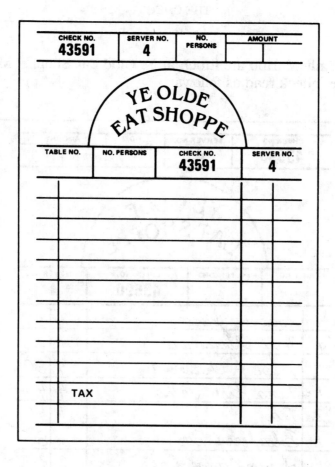

CHECK NO.	SERVER NO.	NO. PERSONS	AMOUNT
43591	4		

YE OLDE
EAT SHOPPE

TABLE NO.	NO. PERSONS	CHECK NO.	SERVER NO.
		43591	4

TAX			

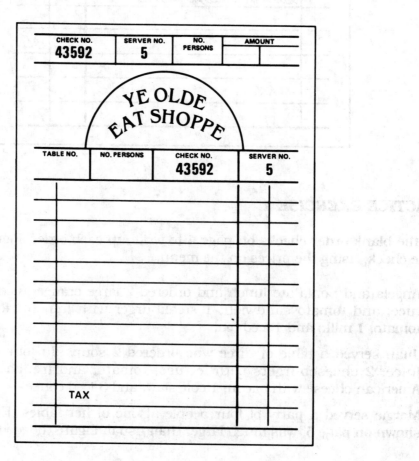

CHECK NO.	SERVER NO.	NO. PERSONS	AMOUNT
43592	5		

YE OLDE
EAT SHOPPE

TABLE NO.	NO. PERSONS	CHECK NO.	SERVER NO.
		43592	5

TAX			

CHECK NO. 43593	SERVER NO. 4	NO. PERSONS	AMOUNT

YE OLDE EAT SHOPPE

TABLE NO.	NO. PERSONS	CHECK NO. 43593	SERVER NO. 4

4	Half Grapefruit		
1	Roast Beef		
2	Turkey		
1	Cheeseburger with F. F., Lettuce + Tomato		
2	Iced Coffee		
2	Iced Tea		
1	Apple Pie		
2	Boston Cream Cake		
1	Choc. Eclair		
	TAX		

THE SALES TAX

Some states charge a sales tax on all purchases, including restaurant bills. The amount of tax varies from state to state. Certain cities charge an additional sales tax. For example, a state may charge a tax of 4% of the bill, and a city or county may charge an additional 4%. In other words, the sales tax in the city or county is 8% of the bill. The state collects 8% and returns 4% to the city or county.

The symbol "%" stands for *percent*. Percent means *hundredths*. Thus 8% means 8 hundredths or 0.08. Since 1 dollar equals 100 cents, an 8% tax means that you pay 8 cents tax for each dollar you spend.

METHOD I

Tax tables are available to all store owners and are usually posted next to each cash register so that the tax can be quickly, easily, and accurately determined for each sale.

ST-110.8 (7/74)

N.Y. State Department of Taxation and Finance • Sales Tax Bureau

8% SALES AND USE TAX COLLECTION CHART **8%**

Amount of Sale	Tax to be Collected	Amount of Sale	Tax to be Collected
$0.01 to $0.10	None	$5.07 to $5.18	$.41
.11 to .17	1¢	5.19 to 5.31	.42
.18 to .29	2¢	5.32 to 5.43	.43
.30 to .42	3¢	5.44 to 5.56	.44
.43 to .54	4¢	5.57 to 5.68	.45
.55 to .67	5¢	5.69 to 5.81	.46
.68 to .79	6¢	5.82 to 5.93	.47
.80 to .92	7¢	5.94 to 6.06	.48
.93 to 1.06	8¢	6.07 to 6.18	.49
1.07 to 1.18	9¢	6.19 to 6.31	.50
1.19 to 1.31	$.10	6.32 to 6.43	.51
1.32 to 1.43	.11	6.44 to 6.56	.52
1.44 to 1.56	.12	6.57 to 6.68	.53
1.57 to 1.68	.13	6.69 to 6.81	.54
1.69 to 1.81	.14	6.82 to 6.93	.55
1.82 to 1.93	.15	6.94 to 7.06	.56
1.94 to 2.06	.16	7.07 to 7.18	.57
2.07 to 2.18	.17	7.19 to 7.31	.58
2.19 to 2.31	.18	7.32 to 7.43	.59
2.32 to 2.43	.19	7.44 to 7.56	.60
2.44 to 2.56	.20	7.57 to 7.68	.61
2.57 to 2.68	.21	7.69 to 7.81	.62
2.69 to 2.81	.22	7.82 to 7.93	.63
2.82 to 2.93	.23	7.94 to 8.06	.64
2.94 to 3.06	.24	8.07 to 8.18	.65
3.07 to 3.18	.25	8.19 to 8.31	.66
3.19 to 3.31	.26	8.32 to 8.43	.67
3.32 to 3.43	.27	8.44 to 8.56	.68
3.44 to 3.56	.28	8.57 to 8.68	.69
3.57 to 3.68	.29	8.69 to 8.81	.70
3.69 to 3.81	.30	8.82 to 8.93	.71
3.82 to 3.93	.31	8.94 to 9.06	.72
3.94 to 4.06	.32	9.07 to 9.18	.73
4.07 to 4.18	.33	9.19 to 9.31	.74
4.19 to 4.31	.34	9.32 to 9.43	.75
4.32 to 4.43	.35	9.44 to 9.56	.76
4.44 to 4.56	.36	9.57 to 9.68	.77
4.57 to 4.68	.37	9.69 to 9.81	.78
4.69 to 4.81	.38	9.82 to 9.93	.79
4.82 to 4.93	.39	9.94 to 10.00	.80
4.94 to 5.06	.40		

On sales over $10.00, compute the tax by multiplying the amount of sale by the applicable tax rate and rounding the result to the nearest whole cent.

METHOD II

Margie doesn't carry a tax table with her, so she must be able to compute the tax. Thus:

$$8\% = \frac{8}{100} = 0.08 \qquad 15\% = \frac{15}{100} = 0.15$$

$$4\frac{1}{2}\% = \frac{4\frac{1}{2}}{100} = 0.04\frac{1}{2} \qquad 3.6\% = \frac{3.6}{100} = 0.036 \quad \text{or} \quad 100\overline{)3.600}$$

$$\begin{array}{r} 0.036 \\ 100\overline{)3.600} \\ \underline{3\ 00x} \\ 600 \\ \underline{600} \end{array}$$

EXAMPLE: What is the sales tax (at 8%) on Tony and Sally Marino's lunch bill? You are looking for 8% of $18.85.

SOLUTION:

$$\begin{array}{r} \$18.85 \\ \times\ 0.08 \\ \hline 1.50(8) \end{array}$$

Round off $1.50(8) to the nearest *cent* by circling the number (8) to the right of the hundredths (0) or *cent* place. Since (8) is more than 5, increase the hundredths place by 1: $1.51.

If the number you circle is less than 5, discard it and keep all the digits to its left. For example:

$1.48(6) = $1.49 \qquad $3.73(4) = $3.73

PRACTICE EXERCISE 2

1. Write the decimal equivalent of each of these percents:

   a. 6%

   b. 22%

   c. 3.2%

   d. $2\frac{1}{2}\%$

   e. 12.6%

   f. 58%

   g. $6\frac{1}{4}\%$

   h. 8.75%

   i. 14.5%

   j. $\frac{1}{2}\%$

2. **a.** Figure an 8% tax for Problems 1–3 in Practice Exercise 1, using both the arithmetic method and the tax table method.

   b. Complete the table below, using an 8% sales tax.

	Tax	
Amount of Check	**Arithmetic Method**	**Tax Table Method**
$7.60		
$9.90		
$3.71		

   c. Find each total bill in Question 2(b), including the tax.

3. Round off each of the following to the nearest cent:

    a. $8.173   f. $6.722
    b. $3.972   g. $8.437
    c. $2.387   h. $3.561
    d. $1.578   i. $5.706
    e. $7.439   j. $4.695

4. Figure an 8% tax on each of the amounts in Problem 3, using both the arithmetic and tax table methods. Add the tax to the rounded amount and find the sum.

5. Using your calculator, figure a $5\frac{1}{2}$% sales tax on each of the following (remember that $5\frac{1}{2}$% is .055). Round off each answer to the nearest cent.

    a. $4.95   f. $10.39
    b. $7.23   g. $12.50
    c. $8.08   h. $15.38
    d. $6.79   i. $19.13
    e. $5.88   j. $25.49

Margie had an order that was greater than $10. The check amounted to $16.95, and the maximum value in the tax table is $10. How can she figure the 8% tax on $16.95 using the tax table?

She knows that $16.95 can be written as $10.00 + $6.95. The tax on each amount can be found in the table, and the two taxes then added. Thus:

For $10.00 the tax is $0.80.
For $6.95   the tax is $0.56.

The total tax for $16.95 is $0.80 + $0.56, or $1.36. This should be the same as finding 8% of $16.95 and then rounding off the answer to the nearest cent:

$16.95
×   0.08
$\overline{}$
$1.35(6)0 = $1.36

PRACTICE EXERCISE 3

1. Figure an 8% tax on each of these amounts, using both methods. Round off each answer to the nearest cent.

    a. $12.43   f. $ 32.90
    b. $18.11   g. $ 73.81
    c. $23.25   h. $ 52.27
    d. $46.56   i. $112.04
    e. $83.35   j. $243.67

2. Add the tax to each amount above, and find each sum.

3. Using your calculator, figure an $8\frac{1}{4}$% tax on each of the amounts in Problem 1 above. Remember that $8\frac{1}{4}$% is equal to 0.0825. Round off each answer to the nearest cent.

THE TIP

Margie and Juan earn an hourly wage for waiting on tables in Ye Olde Eat Shoppe. These wages plus the tips they earn are their total salary.

As the Customer Figures It

Most people figure a tip of approximately 15% of the bill, excluding the tax. The Marinos' bill was

$$\begin{array}{r} \$18.85 \\ +\underline{\quad 1.51} \quad \text{Tax} \\ \$20.36 \end{array}$$

The tip was figured as 15% of $18.85. Therefore:

$$\begin{array}{r} \$18.85 \\ \times \underline{\quad 0.15} \\ 94\ 25 \\ \underline{188\ 5} \\ \$2.82(7)5 \end{array}$$

Rounding off the amount to the nearest cent, you see that the Marinos gave Margie a tip of $2.83 and that their total food bill was

$$\begin{array}{r} \$20.36 \\ +\underline{\quad 2.83} \\ \$23.29 \end{array}$$

The 15% tip can also be computed by finding 10% of the bill, dividing by 2, and adding the two quantities together. Thus: 10% of $10.60 is $1.06; dividing by 2 gives $0.53; and the sum is $1.06 + 0.53 = $1.59, or

$$\begin{array}{r} \$10.60 \\ \times \underline{\quad 0.15} \\ 53\ 00 \\ \underline{1\ 06\ 0} \\ \$1.59(0)0 \end{array}$$

PRACTICE EXERCISE 4

1. a. Figure a tip of 15% for each of the orders in Practice Exercise 1. Round off each answer to the nearest cent.
    b. Figure the sales tax of 8%.
    c. Figure the total amount for food.

2.   Find 15% of each rounded amount in Problem 3 of Practice Exercise 2.

3.   Juan served a party of eight people. The order is written on the check below.

   a. Find the cost of each item.
   b. Find the subtotal.
   c. Figure an 8% sales tax on this subtotal.
   d. Figure a 15% tip on the subtotal.
   e. Find the total amount spent for the meal.

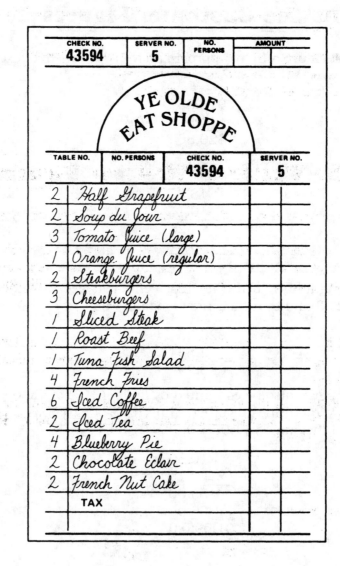

4.   Juan earns $4.95 an hour plus tips. On Monday he worked 7 hours and received $32.53 in tips.

   a. What was his salary for Monday?
   b. What were his total earnings for Monday?

c. How much money would he receive if he earned the same amount for each of the 5 days he worked?

5. Margie earns $4.80 an hour plus tips. On Saturday she worked 8 hours and earned 15% of $615 for tips.

    a. How much money did she earn from salary only?
    b. How much money did she earn from tips only?
    c. What were her total earnings for Saturday?

6. Margie worked a 35-hour week at $4.80 an hour and received 13% of $3215.10 for tips.

    a. How much money did she earn for her weekly salary?
    b. How much money did she receive on tips?
    c. What were her total weekly earnings?

As the Waiter or Waitress Figures It

Restaurants must charge the customer the sales tax set by the city, county, or state.

The size of the tip the customer leaves for service to the waiter or waitress, however, is not fixed by law. The customer is free to leave any amount he or she wishes. Usually, people leave a percentage of the subtotal of the bill.

EXAMPLE: Juan received a tip of $1.35 for an order totaling $9. What percent tip did he receive? You must determine what percent $1.35 is of $9.

SOLUTION: Let's write these amounts as a *ratio* or *fraction* with the $1.35 tip as the numerator and the total bill of $9 as the denominator. Thus:

$$\frac{\$1.35}{\$9}$$

Dividing, we have:

$$9\overline{)1.35} \quad 0.15 = \frac{15}{100} = 15\%$$

$$\begin{array}{r} 0.15 \\ 9\overline{)1.35} \\ \underline{9x} \\ 45 \\ \underline{45} \end{array}$$

Juan received a 15% tip from this diner for the good service he provided.

EXAMPLE: Margie received tips amounting to $79.12 at the end of her working day. If all of her orders or total **receipts** amounted to $635.78, what percentage, to the **nearest tenth**, **did** she receive in tips?

SOLUTION: Write these amounts as a *ratio*:

$$\frac{\$79.12}{\$635.78}$$

and divide to four places past the decimal point:

```
              0.1244           1244
635.78.)79.12.0000     = ───────── =12.44%
        63 57 8xxx           10000
        15 54 20
        12 71 56        Rounding off
         2 82 640       12.4(4)% to the
         2 54 312       nearest tenth
           28 3280      gives 12.4%.
           25 4312
```

Margie received an average of approximately 12.4% in tips from her diners for the service she provided.

PRACTICE EXERCISE 5

1. What percent, to the nearest tenth, did Margie earn for tips on **each of the 6** days listed below?

Day	Total Receipts	Tips Earned	Percent
Monday	$612.97	$73.87	
Tuesday	$648.22	$90.14	
Wednesday	$598.75	$73.90	
Thursday	$609.62	$80.16	
Friday	$703.78	$93.20	
Saturday	$510.48	$71.25	

2. a. If she worked 8 hours each day, earning $4.80 an hour, **what was her** weekly salary?
   b. How much money did she earn in tips for these 6 days?
   c. What were her total weekly earnings in salary and tips combined?
   d. What would her total yearly earnings be if she earned the **same amount** each week for 50 weeks during the year?
   e. What percentage of her yearly earnings were her wages?
   f. What percentage of her yearly earnings were her tips?

3. Juan worked at a small wedding party for 32 people at $23.75 per person. He was promised a 15% tip for his service. He hired Angela to help clear off the dishes and agreed to pay her 15% of the tip he received.

   a. How much did the party cost for the 32 people at $23.75 per person?
   b. What was the amount of Juan's tip?
   c. How much money did Angela earn?
   d. How much money did Juan earn after sharing his tip with Angela?
   e. What was the cost of the entire wedding party, including Juan's tip and an 8% sales tax on the subtotal?

MAKING CHANGE

Charlotte is the cashier at Ye Olde Eat Shoppe. She checks each of the bills presented to her before giving any change to the diner. Since she is always at the cash register, she uses the tax table posted there as her guide, as well as a copy of the menu. She must know how to add very efficiently and quite rapidly. Charlotte always tries to give the diner the smallest number of bills and the fewest coins when making change.

The Marinos' bill totaled $20.36, including the tax, and they gave Charlotte a $20 and a $10 bill to pay for it. What bills and coins should Charlotte return to them in change if she uses as few of each as possible? The correct change can be found using either of the two methods described below.

METHOD I

The most common method is to subtract, but it is slow and not very efficient.

$$
\begin{array}{r}
9\ 91 \\
\$\cancel{3}\cancel{0}.\cancel{0}0 \\
-20.36 \\
\hline
\$\ 9.64
\end{array}
$$

Charlotte would give a $5 bill, four $1 bills, and a half dollar, a dime, and four pennies.

METHOD II

Another method, used by most cashiers, is to start with the sum on the bill and keep adding to it until the amount the diner paid is reached. In this case, you would start with $20.36 and add until you arrive at $30, using the coins at your disposal, for instance,

$$\underbrace{\$20.36 + 0.04 + 0.10 + 0.50 + 1 + 1 + 1 + 1 + 5} = \$30$$

Change = $9.64

Another time Charlotte receives a $20 bill to pay a check of $7.52. What is the best way of making change? Charlotte says, $7.52, 7.53, 7.54, 7.55, 7.65, 7.75, 8, 9, 10, 20, or

$$\underbrace{\$7.52 + 0.01 + 0.01 + 0.01 + 0.10 + 0.10 + 0.25 + 1 + 1 + 10} = \$20$$

Change = $12.48

or

$$
\begin{array}{r}
19\ 9_1 \\
\$\cancel{2}\cancel{0}.\cancel{0}0 \\
-\ 7.52 \\
\hline
\$12.48
\end{array}
$$

PRACTICE EXERCISE 6

1. How can Charlotte give change with the smallest number of bills and the fewest coins for each of the following transactions? The first one has been done for you as an example.

	Total Check	Amount Received	Bills ($)				Coins (¢)					Change
			20	10	5	1	50	25	10	5	1	
a.	$ 2.17	$ 3					1	1		1	3	$0.83
b.	$ 3.29	$ 5										
c.	$12.63	$20										
d.	$ 7.79	$10										
e.	$13.24	$15										
f.	$18.23	$20										
g.	$ 1.19	$10										
h.	$21.10	$25										
i.	$45.46	$50										
j.	$27.93	$30										

2. Charlotte must figure the wages earned for Juan, Margie, and the other restaurant employees at the end of each week. Juan works 40 hours a week at $4.95 an hour, while Margie works 40 hours a week and earns $4.80 an hour.

    a. How much does Juan earn weekly?
    b. How much does Margie earn weekly?
    c. Using the same currency breakdown as in the table in Problem 1, how did Charlotte pay each of them with the smallest number of bills and the fewest coins?

SALARY DEDUCTIONS

Margie earned wages of $4.80 an hour for 40 hours of work. She expected to receive wages totaling $192.00 and was surprised to find quite a bit less than that amount in her pay envelope. Why did this occur? There are certain taxes to be paid to the federal, state, and sometimes the city government, which are deducted from everyone's wages.

Not everyone pays the same amount of taxes. Your payments depend on two things:

1. The amount of money you earn.

2. The number of people who depend on you for support. They are called *dependents*, *exemptions*, or *allowances*.

Your employer deducts these taxes from your wages and pays them for you to the proper sources. The employer knows the total amount you earn each week, and you tell him or her the number of dependents you have when you are hired and complete a W-4 form. A sample of this form is shown below.

Form **W-4** Department of the Treasury Internal Revenue Service	**Employee's Withholding Allowance Certificate** ▶ For Privacy Act and Paperwork Reduction Act Notice, see reverse.		OMB No. 1545-0010 19**93**
1 Type or print your first name and middle initial *MArgie*	Last name *Romano*		2 Your social security number *075-58-4097*
Home address (number and street or rural route) *42 East 14 Street*	3 ☒ Single ☐ Married ☐ Married, but withhold at higher Single rate. **Note:** If married, but legally separated, or spouse is a nonresident alien, check the Single box.		
City or town, state, and ZIP code *New York, New York 10003*	4 If your last name differs from that on your social security card, check here and call 1-800-772-1213 for more information · · · ▶ ☐		

5 Total number of allowances you are claiming (from line G above or from the worksheets on page 2 if they apply) . **5** *1*
6 Additional amount, if any, you want withheld from each paycheck . . . . **6** $ —
7 I claim exemption from withholding for 1993 and I certify that I meet **ALL** of the following conditions for exemption:
  ● Last year I had a right to a refund of **ALL** Federal income tax withheld because I had **NO** tax liability; **AND**
  ● This year I expect a refund of **ALL** Federal income tax withheld because I expect to have **NO** tax liability; **AND**
  ● This year if my income exceeds $600 and includes nonwage income, another person cannot claim me as a dependent.
  If you meet all of the above conditions, enter "EXEMPT" here · · · · · · · · · ▶ **7**

Under penalties of perjury, I certify that I am entitled to the number of withholding allowances claimed on this certificate or entitled to claim exempt status.

Employee's signature ▶ *Margie Romano*   Date ▶ *June 29* , 19 *94*

8 Employer's name and address (Employer: Complete 8 and 10 only if sending to the IRS) | 9 Office code (optional) | 10 Employer identification number

Margie is unmarried, and no one else is dependent on her for support. She indicates these facts on the form by checking the "Single" box and claiming one allowance.

Federal Tax

There are two methods by which Charlotte can figure the federal income withholding tax for each employee. Each method depends on her ability to read tables and do the correct mathematical computations.

METHOD I

Charlotte computed Margie's withholding tax using the percentage method in the following way:

Use these steps to figure the income tax to withhold under the percentage method:

1. Multiply one withholding allowance (see table below) by the number of allowances the employee claims.

2. Subtract that amount from the employee's wages.

3. Determine amount to withhold from appropriate table.

Percentage Method—Amount for One Withholding Allowance

Payroll Period	One withholding allowance
Weekly	$45.19
Biweekly	90.38
Semimonthly	97.92
Monthly	195.83
Quarterly	587.50
Semiannually	1,175.00
Annually	2,350.00
Daily or miscellaneous (each day of the payroll period)	9.04

Margie earned $192.00 this week and claimed one allowance. Since the payroll period is weekly, Charlotte used the figure $45.19 and then followed directions 1., 2., and 3. above. From 1. she multiplied $45.19 by one allowance, which equals $45.19. Following direction 2., she subtracted $45.19 from $192.00:

$$\begin{array}{r} {}^{8\;1}\;{}^{9\;1} \\ \$19\cancel{2}.\cancel{0}0 \\ -\;45.19 \\ \hline \$146.81 \end{array}$$

She knew that Margie is single, and following direction 3., she consulted the following table:

Table for Percentage Method of Withholding

TABLE 1—WEEKLY Payroll Period

(a) SINGLE person (including head of household)—			(b) MARRIED person—		
If the amount of wages (after subtracting withholding allowances) is:	The amount of income tax to withhold is:		If the amount of wages (after subtracting withholding allowances) is:	The amount of income tax to withhold is:	
Not over $49.	$0		Not over $119	$0	
Over— **But not over—**		of excess over—	**Over—** **But not over—**		of excess over—
$49 —$451 . . .	15%	—$49	$119 —$784 . . .	15%	—$119
$451 —$942 . . .	$60.30 plus 28%	—$451	$784 —$1,563 . . .	$99.75 plus 28%	—$784
$942	$197.78 plus 31%	—$942	$1,563	$317.87 plus 31%	—$1,563

Margie's earnings of $146.81 fall in the table between $49 and $451. Her tax is 15% of all moneys earned above $49.

Charlotte subtracted $146.81 − 49 = $97.81, which represents all moneys earned above $49. She took 15% of that amount:

$$15\% = \frac{15}{100} = 0.15 \qquad \text{Thus:} \qquad \begin{array}{r} \$97.81 \\ \times\ 0.15 \\ \hline 489\ 05 \\ 978\ 1 \\ \hline \$14.67\ 15 = \$14.67 \end{array}$$

Margie's tax is $14.67.

METHOD II

Another way to figure the withholding tax is to use the tax tables provided by the federal government. Portions of these tables are reproduced on pages 18–21.

Charlotte looked up Margie's weekly earnings of $192.00 in the table headed "Single Persons—Weekly Payroll Period." She found that the amount was at least $190 and not more than $195. In the column for one allowance on the right, the tax to be withheld is given as $15.00. As you see, the tax differs according to the method used.

Both methods provide only approximations of what the weekly tax deduction should be. At the end of the year, when you file your federal income tax return, your yearly taxes are determined. If the taxes withheld were more than the amount required, you are entitled to a refund. If your withholding taxes were less, you must pay the difference. Regardless of the method used during the year, you end up in the same place at the end of the year.

FEDERAL TAX WITHHOLDING
SINGLE Persons—WEEKLY Payroll Period

If the wages are—		And the number of withholding allowances claimed is—										
At least	But less than	0	1	2	3	4	5	6	7	8	9	10
		The amount of income tax to be withheld is—										
$0	$50	$0	$0	$0	$0	$0	$0	$0	$0	$0	$0	$0
50	55	1	0	0	0	0	0	0	0	0	0	0
55	60	1	0	0	0	0	0	0	0	0	0	0
60	65	2	0	0	0	0	0	0	0	0	0	0
65	70	3	0	0	0	0	0	0	0	0	0	0
70	75	4	0	0	0	0	0	0	0	0	0	0
75	80	4	0	0	0	0	0	0	0	0	0	0
80	85	5	0	0	0	0	0	0	0	0	0	0
85	90	6	0	0	0	0	0	0	0	0	0	0
90	95	7	0	0	0	0	0	0	0	0	0	0
95	100	7	1	0	0	0	0	0	0	0	0	0
100	105	8	1	0	0	0	0	0	0	0	0	0
105	110	9	2	0	0	0	0	0	0	0	0	0
110	115	10	3	0	0	0	0	0	0	0	0	0
115	120	10	4	0	0	0	0	0	0	0	0	0
120	125	11	4	0	0	0	0	0	0	0	0	0
125	130	12	5	0	0	0	0	0	0	0	0	0
130	135	13	6	0	0	0	0	0	0	0	0	0
135	140	13	7	0	0	0	0	0	0	0	0	0
140	145	14	7	1	0	0	0	0	0	0	0	0
145	150	15	8	1	0	0	0	0	0	0	0	0
150	155	16	9	2	0	0	0	0	0	0	0	0
155	160	16	10	3	0	0	0	0	0	0	0	0
160	165	17	10	4	0	0	0	0	0	0	0	0
165	170	18	11	4	0	0	0	0	0	0	0	0
170	175	19	12	5	0	0	0	0	0	0	0	0
175	180	19	13	6	0	0	0	0	0	0	0	0
180	185	20	13	7	0	0	0	0	0	0	0	0
185	190	21	14	7	1	0	0	0	0	0	0	0
190	195	22	15	8	1	0	0	0	0	0	0	0
195	200	22	16	9	2	0	0	0	0	0	0	0
200	210	23	17	10	3	0	0	0	0	0	0	0
210	220	25	18	11	5	0	0	0	0	0	0	0
220	230	26	20	13	6	0	0	0	0	0	0	0
230	240	28	21	14	8	1	0	0	0	0	0	0
240	250	29	23	16	9	2	0	0	0	0	0	0
250	260	31	24	17	11	4	0	0	0	0	0	0
260	270	32	26	19	12	5	0	0	0	0	0	0
270	280	34	27	20	14	7	0	0	0	0	0	0
280	290	35	29	22	15	8	2	0	0	0	0	0
290	300	37	30	23	17	10	3	0	0	0	0	0
300	310	38	32	25	18	11	5	0	0	0	0	0
310	320	40	33	26	20	13	6	0	0	0	0	0
320	330	41	35	28	21	14	8	1	0	0	0	0
330	340	43	36	29	23	16	9	2	0	0	0	0
340	350	44	38	31	24	17	11	4	0	0	0	0
350	360	46	39	32	26	19	12	5	0	0	0	0
360	370	47	41	34	27	20	14	7	0	0	0	0
370	380	49	42	35	29	22	15	8	2	0	0	0
380	390	50	44	37	30	23	17	10	3	0	0	0
390	400	52	45	38	32	25	18	11	5	0	0	0
400	410	53	47	40	33	26	20	13	6	0	0	0
410	420	55	48	41	35	28	21	14	8	1	0	0
420	430	56	50	43	36	29	23	16	9	2	0	0
430	440	58	51	44	38	31	24	17	11	4	0	0
440	450	59	53	46	39	32	26	19	12	5	0	0
450	460	61	54	47	41	34	27	20	14	7	0	0
460	470	64	56	49	42	35	29	22	15	8	1	0
470	480	67	57	50	44	37	30	23	17	10	3	0
480	490	70	59	52	45	38	32	25	18	11	4	0
490	500	73	60	53	47	40	33	26	20	13	6	0
500	510	75	63	55	48	41	35	28	21	14	7	1
510	520	78	66	56	50	43	36	29	23	16	9	2
520	530	81	68	58	51	44	38	31	24	17	10	4
530	540	84	71	59	53	46	39	32	26	19	12	5
540	550	87	74	61	54	47	41	34	27	20	13	7
550	560	89	77	64	56	49	42	35	29	22	15	8
560	570	92	80	67	57	50	44	37	30	23	16	10
570	580	95	82	70	59	52	45	38	32	25	18	11
580	590	98	85	73	60	53	47	40	33	26	19	13

(Continued on next page)

FEDERAL TAX WITHHOLDING
SINGLE Persons—WEEKLY Payroll Period

If the wages are—		And the number of withholding allowances claimed is—										
At least	But less than	0	1	2	3	4	5	6	7	8	9	10
		The amount of income tax to be withheld is—										
$590	$600	$101	$88	$75	$63	$55	$48	$41	$35	$28	$21	$14
600	610	103	91	78	66	56	50	43	36	29	22	16
610	620	106	94	81	68	58	51	44	38	31	24	17
620	630	109	96	84	71	59	53	46	39	32	25	19
630	640	112	99	87	74	61	54	47	41	34	27	20
640	650	115	102	89	77	64	56	49	42	35	28	22
650	660	117	105	92	80	67	57	50	44	37	30	23
660	670	120	108	95	82	70	59	52	45	38	31	25
670	680	123	110	98	85	72	60	53	47	40	33	26
680	690	126	113	101	88	75	63	55	48	41	34	28
690	700	129	116	103	91	78	65	56	50	43	36	29
700	710	131	119	106	94	81	68	58	51	44	37	31
710	720	134	122	109	96	84	71	59	53	46	39	32
720	730	137	124	112	99	86	74	61	54	47	40	34
730	740	140	127	115	102	89	77	64	56	49	42	35
740	750	143	130	117	105	92	79	67	57	50	43	37
750	760	145	133	120	108	95	82	70	59	52	45	38
760	770	148	136	123	110	98	85	72	60	53	46	40
770	780	151	138	126	113	100	88	75	63	55	48	41
780	790	154	141	129	116	103	91	78	65	56	49	43
790	800	157	144	131	119	106	93	81	68	58	51	44
800	810	159	147	134	122	109	96	84	71	59	52	46
810	820	162	150	137	124	112	99	86	74	61	54	47
820	830	165	152	140	127	114	102	89	77	64	55	49
830	840	168	155	143	130	117	105	92	79	67	57	50
840	850	171	158	145	133	120	107	95	82	69	58	52
850	860	173	161	148	136	123	110	98	85	72	60	53
860	870	176	164	151	138	126	113	100	88	75	62	55
870	880	179	166	154	141	128	116	103	91	78	65	56
880	890	182	169	157	144	131	119	106	93	81	68	58
890	900	185	172	159	147	134	121	109	96	83	71	59
900	910	187	175	162	150	137	124	112	99	86	74	61
910	920	190	178	165	152	140	127	114	102	89	76	64
920	930	193	180	168	155	142	130	117	105	92	79	67
930	940	196	183	171	158	145	133	120	107	95	82	69
940	950	199	186	173	161	148	135	123	110	97	85	72
950	960	202	189	176	164	151	138	126	113	100	88	75
960	970	205	192	179	166	154	141	128	116	103	90	78
970	980	208	194	182	169	156	144	131	119	106	93	81
980	990	211	197	185	172	159	147	134	121	109	96	83
990	1,000	214	200	187	175	162	149	137	124	111	99	86
1,000	1,010	217	203	190	178	165	152	140	127	114	102	89
1,010	1,020	220	206	193	180	168	155	142	130	117	104	92
1,020	1,030	224	210	196	183	170	158	145	133	120	107	95
1,030	1,040	227	213	199	186	173	161	148	135	123	110	97
1,040	1,050	230	216	202	189	176	163	151	138	125	113	100
1,050	1,060	233	219	205	192	179	166	154	141	128	116	103
1,060	1,070	236	222	208	194	182	169	156	144	131	118	106
1,070	1,080	239	225	211	197	184	172	159	147	134	121	109
1,080	1,090	242	228	214	200	187	175	162	149	137	124	111
1,090	1,100	245	231	217	203	190	177	165	152	139	127	114
1,100	1,110	248	234	220	206	193	180	168	155	142	130	117
1,110	1,120	251	237	223	209	196	183	170	158	145	132	120
1,120	1,130	255	241	227	213	199	186	173	161	148	135	123
1,130	1,140	258	244	230	216	202	189	176	163	151	138	125
1,140	1,150	261	247	233	219	205	191	179	166	153	141	128
1,150	1,160	264	250	236	222	208	194	182	169	156	144	131
1,160	1,170	267	253	239	225	211	197	184	172	159	146	134
1,170	1,180	270	256	242	228	214	200	187	175	162	149	137
1,180	1,190	273	259	245	231	217	203	190	177	165	152	139
1,190	1,200	276	262	248	234	220	206	193	180	167	155	142
1,200	1,210	279	265	251	237	223	209	196	183	170	158	145
1,210	1,220	282	268	254	240	226	212	198	186	173	160	148
1,220	1,230	286	272	258	244	230	216	202	189	176	163	151
1,230	1,240	289	275	261	247	233	219	205	191	179	166	153
1,240	1,250	292	278	264	250	236	222	208	194	181	169	156

$1,250 and over   Use Table 1(a) for a **SINGLE person** on page 17.

FEDERAL TAX WITHHOLDING

MARRIED Persons—WEEKLY Payroll Period

If the wages are—		And the number of withholding allowances claimed is—										
At least	But less than	0	1	2	3	4	5	6	7	8	9	10
		The amount of income tax to be withheld is—										
$0	$125	$0	$0	$0	$0	$0	$0	$0	$0	$0	$0	$0
125	130	1	0	0	0	0	0	0	0	0	0	0
130	135	2	0	0	0	0	0	0	0	0	0	0
135	140	3	0	0	0	0	0	0	0	0	0	0
140	145	3	0	0	0	0	0	0	0	0	0	0
145	150	4	0	0	0	0	0	0	0	0	0	0
150	155	5	0	0	0	0	0	0	0	0	0	0
155	160	6	0	0	0	0	0	0	0	0	0	0
160	165	6	0	0	0	0	0	0	0	0	0	0
165	170	7	0	0	0	0	0	0	0	0	0	0
170	175	8	1	0	0	0	0	0	0	0	0	0
175	180	9	2	0	0	0	0	0	0	0	0	0
180	185	9	3	0	0	0	0	0	0	0	0	0
185	190	10	3	0	0	0	0	0	0	0	0	0
190	195	11	4	0	0	0	0	0	0	0	0	0
195	200	12	5	0	0	0	0	0	0	0	0	0
200	210	13	6	0	0	0	0	0	0	0	0	0
210	220	14	8	1	0	0	0	0	0	0	0	0
220	230	16	9	2	0	0	0	0	0	0	0	0
230	240	17	11	4	0	0	0	0	0	0	0	0
240	250	19	12	5	0	0	0	0	0	0	0	0
250	260	20	14	7	0	0	0	0	0	0	0	0
260	270	22	15	8	2	0	0	0	0	0	0	0
270	280	23	17	10	3	0	0	0	0	0	0	0
280	290	25	18	11	5	0	0	0	0	0	0	0
290	300	26	20	13	6	0	0	0	0	0	0	0
300	310	28	21	14	8	1	0	0	0	0	0	0
310	320	29	23	16	9	2	0	0	0	0	0	0
320	330	31	24	17	11	4	0	0	0	0	0	0
330	340	32	26	19	12	5	0	0	0	0	0	0
340	350	34	27	20	14	7	0	0	0	0	0	0
350	360	35	29	22	15	8	1	0	0	0	0	0
360	370	37	30	23	17	10	3	0	0	0	0	0
370	380	38	32	25	18	11	4	0	0	0	0	0
380	390	40	33	26	20	13	6	0	0	0	0	0
390	400	41	35	28	21	14	7	1	0	0	0	0
400	410	43	36	29	23	16	9	2	0	0	0	0
410	420	44	38	31	24	17	10	4	0	0	0	0
420	430	46	39	32	26	19	12	5	0	0	0	0
430	440	47	41	34	27	20	13	7	0	0	0	0
440	450	49	42	35	29	22	15	8	1	0	0	0
450	460	50	44	37	30	23	16	10	3	0	0	0
460	470	52	45	38	32	25	18	11	4	0	0	0
470	480	53	47	40	33	26	19	13	6	0	0	0
480	490	55	48	41	35	28	21	14	7	1	0	0
490	500	56	50	43	36	29	22	16	9	2	0	0
500	510	58	51	44	38	31	24	17	10	4	0	0
510	520	59	53	46	39	32	25	19	12	5	0	0
520	530	61	54	47	41	34	27	20	13	7	0	0
530	540	62	56	49	42	35	28	22	15	8	1	0
540	550	64	57	50	44	37	30	23	16	10	3	0
550	560	65	59	52	45	38	31	25	18	11	4	0
560	570	67	60	53	47	40	33	26	19	13	6	0
570	580	68	62	55	48	41	34	28	21	14	7	1
580	590	70	63	56	50	43	36	29	22	16	9	2
590	600	71	65	58	51	44	37	31	24	17	10	4
600	610	73	66	59	53	46	39	32	25	19	12	5
610	620	74	68	61	54	47	40	34	27	20	13	7
620	630	76	69	62	56	49	42	35	28	22	15	8
630	640	77	71	64	57	50	43	37	30	23	16	10
640	650	79	72	65	59	52	45	38	31	25	18	11
650	660	80	74	67	60	53	46	40	33	26	19	13
660	670	82	75	68	62	55	48	41	34	28	21	14
670	680	83	77	70	63	56	49	43	36	29	22	16
680	690	85	78	71	65	58	51	44	37	31	24	17
690	700	86	80	73	66	59	52	46	39	32	25	19
700	710	88	81	74	68	61	54	47	40	34	27	20
710	720	89	83	76	69	62	55	49	42	35	28	22
720	730	91	84	77	71	64	57	50	43	37	30	23
730	740	92	86	79	72	65	58	52	45	38	31	25

(Continued on next page)

FEDERAL TAX WITHHOLDING
MARRIED Persons — WEEKLY Payroll Period

If the wages are—		And the number of withholding allowances claimed is—										
At least	But less than	0	1	2	3	4	5	6	7	8	9	10
		The amount of income tax to be withheld is—										
$740	$750	$94	$87	$80	$74	$67	$60	$53	$46	$40	$33	$26
750	760	95	89	82	75	68	61	55	48	41	34	28
760	770	97	90	83	77	70	63	56	49	43	36	29
770	780	98	92	85	78	71	64	58	51	44	37	31
780	790	100	93	86	80	73	66	59	52	46	39	32
790	800	103	95	88	81	74	67	61	54	47	40	34
800	810	106	96	89	83	76	69	62	55	49	42	35
810	820	108	98	91	84	77	70	64	57	50	43	37
820	830	111	99	92	86	79	72	65	58	52	45	38
830	840	114	101	94	87	80	73	67	60	53	46	40
840	850	117	104	95	89	82	75	68	61	55	48	41
850	860	120	107	97	90	83	76	70	63	56	49	43
860	870	122	110	98	92	85	78	71	64	58	51	44
870	880	125	113	100	93	86	79	73	66	59	52	46
880	890	128	115	103	95	88	81	74	67	61	54	47
890	900	131	118	106	96	89	82	76	69	62	55	49
900	910	134	121	108	98	91	84	77	70	64	57	50
910	920	136	124	111	99	92	85	79	72	65	58	52
920	930	139	127	114	101	94	87	80	73	67	60	53
930	940	142	129	117	104	95	88	82	75	68	61	55
940	950	145	132	120	107	97	90	83	76	70	63	56
950	960	148	135	122	110	98	91	85	78	71	64	58
960	970	150	138	125	112	100	93	86	79	73	66	59
970	980	153	141	128	115	103	94	88	81	74	67	61
980	990	156	143	131	118	105	96	89	82	76	69	62
990	1,000	159	146	134	121	108	97	91	84	77	70	64
1,000	1,010	162	149	136	124	111	99	92	85	79	72	65
1,010	1,020	164	152	139	126	114	101	94	87	80	73	67
1,020	1,030	167	155	142	129	117	104	95	88	82	75	68
1,030	1,040	170	157	145	132	119	107	97	90	83	76	70
1,040	1,050	173	160	148	135	122	110	98	91	85	78	71
1,050	1,060	176	163	150	138	125	112	100	93	86	79	73
1,060	1,070	178	166	153	140	128	115	103	94	88	81	74
1,070	1,080	181	169	156	143	131	118	105	96	89	82	76
1,080	1,090	184	171	159	146	133	121	108	97	91	84	77
1,090	1,100	187	174	162	149	136	124	111	99	92	85	79
1,100	1,110	190	177	164	152	139	126	114	101	94	87	80
1,110	1,120	192	180	167	154	142	129	117	104	95	88	82
1,120	1,130	195	183	170	157	145	132	119	107	97	90	83
1,130	1,140	198	185	173	160	147	135	122	109	98	91	85
1,140	1,150	201	188	176	163	150	138	125	112	100	93	86
1,150	1,160	204	191	178	166	153	140	128	115	102	94	88
1,160	1,170	206	194	181	168	156	143	131	118	105	96	89
1,170	1,180	209	197	184	171	159	146	133	121	108	97	91
1,180	1,190	212	199	187	174	161	149	136	123	111	99	92
1,190	1,200	215	202	190	177	164	152	139	126	114	101	94
1,200	1,210	218	205	192	180	167	154	142	129	116	104	95
1,210	1,220	220	208	195	182	170	157	145	132	119	107	97
1,220	1,230	223	211	198	185	173	160	147	135	122	109	98
1,230	1,240	226	213	201	188	175	163	150	137	125	112	100
1,240	1,250	229	216	204	191	178	166	153	140	128	115	102
1,250	1,260	232	219	206	194	181	168	156	143	130	118	105
1,260	1,270	234	222	209	196	184	171	159	146	133	121	108
1,270	1,280	237	225	212	199	187	174	161	149	136	123	111
1,280	1,290	240	227	215	202	189	177	164	151	139	126	114
1,290	1,300	243	230	218	205	192	180	167	154	142	129	116
1,300	1,310	246	233	220	208	195	182	170	157	144	132	119
1,310	1,320	248	236	223	210	198	185	173	160	147	135	122
1,320	1,330	251	239	226	213	201	188	175	163	150	137	125
1,330	1,340	254	241	229	216	203	191	178	165	153	140	128
1,340	1,350	257	244	232	219	206	194	181	168	156	143	130
1,350	1,360	260	247	234	222	209	196	184	171	158	146	133
1,360	1,370	262	250	237	224	212	199	187	174	161	149	136
1,370	1,380	265	253	240	227	215	202	189	177	164	151	139
1,380	1,390	268	255	243	230	217	205	192	179	167	154	142
1,390	1,400	271	258	246	233	220	208	195	182	170	157	144

$1,400 and over   Use Table 1(b) for a **MARRIED person** on page 17.

PRACTICE EXERCISE 7

1. Juan is single with two dependents. He worked 40 hours at $4.95 an hour.

    a. What was his weekly wage?
    b. Using the percentage method, find the amount of his withholding tax.
    c. Using the tax tables, find the amount of his withholding tax.
    d. Which amount is smaller? What is the difference?

2. Charlotte's rate of pay is higher since she doesn't receive any tips. Charlotte is married with two dependents. She worked 40 hours at $8.35 an hour.

    a. What was her weekly wage?
    b. Using the percentage method, find the amount of her withholding tax.
    c. Using the tax tables, find the amount of her withholding tax.
    d. Which amount is larger? By how much?

3. There are other employees working in Ye Olde Eat Shoppe. Can you help Charlotte compute their wages and taxes?

Name	Job	Hours Worked	Hourly Rate of Pay	Wages	Number of Dependents	Marriage Status	Tax Percentage Method	Tax Table Method
Alice	Chef	40	$11.75		2	Married		
Ivan	Baker	40	10.25		3	Married		
Andrew	Dish-washer	42	5.15		1	Single		
Eric	Busboy	42	4.85		1	Single		
Bob	Busboy	41	4.80		2	Single		

State Tax

Besides deducting federal taxes from each employee's wages, Charlotte must also deduct state taxes. Here, too, different methods are used to compute the tax. Charlotte showed the two methods in the following way.

Margie's gross wages are $192.00 with a withholding tax of $15.00

METHOD I

In the "Weekly Withholding Tax Table" on page 23, $192.00 is at least $190 but less than $200. Charlotte found that the state tax for a person with one dependent, or exemption, is $3.20.

Weekly Withholding Tax Table

Method I

WAGES		EXEMPTIONS CLAIMED										
At Least	But Less Than	0	1	2	3	4	5	6	7	8	9	10 or more
		TAX TO BE WITHHELD										
$ 130	$ 135	$ 1.50	$ .70									
135	140	1.70	.90	$ .10								
140	145	1.90	1.10	.30								
145	150	2.10	1.30	.50								
150	160	2.40	1.60	.80								
160	170	2.80	2.00	1.20	$ .40							
170	180	3.20	2.40	1.60	.80	$ .10						
180	190	3.60	2.80	2.00	1.20	.50						
190	200	4.00	3.20	2.40	1.60	.90	$ .10					
200	210	4.40	3.60	2.80	2.00	1.30	.50					
210	220	4.90	4.00	3.20	2.40	1.70	.90	$ .10				
220	230	5.40	4.40	3.60	2.80	2.10	1.30	.50				
230	240	5.90	4.90	4.00	3.20	2.50	1.70	.90	$ .20			
240	250	6.40	5.40	4.50	3.60	2.90	2.10	1.30	.60			
250	260	6.90	5.90	5.00	4.00	3.30	2.50	1.70	1.00	$ .20		
260	270	7.50	6.40	5.50	4.50	3.70	2.90	2.10	1.40	.60		
270	280	8.10	7.00	6.00	5.00	4.10	3.30	2.50	1.80	1.00	$ .20	
280	290	8.70	7.60	6.50	5.50	4.50	3.70	2.90	2.20	1.40	.60	
290	300	9.30	8.20	7.00	6.00	5.00	4.10	3.30	2.60	1.80	1.00	$ .30
300	310	9.90	8.80	7.60	6.50	5.50	4.60	3.70	3.00	2.20	1.40	.70
310	320	10.60	9.40	8.20	7.10	6.00	5.10	4.10	3.40	2.60	1.80	1.10
320	330	11.30	10.00	8.80	7.70	6.50	5.60	4.60	3.80	3.00	2.20	1.50
330	340	12.00	10.70	9.40	8.30	7.10	6.10	5.10	4.20	3.40	2.60	1.90
340	350	12.70	11.40	10.00	8.90	7.70	6.60	5.60	4.70	3.80	3.00	2.30
350	360	13.50	12.10	10.70	9.50	8.30	7.20	6.10	5.20	4.20	3.40	2.70
360	370	14.30	12.80	11.40	10.10	8.90	7.80	6.60	5.70	4.70	3.80	3.10
370	380	15.10	13.50	12.10	10.80	9.50	8.40	7.20	6.20	5.20	4.20	3.50
380	390	15.80	14.30	12.80	11.50	10.10	9.00	7.80	6.70	5.70	4.70	3.90
390	400	16.60	15.10	13.60	12.20	10.80	9.60	8.40	7.30	6.20	5.20	4.30
400	410	17.40	15.90	14.40	12.90	11.50	10.20	9.00	7.90	6.70	5.70	4.80
410	420	18.20	16.70	15.20	13.70	12.20	10.90	9.60	8.50	7.30	6.20	5.30
420	430	19.00	17.50	16.00	14.50	12.90	11.60	10.20	9.10	7.90	6.80	5.80
430	440	19.80	18.30	16.80	15.20	13.70	12.30	10.90	9.70	8.50	7.40	6.30
440	450	20.60	19.10	17.50	16.00	14.50	13.00	11.60	10.30	9.10	8.00	6.80
450	460	21.40	19.80	18.30	16.80	15.30	13.80	12.30	11.00	9.70	8.60	7.40
460	470	22.10	20.60	19.10	17.60	16.10	14.60	13.10	11.70	10.30	9.20	8.00
470	480	22.90	21.40	19.90	18.40	16.90	15.40	13.80	12.40	11.00	9.80	8.60
480	490	23.70	22.20	20.70	19.20	17.70	16.20	14.60	13.10	11.70	10.40	9.20
490	500	24.50	23.00	21.50	20.00	18.50	16.90	15.40	13.90	12.40	11.10	9.80
500	510	25.30	23.80	22.30	20.80	19.20	17.70	16.20	14.70	13.20	11.80	10.40
510	520	26.10	24.60	23.10	21.50	20.00	18.50	17.00	15.50	14.00	12.50	11.10
520	530	26.90	25.40	23.80	22.30	20.80	19.30	17.80	16.30	14.80	13.20	11.80
530	540	27.70	26.10	24.60	23.10	21.60	20.10	18.60	17.10	15.50	14.00	12.50
540	550	28.40	26.90	25.40	23.90	22.40	20.90	19.40	17.80	16.30	14.80	13.30
550	560	29.20	27.70	26.20	24.70	23.20	21.70	20.10	18.60	17.10	15.60	14.10
560	570	30.00	28.50	27.00	25.50	24.00	22.50	20.90	19.40	17.90	16.40	14.90
570	580	30.80	29.30	27.80	26.30	24.80	23.20	21.70	20.20	18.70	17.20	15.70
580	590	31.60	30.10	28.60	27.10	25.50	24.00	22.50	21.00	19.50	18.00	16.50
590	600	32.40	30.90	29.40	27.80	26.30	24.80	23.30	21.80	20.30	18.80	17.20
600	610	33.20	31.70	30.10	28.60	27.10	25.60	24.10	22.60	21.10	19.50	18.00
610	620	34.00	32.40	30.90	29.40	27.90	26.40	24.90	23.40	21.80	20.30	18.80
620	630	34.70	33.20	31.70	30.20	28.70	27.20	25.70	24.10	22.60	21.10	19.60
630	640	35.50	34.00	32.50	31.00	29.50	28.00	26.40	24.90	23.40	21.90	20.40
640	650	36.30	34.80	33.30	31.80	30.30	28.80	27.20	25.70	24.20	22.70	21.20
$650 & OVER		Use **Method II**, "Exact Calculation Method", on page 24.										

WAGES		EXEMPTIONS CLAIMED				
At Least	But Less Than	0	1	2	3	4
		TAX TO BE WITHHELD				
$ 0	$ 18	$ .00				
18	36	.00				
36	50	.00				
50	60	.00				
60	68	.00				
68	74	.00				
74	78	.00				
78	80	.00				
80	82	.00				
82	84	.00				
84	86	.00				
86	88	.00				
88	90	.00				
90	92	.00				
92	94	.00				
94	96	.00				
96	98	.00				
98	100	.10				
100	105	.30				
105	110	.50				
110	115	.70				
115	120	.90	$ .10			
120	125	1.10	.30			
125	130	1.30	.50			

METHOD II

Tables A and B below are reproduced from the New York State tax booklet.

Method II

This method is based upon applying a given percentage to the portion of the wages (after deductions and exemptions) which falls within a wage bracket and adding to this product the given accumulated tax for all lower tax brackets. After subtracting the amount of deductions (from Table A) and the amount of exemptions (from Table B) below, the following table is used.

WEEKLY PAYROLL

Line No.	If wages are (after deductions and exemptions)		The AMOUNT to be WITHHELD is the SUM of:				
	At Least	But less than	This Amount (tax-lower brackets)	PLUS	This Percent	-OF	Excess of Wages (after deductions and exemptions) over this amount
	Col. 1	Col. 2	Col. 3		Col. 4		Col. 5
1	$ 0	$106	$0.00	PLUS	4%	EXCESS OVER	$ 0
2	106	154	4.23	"	5%	" "	106
3	154	212	6.63	"	6%	" "	154
4	212	250	10.10	"	7%	" "	212
5	250	1,731	12.79	"	7.875%	" "	250
6	1,731	1,923	129.40	"	9.315%	" "	1,731
7	1,923	2,885	147.33	"	9.815%	" "	1,923
8	2,885	AND UP	241.71	"	8.375%	" "	2,885

EXACT CALCULATION METHOD

The steps in computing the amount of tax to be withheld are as follows:

Step (1) Determine the amount of deduction allowance (from Table A).
  (2) Multiply the amount of one exemption (from Table B) by the number of exemptions claimed.
  (3) Add the amount of deductions obtained in Step (1) to the amount of exemptions obtained in Step (2).
  (4) Subtract the amount of deductions and exemptions obtained in Step (3) from the employee's gross wages.
  (5) Using the proper table in the Table II series (depending on the particular payroll period), find the applicable line on which the wages after deductions and exemptions in Step (4) are located in columns 1 and 2.
  (6) Subtract the amount in column 5 of this line from the amount of wages after deductions and exemptions in Step (4).
  (7) Multiply the remainder obtained in Step (6) by the percentage in column 4 of this line.
  (8) Add the product in Step (7) to the amount in column 3 of the applicable line. This is the New York State tax to be withheld for the particular payroll period.

EXAMPLE 1: (weekly payroll)
Weekly gross pay of $400, single with 3 exemptions claimed:
  (1) Deduction allowance (from Table A) = $96.15
  (2) $19.25 (from Table B) × 3 = $57.75
  (3) $96.15 + $57.75 = $153.90
  (4) $400 − $153.90 = $246.10
  (5) Line 4 of Table II A is applicable ($246.10 is between $212 and $250)
  (6) $246.10 − $212 = $34.10
  (7) $34.10 × 7% (.07) = $2.39
  (8) $2.39 + $10.10 = $12.49 (New York State tax to be withheld)

TABLE A Deduction Allowance Table

NEW YORK STATE, CITY OF NEW YORK AND CITY OF YONKERS

	Payroll Period					
	Weekly	Biweekly	Semimonthly	Monthly	Daily	Annual
Single	$96.15	$192.30	$208.35	$416.70	$19.25	$ 5,000.00
Married	105.75	211.50	229.15	458.30	21.15	5,500.00

TABLE B Exemption Allowance Table

Based on a full year exemption of $1,000.00

Payroll Period	Amount of one exemption
Weekly	$ 19.25
Biweekly	38.50
Semimonthly	41.65
Monthly	83.30
Quarterly	250.00
Semiannual	500.00
Annual	1,000.00
Daily or miscellaneous	3.85

The eight steps to be followed are listed to the right of Table IIA on page 24:

(1) Charlotte used Table A and a weekly payroll period and found that the deduction allowance is $96.15.

(2) From Table B: The payroll period is weekly, so $19.25 \times 1 = $19.25.

(3) Step (1) + Step (2) = $96.15 + $19.25 = $115.40.

(4) $192.00   Gross wages
     −115.40   Deductions and exemptions
     $ 76.60

(5) $76.60 is less than $106 and is therefore located on line 1 in Table IIA. The tax is 4% of the excess over 0.

(6) $76.60 \times 4\% = $76.60
                      $\times$ 0.04
                      $3.064 = $3.06

(7) Thus the total tax is $3.06.

Method I is certainly the simpler method and probably is used more often by employers. Margie's earnings are now $192.00 less $15.00 for federal tax and $3.20 for state tax.

PRACTICE EXERCISE 8

Compute the state tax for each employee (Alice, Ivan, Andrew, Eric, and Bob) using both Methods I and II illustrated above. The information needed is given on pages 23 and 24.

Social Security Tax

Social Security deductions are taken from your salary during each working year. Old-age, disability, and death benefits can be received by you or your beneficiaries from the payments you make to the Social Security plan. You and your employer pay equal shares to the federal government, which runs the Social Security agency. Currently, your payroll deductions for Social Security amount to 6.2% of your salary, up to a maximum of $3571.20. In addition, there is a 1.45%

tax for Medicare, making the total deduction 6.2% + 1.45% = 7.65%. Your contribution is matched by your employer. Margie must pay a tax of 7.65% on her salary of $192.00. Thus:

$$7.65\% = \frac{7.65}{100} = 100\overline{)7.6500}$$

$$
\begin{array}{r}
0.0765 \\
100\overline{)7.6500} \\
7\,00xx \\
650 \\
600 \\
500 \\
500 \\
\end{array}
$$

and

$$
\begin{array}{r}
\$192.00 \\
\times\,0.07\ 65 \\
\hline
960\ 00 \\
1\,1520\ 0 \\
134400 \\
\hline
14.6880\ 00 = \$14.69 \\
\end{array}
$$

Therefore Margie earns a salary of $192.00 less the following amounts:

Taxes = $15.00 + $3.20 = $18.20   Social Security/Medicare = $14.69

$$
\begin{array}{r}
\$18.20 \\
+14.69 \\
\hline
\$32.89 \\
\end{array}
$$

$$
\begin{array}{r}
\$192.00 \\
-\ 32.89 \\
\hline
\$159.11 \\
\end{array}
$$
    Margie's take-home pay

PRACTICE EXERCISE 9

For each employee (Alice, Ivan, Andrew, Eric, and Bob) do the following:

1. Compute the Social Security/Medicare deduction.

2. Figure the total deduction: federal tax, using the tables; state tax, using Method I; and Social Security/Medicare.

3. Figure the take-home pay.

EXTENDING RECIPES

Many recipes are calculated to feed 4, 6, or 8 people. Alice, the chef, when cooking for many customers, must learn to increase or decrease the amounts in a recipe so that the new quantities are consistent with the original ones.

Suppose that a recipe is written to feed 4 people and Alice wants to make the same recipe for 20 people. What should she do? You figured it correctly: she must multiply each quantity in the recipe by 5. If 2 eggs are required in the original recipe, Alice must put in 10 eggs when preparing it for 20 people. If a half cup of water is required, she must use $5 \times \frac{1}{2}$ or

$$\frac{5}{1} \times \frac{1}{2} = \frac{5}{2} = \frac{2}{2} + \frac{2}{2} + \frac{1}{2} = 1 + 1 + \frac{1}{2} = 2\frac{1}{2} \text{ cups of water for the new recipe}$$

Another way of doing the same problem is to write and solve a proportion. A *proportion* is two equal ratios; $\frac{2}{4}$ is a ratio and so is $\frac{3}{6}$. Since $\frac{2}{4} = \frac{3}{6}$, we say that these two ratios form a proportion. The 2 and 6 are called the *extremes*, and the 4 and 3 are called the *means*, of the proportion. If four terms form a proportion, then the product of the means is equal to the product of the extremes. Thus:

$$2 \times 6 = 3 \times 4.$$

A proportion can help Alice do the mathematics required to increase or decrease recipe amounts. If $\frac{1}{2}$ cup of water is needed in a recipe for 4 people, how much water is required for 20 people? Writing this as a proportion, we obtain

$$\frac{\frac{1}{2} \text{ cup}}{4 \text{ people}} = \frac{x \text{ cups}}{20 \text{ people}}$$

We can then solve this *equation* by finding the product of the means, $4 \cdot x$, and setting it equal to the product of the extremes, $\frac{1}{2} \cdot 20$. Thus:

$$4 \cdot x = \frac{1}{2} \cdot 20$$

$$4x = 10$$

This *equation* is solved by undoing the mathematical process that combines the 4 and the x. Since they are multiplied together, we must divide by 4 if we wish to undo the operation. Thus: $4x = 10$, and dividing both sides of the equation by 4 (this is necessary if we are to maintain the equality of both quantities) gives

$$\frac{4x}{4} = \frac{10}{4}$$

$$x = \frac{\overset{5}{\cancel{10}}}{\underset{2}{\cancel{4}}} = \frac{5}{2} = \frac{2}{2} + \frac{2}{2} + \frac{1}{2}$$

$$x = 1 + 1 + \frac{1}{2}$$

$$x = 2\frac{1}{2} \text{ cups}$$

EXAMPLE: A recipe is written to serve 8 people and requires $\frac{3}{4}$ tablespoon of olive oil. How many tablespoons will be required if the recipe is increased for 60 people?

SOLUTION: We write the proportion:

$$\frac{\frac{3}{4} \text{ Tbs.}}{8 \text{ people}} = \frac{x \text{ Tbs.}}{60 \text{ people}}$$

$$8x = \frac{3}{4} \cdot 60$$

$$8x = \frac{3}{\cancel{4}} \cdot \frac{\cancel{60}^{15}}{1}$$

$$\frac{8x}{8} = \frac{45}{8} \qquad \text{or} \qquad \frac{8x}{8} = \frac{45}{8}$$

$$x = 8\overline{)45} \quad \left\{ \begin{array}{l} \text{This is a} \\ \text{shorter} \\ \text{method} \\ \text{used to} \\ \text{change an} \\ \text{improper} \\ \text{fraction} \\ \text{into a} \\ \text{mixed} \\ \text{number.} \end{array} \right\}$$

$$\begin{array}{r} 40 \\ \overline{5} \end{array}$$

$$x = 5\frac{5}{8} \text{ Tbs.}$$

$$x = \frac{8}{8} + \frac{8}{8} + \frac{8}{8} + \frac{8}{8} + \frac{8}{8} + \frac{5}{8}$$

$$x = 1 + 1 + 1 + 1 + 1 + \frac{5}{8}$$

$$x = 5\frac{5}{8} \text{ Tbs.}$$

PRACTICE EXERCISE 10

1. Change each of these improper fractions to a mixed number:

a. $\dfrac{19}{6}$ f. $\dfrac{7}{2}$

b. $\dfrac{14}{3}$ g. $\dfrac{27}{4}$

c. $\dfrac{25}{8}$ h. $\dfrac{16}{7}$

d. $\dfrac{17}{5}$ i. $\dfrac{35}{32}$

e. $\dfrac{22}{9}$ j. $\dfrac{23}{16}$

2. Solve the following proportions for *x*:

a. $\dfrac{5}{3} = \dfrac{x}{6}$

d. $\dfrac{4}{x} = \dfrac{2}{\frac{1}{2}}$

b. $\dfrac{x}{4} = \dfrac{2}{8}$

e. $\dfrac{5}{6} = \dfrac{x}{30}$

c. $\dfrac{x}{6} = \dfrac{\frac{1}{2}}{2}$

f. $\dfrac{\frac{1}{2}}{8} = \dfrac{x}{40}$

3. Compute each of the quantities listed in this recipe, which feeds 6 people, needed to feed 40 people. The first proportion is done for you as an example.

EGGPLANT and RICE PROVENÇAL

6 *people*	40 *people*
a. large eggplants, about 2 pounds	$\dfrac{2 \text{ pounds}}{6 \text{ people}} = \dfrac{x \text{ pounds}}{40 \text{ people}}$

b. 4 tablespoons olive oil
c. 3 cups chopped onions
d. 1 green pepper
e. 2 cloves garlic
f. 1 teaspoon thyme
g. 1 bay leaf
h. $3\dfrac{1}{4}$ cups chicken broth
i. $\dfrac{1}{2}$ cup Parmesan cheese
j. 2 tablespoons butter

FIGURING THE AREAS AND VOLUMES OF BOWLS AND PANS

Ivan is the baker in the restaurant, and when he increases cake and pie recipes, he must compute in the same way Alice, the chef, does. In addition, he must mix the ingredients of a cake in a large bowl and then transfer the contents into smaller round, rectangular, or square baking pans. Ivan must figure out how many individual cakes he can make from the contents of the large mixing bowl. To do this, he must know how to compute the areas of rectangles, squares, and circles. He must also know how to find the volumes of rectangular prisms, square prisms, and cylinders.

A *rectangular prism* is a three-dimensional solid with a length (*l*), width (*w*), and height (*h*). It is pictured below.

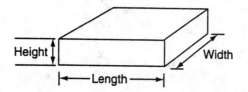

The *volume* of this figure means the number of cubic units that fit inside it. Cubic units may be cubic inches, cubic feet, etc. A cubic inch has a length, width, and height each equal to 1 inch. A cubic foot has dimensions each equal to 1 foot. A cubic inch is drawn below.

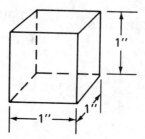

To find the volume of a rectangular prism, you must first find the area of the base of the figure and then multiply it by the height. *Area* means the number of square units that fit into the figure. Square units may be square inches, square feet, etc. The diagram below pictures a rectangle filled with square units.

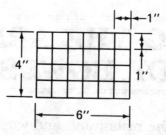

As you see, there are 24 square inches in this rectangle, and therefore its area is 24 sq. in.

If these were the dimensions of a rectangular solid whose height was 2 in., then the volume would be the area of the base times the height. Written as a formula or equation, it becomes

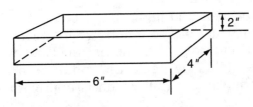

$V = B \cdot h$
$V = 24 \cdot 2$
$V = 48$ cu. in.

EXAMPLE: Find the volume of a rectangular prism whose base measures 18 cm by 12 cm and whose height is 8 cm.

SOLUTION: The area of the base is $18 \cdot 12$, or 216 cm^2. Multiplying by 8 cm results in a volume of 1728 cm^3. This problem could also be solved by using the formula $V = l \cdot w \cdot h$.

When the result is written as 216 cm^2 or 1728 cm^3, the little raised 2 and 3 are called *exponents* and mean that a value is a *factor* more than once. For instance, 6^2 means that 6 is a factor twice, and to find the value we must multiply 6 by 6. Thus: $6^2 = 36$, and 8^3 means $8 \cdot 8 \cdot 8 = 512$. Exponents may be numbers other than 2 or 3, but they are always treated in the same way as shown above.

If the length and width of the base have the same dimensions, the figure is called a *square prism*. The volume of a square prism is found in the same manner as is the volume of a rectangular prism.

EXAMPLE: Find the volume of a square prism whose base dimensions are both 8 cm and whose height is 3 cm.

SOLUTION: The area of the base is $8 \cdot 8$ or $8^2 = 64$, and multiplying this by the height, 3, gives 192 cm^3. Using the formula $V = l \cdot w \cdot h$, we obtain

$$V = 8 \cdot 8 \cdot 3$$
$$V = 192 \text{ cm}^3$$

Ivan has mixed a large recipe for cakes in a container shaped like this:

This is called a *cylinder*, and its base is a circle. We find the volume of this figure in the same manner as for the others we just completed; that is, we find the area of the base and multiply it by the height. How do we find the area of a circle? We multiply the radius by itself (the radius is a factor twice) and then multiply the result by the value π (pi), which is approximately equal to 3.14, or $\frac{22}{7}$. Writing the area as a formula, we have $A = \pi r^2$.

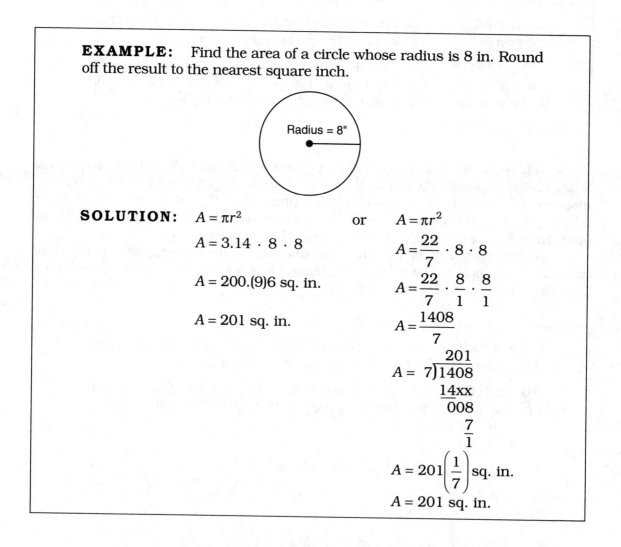

EXAMPLE: Find the area of a circle whose radius is 8 in. Round off the result to the nearest square inch.

Radius = 8"

SOLUTION: $A = \pi r^2$ or $A = \pi r^2$

$A = 3.14 \cdot 8 \cdot 8$ $A = \frac{22}{7} \cdot 8 \cdot 8$

$A = 200.(9)6$ sq. in. $A = \frac{22}{7} \cdot \frac{8}{1} \cdot \frac{8}{1}$

$A = 201$ sq. in. $A = \frac{1408}{7}$

$$A = 7\overline{)1408}$$
with quotient 201, $\frac{14xx}{008}$, $\frac{7}{1}$

$A = 201\left(\frac{1}{7}\right)$ sq. in.

$A = 201$ sq. in.

Ivan has mixed his cake in a cylindrical pan, the radius of whose base is 8 in. The height of the mix is 6 in. What is the volume of the mix?

The volume is the area of the base times its height, or $V = B \cdot h$.

$V = \pi \cdot r^2 \cdot h$

$V = 3.14 \cdot 8 \cdot 8 \cdot 6$

$V = 3.14 \cdot 384$

$V = 1205.76$ cu. in.

$V = 1206$ cu. in.

$$
\begin{array}{r}
384 \\
\times\, 3.14 \\
\hline
15\ 36 \\
38\ 4 \\
1152 \\
\hline
1205.76
\end{array}
$$

If Ivan has many cake pans measuring $12 \times 15 \times 2\frac{1}{2}$ in. and wants to fill them to a height of 2 in., how many pans can he fill from the mix computed above?
Each rectangular pan has a volume of

$$V = 12 \cdot 15 \cdot 2 \quad \text{or} \quad V = 360 \text{ cu. in.}$$

(*Remember:* Fill them only to a height of 2 in.)

Each pan requires 360 cu. in. and Ivan has approximately 1206 cu. in. He can divide and see that

$$
\begin{array}{r}
3 \\
360\overline{)1206} \\
1080 \\
\hline
126
\end{array}
$$

He can fill 3 pans with his mix and still have some left over.

If he has another cake pan measuring $9 \times 13 \times 2\frac{1}{2}$ in., can he fill it to a height of 2 in. with the remaining cake mix?
The volume of this smaller cake pan is

$$V = 9 \cdot 13 \cdot 2 \quad \text{or} \quad V = 234 \text{ cu. in.}$$

After filling 3 cake pans, he has used 1080 cu. in., which leaves him

$$
\begin{array}{r}
1206 \\
-1080 \\
\hline
126
\end{array}
$$

126 cu. in., not enough to fill this cake pan

Ivan decides to put the remaining mix into a cylindrical cake pan with a 4-in. radius as a base and a height of $2\frac{1}{2}$ in. Can this be filled with the remaining cake mix?

$$V = \pi \cdot r^2 \cdot h$$

$$V = \frac{22}{7} \cdot 4 \cdot 4 \cdot 2\frac{1}{2} \quad \left(2\frac{1}{2} = \frac{2 \times 2 + 1}{2} = \frac{5}{2} \right)$$

$$V = \frac{\overset{11}{\cancel{22}}}{7} \cdot \frac{4}{1} \cdot \frac{4}{1} \cdot \frac{5}{\underset{1}{\cancel{2}}}$$

$$V = \frac{880}{7}$$

$$
\begin{array}{r}
125 \\
V = 7\overline{)880} \\
\underline{7\text{xx}} \\
18 \\
\underline{14} \\
40 \\
\underline{35} \\
5
\end{array}
$$

$$V = 125\frac{5}{7} \quad \text{or} \quad V = 126 \text{ cu. in.}$$

Therefore Ivan can fill 3 rectangular cake pans and this cylindrical cake pan with the cake ingredients.

PRACTICE EXERCISE 11

1. Find the areas of the rectangles whose dimensions are listed below. Remember that all dimensions must be alike before multiplying.

    a. $l = 14''$, $w = 5''$

    b. $l = 25$ m, $w = 4$ m

    c. $l = 2'$, $w = 6''$

    d. $l = 3\frac{1}{2}$ cm, $w = 8$ cm

    e. $l = 12''$, $w = 2\frac{1}{4}''$

    f. $l = 3.8''$, $w = 2.5''$

    g. $l = 6.5''$, $w = 3.5''$

    h. $l = 6\frac{1}{4}$ dm, $w = 7\frac{1}{2}$ dm

    i. $l = 3'4''$, $w = 2'6''$

    j. $l = 3.6''$, $w = 2\frac{1}{2}''$

2. Find the value of each of the following expressions:

    a. 7^2

    b. 3^4 (*Hint:* $3 \cdot 3 \cdot 3 \cdot 3$.)

    c. 2^5

    d. 10^2

    e. 13^2

    f. 5^3

    g. $\left(\frac{1}{2}\right)^4$

    h. $(3.2)^2$

    i. $(5.17)^2$

    j. 12^3

3. Find the volume of each of these rectangular prisms. (Be careful with the different units of measurement.)

    a. $l = 9$ cm, $w = 5$ cm, $h = 2$ cm

    b. $l = 12''$, $w = 10''$, $h = 3''$

    c. $l = 2'$, $w = 1\frac{1}{2}'$, $h = \frac{1}{2}'$

    d. $l = 2\frac{1}{2}'$, $w = 1\frac{1}{4}'$, $h = \frac{1}{2}'$

    e. $l = 9$ m, $w = 6$ m, $h = 0.5$ m

    f. $l = 2'3''$, $w = 1'4''$, $h = 6''$

    g. $l = 3.2$ dm, $w = 2.3$ dm, $h = 1.5$ dm

    h. $l = 4'2''$, $w = 10''$, $h = \frac{1}{4}'$

    i. $l = 7''$, $w = 3.1''$, $h = 8.9''$

4. Find the volume of each of these square prisms:

    a. $l = w = 8$ m, $h = 3$ m

    b. $l = w = 2'$, $h = \frac{1}{2}'$

    c. $l = w = 2'3''$, $h = \frac{1}{2}'$

    d. $l = w = 1'6''$, $h = 4''$

    e. $l = w = 3.5$ cm, $h = 1.3$ cm

5. Find the area of each of these circles. Use $\pi = 3.14$, and round off the result to the nearest square unit.

    a. $r = 6''$

    b. $r = 3$ cm

    c. $r = 3.2'$

    d. $r = 5.1''$

    e. $r = 2\dfrac{1}{2}''$

    f. $d = 8$ dm   (d = diameter, a line in the circle passing through the center. The radius is one-half the diameter.)

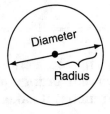

6. Find the area of each of these circles. Use $\pi = \dfrac{22}{7}$, and round off the result to the nearest square unit.

    a. $r = 7''$ **d.** $d = 14$ m

    b. $r = 1\dfrac{1}{2}''$ **e.** $r = 5\dfrac{1}{4}$ cm

    c. $d = 6$ cm   **f.** $r = 14''$

7. Find the volume of each of these cylinders. Round off the result to the nearest cubic unit.

    a. $r = 6$ m, $h = 2$ m, $\pi = 3.14$

    b. $r = 3''$, $h = 8''$, $\pi = 3.14$

    c. $r = 8''$, $h = 1.4''$, $\pi = \dfrac{22}{7}$

    d. $r = 2'3''$, $h = 7''$, $\pi = \dfrac{22}{7}$

    e. $d = 1'4''$, $h = \dfrac{1}{2}'$, $\pi = 3.14$

    f. $d = 3'6''$, $h = 8''$, $\pi = \dfrac{22}{7}$

    g. $r = 3\dfrac{1}{4}$ cm, $h = 3\dfrac{1}{2}$ cm, $\pi = \dfrac{22}{7}$

8. Ivan has mixed cake ingredients in a large, cylindrical bowl whose radius is 12″ and whose height is 8″. He has large cake pans measuring 12″ × 18″ × 3″ and smaller ones measuring 9″ × 13″ × 3″ and 8″ × 8″ × 3″. How many large, medium, and small cake pans can be filled to a height of 2″ from the mixed cake ingredients?

9. Using the same cake pans as in Problem 8, find how many large, medium, and small cake pans can be filled with 1291 cu. in. of cake ingredients.

10. Calculator exercise: Do each of the problems in Question 2 on your calculator.

11. Calculator exercise: Do each of the problems in Question 5 on your calculator.

EXAM TIME

1. Using the menu from Ye Olde Eat Shoppe, find the amount of this check: a regular orange juice, a fried egg sandwich with bacon, cole slaw, tea, and French nut cake.

2. Find the decimal equivalent of $6\frac{1}{2}\%$.

3. Find the sales tax of 8% on a purchase of $25.46. Round off to the nearest cent.

4. The charge in a restaurant was $23.85. What amount should be left for a 15% tip?

5. If I earn $26.02 in tips from total receipts of $173.25, what percent tips did I receive?

6. What is the most efficient manner of returning change to a customer whose bill was $7.84 and who paid with a $10 bill?

7. Jim is married, earns $523 a week, and has three dependents. Using the table on page 20, find his tax.

8. Find the Social Security/Medicare deduction from a weekly salary of $330 if 7.65% is deducted.

9. If 4 Tbs. of oil are needed in a recipe for 6 people, how many tablespoons are needed in the same recipe for 15 people?

10. Change $\frac{35}{6}$ to a mixed number.

11. Change $3\frac{1}{2}$ to an improper fraction.

12. Find the area of a rectangle whose dimensions are 15 cm by 12 cm.

13. Find the area of a circle whose radius is 3″. (Use $\pi = 3.14$.) Round off to the nearest square inch.

14. Find the volume of a rectangular prism whose dimensions are 8″ × 12″ × 4″.

15. Find the volume of a square prism if the measurement of its base is 14″ and its height is 6″.

16. Find the volume of a cylinder whose radius is 7″ and whose height is 7″. $\left(\text{Use } \pi = \dfrac{22}{7}.\right)$

17. Find the value of 7^3.

18. Find the area of a circle, to the nearest square inch, whose radius is 3.1″. (Use $\pi = 3.14$.)

Now Check Your Answers

Now that you have completed the exam, check your answers against the correct ones, which follow the answers to the practice exercises below.

ANSWERS TO PRACTICE EXERCISES AND EXAM

PRACTICE EXERCISE 1

1. $14.00   **2.** $26.30   **3.** $40.85

PRACTICE EXERCISE 2

1. a. 0.06   **b.** 0.22   **c.** 0.032   **d.** $0.02\dfrac{1}{2}$ or 0.025   **e.** 0.126

f. 0.58   **g.** $0.06\dfrac{1}{4}$ or 0.062   **h.** 0.875   **i.** 0.145

j. $0.00\dfrac{1}{2}$ or 0.005

2. a. $1.12, $0.56 + 0.56; $2.10, (2 × $0.80) + 0.50; $3.27, (4 × $0.80) + 0.07

b.

Amount	Arithmetic Method	Tax Table Method
$7.60	$0.61	$0.61
9.90	0.79	0.79
3.71	0.30	0.30

c. $8.21, $10.69, $4.01

3. a. $8.17   **b.** $3.97   **c.** $2.39   **d.** $1.58   **e.** $7.44
f. $6.72   **g.** $8.44   **h.** $3.56   **i.** $5.71   **j.** $4.70

4.

	Amount	Arithmetic Method	Tax Table Method	Total
a.	$8.17	$0.65	$0.65	$8.82
b.	3.97	0.32	0.32	4.29
c.	2.39	0.19	0.19	2.58
d.	1.58	0.13	0.13	1.71
e.	7.44	0.60	0.60	8.04
f.	6.72	0.54	0.54	7.26
g.	8.44	0.68	0.68	9.12
h.	3.56	0.28	0.28	3.84
i.	5.71	0.46	0.46	6.17
j.	4.70	0.38	0.38	5.08

5. a. $0.27   b. $0.40   c. $0.44   d. $0.82   e. $0.32
   f. $0.57   g. $0.69   h. $0.85   i. $1.05   j. $1.40

PRACTICE EXERCISE 3

		Amount	Arithmetic	Tax Table	Sum
1.,2.	a.	$ 12.43	$ 0.99	$ 0.80 + 0.19	$ 13.42
	b.	18.11	1.45	0.80 + 0.65	19.56
	c.	23.25	1.86	1.60 + 0.26	25.11
	d.	46.56	3.72	3.20 + 0.52	50.28
	e.	83.35	6.67	6.40 + 0.27	90.02
	f.	32.90	2.63	2.40 + 0.23	35.53
	g.	73.81	5.90	5.60 + 0.30	79.71
	h.	52.27	4.18	4.00 + 0.18	56.45
	i.	112.04	8.96	8.80 + 0.16	121.00
	j.	243.67	19.49	19.20 + 0.29	263.16

3. a. $1.03   b. $1.49   c. $1.92   d. $3.84   e. $6.88
   f. $2.71   g. $6.09   h. $4.31   i. $9.24   j. $20.10

PRACTICE EXERCISE 4

	Amount	Sales Tax	Tip	Total
1.	$14.00	$1.12	$2.10	$17.22
	26.30	2.10	3.95	32.35
	40.85	3.27	6.13	50.25

2. a. $1.23   b. 0.60   c. 0.36   d. 0.24   e. 1.12   f. 1.01
   g. 1.26   h. 0.53   i. 0.86   j. 0.71

3.
2 Half-Grapefruit	$2.50
2 Soup du Jour	4.80
3 Tomato Juice (large)	4.50
1 Orange Juice (regular)	1.00
2 Steakburgers	6.30
3 Cheeseburgers	10.65
1 Sliced Steak	6.40
1 Roast Beef	5.65
1 Tuna Fish Salad	4.40
4 French Fries	6.60

6 Iced Coffee	$6.00
2 Iced Tea	2.00
4 Blueberry Pie	7.40
2 Chocolate Eclair	3.50
2 French Nut Cake	5.90
Subtotal	$77.60
Sales tax	6.21
Tip	11.64
Total	$95.45

4. a. $34.65   b. $67.18   c. $335.90

5. a. $38.40   b. $92.25   c. $130.65

6. a. $168.00   b. $417.96   c. $585.96

PRACTICE EXERCISE 5

1.
Monday	12.1%	Thursday	13.1%
Tuesday	13.9%	Friday	13.2%
Wednesday	12.3%	Saturday	14.0%

2. a. $230.40   b. $482.52   c. $712.92
    d. $35,646.00   e. 32%   f. 68%

3. a. $760.00   b. $114.00   c. $17.10
    d. $96.90   e. $760.00 + 114.00 + 60.80 = $934.80

PRACTICE EXERCISE 6

		Bills ($)		Coins (¢)				
	Change	5	1	50	25	10	5	1
1. a. $ 3.00 − 2.17 = 0.83				1	1		1	3
b. 5.00 − 3.29 = 1.71			1	1		2		1
c. 20.00 − 12.63 = 7.37		1	2		1	1		2
d. 10.00 − 7.79 = 2.21			2			2		1
e. 15.00 − 13.24 = 1.76			1	1	1			1
f. 20.00 − 18.23 = 1.77			1	1	1			2
g. 10.00 − 1.19 = 8.81		1	3	1	1		1	1
h. 25.00 − 21.10 = 3.90			3	1	1	1	1	
i. 50.00 − 45.46 = 4.54			4	1				4
j. 30.00 − 27.93 = 2.07			2				1	2

2. a. $198.00
    b. $196.00

c. Juan: 9—$20 bills   Margie: 9—$20 bills
          1—$10 bill   1—$10 bill
          1—$5 bill   1—$5 bill
          3—$1 bills   1—$1 bill

PRACTICE EXERCISE 7

1.   a. $198.00

     b. $45.19 × 2 = $90.38 $198.00 $107.62 $58.62 × .15 = $8.79

$$\begin{array}{r} \$198.00 \\ -\ 90.38 \\ \hline \$107.62 \end{array} \qquad \begin{array}{r} \$107.62 \\ -\ 49.00 \\ \hline \$\ 58.62 \end{array}$$

     c. $9.00   d.   (c): percentage method $0.21 less

2.   a. $334.00
     b. $18.69
     c. $19.00
     d.   (c): table method $0.31 more

3.

	Wages	% Method	Tax Table
Alice	$470.00	$39.09	$40.00
Ivan	410.00	23.31	24.00
Andrew	216.30	18.32	18.00
Eric	203.70	16.43	17.00
Bob	196.80	8.61	9.00

PRACTICE EXERCISE 8

	Method I	Method II
Alice	$18.75	$19.10
Ivan	12.52	12.90
Andrew	4.04	4.00
Eric	3.53	3.60
Bob	2.49	2.40

PRACTICE EXERCISE 9

1.   Alice $35.96   Eric $15.58
     Ivan   31.37   Bob   15.06
     Andrew   16.55

2.

	Federal Tax	State Tax	Social Security	Total Deduction
Alice	$40.00	$18.75	$35.96	$94.71
Ivan	24.00	12.52	31.37	57.89
Andrew	18.00	4.04	16.55	38.59
Eric	17.00	3.53	15.88	36.11
Bob	9.00	2.49	15.06	26.55

3. Alice $470.00 – 94.71 = $375.29
    Ivan   410.00 – 57.89 = 352.11
    Andrew   216.30 – 38.59 = 177.71
    Eric   203.70 – 36.11 = 167.59
    Bob   196.80 – 26.55 = 170.25

PRACTICE EXERCISE 10

1. a. $3\frac{1}{6}$ b. $4\frac{2}{3}$ c. $3\frac{1}{8}$ d. $3\frac{2}{5}$ e. $2\frac{4}{9}$

    f. $3\frac{1}{2}$ g. $6\frac{3}{4}$ h. $2\frac{2}{7}$ i. $1\frac{3}{32}$ j. $1\frac{7}{16}$

2. a. $x = 10$ d. $x = 1$
    b. $x = 1$ e. $x = 0.25$
    c. $x = \frac{3}{2}$ f. $x = \frac{5}{2}$

3. a. $x = 13\frac{1}{3}$ b. $x = 26\frac{2}{3}$ c. $x = 20$ d. $x = 6\frac{2}{3}$ e. $x = 13\frac{1}{3}$

    f. $x = 6\frac{2}{3}$ g. $x = 6\frac{2}{3}$ h. $x = 21\frac{2}{3}$ i. $x = 3\frac{1}{3}$ j. $x = 13\frac{1}{3}$

PRACTICE EXERCISE 11

1. a. 70 sq. in.   b. 100 m^2 c. 144 sq. in. or 1 sq. ft.
    d. 28 cm^2 e. 27 sq. in.   f. 9.5 sq. in.   g. 22.75 sq. in.
    h. $46\frac{7}{8}$dm^2 i. 1200 sq. in. or $8\frac{1}{3}$ sq. ft.   j. 9 sq. in.

2. a. 49   b. 81   c. 32   d. 100   e. 169   f. 125   g. $\frac{1}{16}$

    h. 10.24   i. 26.7289   j. 1728

3. a. 90 cm^3 b. 360 cu. in.   c. $1\frac{1}{2}$ cu. ft.   d. $1\frac{9}{16}$ cu. ft.

    e. 27 m^3 f. 2592 cu. in. or $1\frac{1}{2}$ cu. ft.

    g. 11.04 dm^3 h. 1500 cu. in. or $\frac{125}{144}$ cu. ft.   i. 193.13 cu. in.

4. a. 192 m^3
    b. 2 ft.^{3}
    c. 4374 in.^{3} or 2.53125 ft.^{3}
    d. 1296 m^3
    e. 15.925 cm^3

5. a. 113 sq. in.   b. 28 cm^2 c. 32 sq. ft.
    d. 82 sq. in.   e. 20 sq. in.   f. 50 dm^2

6. a. 154 in.^{2}　　d. 616 m^2
　 b. 7 m^2　e. 87 cm^2
　 c. 113 cm^2　f. 616 m^2

7. a. 226 m^3　b. 226 cu. in.　　c. 282 cu. in.
　 d. 16,038 cu. in. or 9 cu. ft.　　e. 1206 cu. in. or 1 cu. ft.
　 f. 6 cu. ft. or 11,088 cu. in.　　g. 116 cm^3

8. 8 large, 1 medium, 1 small

9. 2 large, 1 medium, 1 small

EXAM TIME

1. $9.85　　2. $0.06\frac{1}{2}$ or 0.065　　3. $2.04　　4. $3.58　　5. 15%

6. One 1¢, one 5¢, one 10¢, two $1, one $5.　　7. $41.00　　8. $25.25

9. 10 Tbs.　　10. $5\frac{5}{6}$　　11. $\frac{7}{2}$　　12. 180 cm^2　　13. 28 sq. in.

14. 384 cu. in.　　15. 1176 cu. in.　　16. 1078 cu. in.　　17. 343

18. 30 sq. in.

How Well Did You Do?

0–11　　**Poor.** Reread the unit, redo all the practice exercises, and retake the exam.

12, 13　　**Fair.** Reread the sections dealing with the problems you got wrong. Redo those practice exercises, and retake the exam.

14–16　　**Good.** Review the problems you got wrong. Redo them correctly.

17, 18　　**Very good!** Continue on to the next unit. You're on your way.

THE WHEELS-AND-DEALS AUTOMOBILE SALESROOM

The Wheels-and-Deals Automobile Sales Company has rented space in a new location. The owners have decided to renovate the building and refurnish the office.

WORKING ON THE INSIDE

Finding the Perimeter

They will start the renovation by installing molding and painting the rooms. A number of painters are invited to inspect the building and give their estimates for the cost of renovation.

Mr. Ramirez is one of the persons asked to estimate the cost of the molding and painting job. He must know how to measure with a ruler and to use these measurements to find the perimeters and the areas of the walls and ceilings.

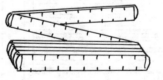

Most rooms have *rectangular* walls and ceilings. A rectangle is pictured below.

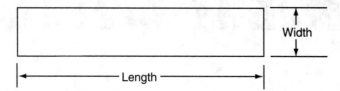

We usually refer to the longer dimension as the *length* of the rectangle and the shorter one as the *width*. The two lengths are *parallel* to each other, as are the two widths, which means they will never meet no matter how far they are extended. The length and width are *perpendicular* to each other, which means the angle formed at their intersection is 90° or a right angle.

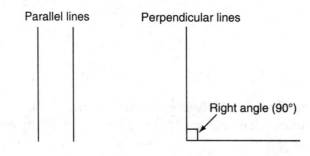

The *perimeter* of a closed geometric figure is found by adding the measures of the sides of the figure. The measures must all be expressed in the same unit: inches, feet, yards, etc.

The perimeter (*P*) of the rectangle below is

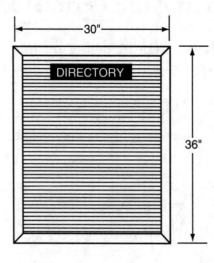

$$P = 30'' + 36'' + 30'' + 36''$$
$$P = 132''$$

EXAMPLE: Mr. Ramirez must purchase and install a molding around the directory sign shown on page 44. If the molding costs $1.75 a foot and installing it costs an additional $1.10 a foot, what will be the total cost, including installation?

SOLUTION: Since 12" = 1', he divides:

$$\begin{array}{r} 11' \\ 12\overline{)132} \\ \underline{12} \\ 12 \\ \underline{12} \end{array}$$

and finds that a perimeter of 132" is equivalent to 11'.
        If the price of the molding is $1.75 a foot and the cost of installing it is $1.10 a foot, the total cost is

$$\begin{array}{r} \$1.75 \\ +\ 1.10 \\ \hline \$2.85 \end{array} \quad \text{a foot} \qquad \text{and} \qquad \begin{array}{r} \$2.85 \\ \times\quad 11 \\ \hline 2\ 85 \\ 28\ 5 \\ \hline \$31.35 \end{array}$$

PRACTICE EXERCISE 1

1. Find the perimeter of a rectangle whose dimensions are 14" by 19".

2. Find the perimeter of a triangle whose sides measure 18', 14', and 22'.

3. Find the cost of 38' of molding if it costs $1.42 a foot.

4. Find the cost of 138" of molding if it costs $1.36 a foot.

5. How many feet of molding are required if it is to be placed around the floor of a room whose dimensions are 7'6" by 10'6"? (Molding is not needed for two doorways in the room whose widths are each 2'6".)

6. Find the perimeter of a triangle whose sides are 6.2", 7.5", and 8.9".

7. Find the perimeter of a rectangle whose dimensions are $9\frac{1}{2}'$ by $10\frac{2}{3}'$.

8. Find the cost of painting a fence 30' long if labor costs $20.10 for each 10' of fencing.

9. The dimensions of the room drawn below are $14' \times 11' \times 8'$. If molding is placed along the 12 edges of the room, how many feet of molding are required?

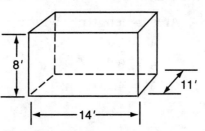

Finding the Area

The area of a wall is very important to a painter. *Area* is the amount of surface in a closed geometric figure. The unit used to measure area is the square unit: the square inch, square foot, square yard, etc.

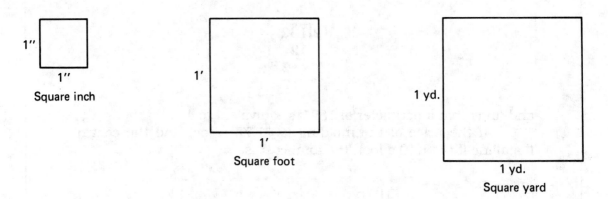

The area of a rectangle whose dimensions are 6″ by 7″:

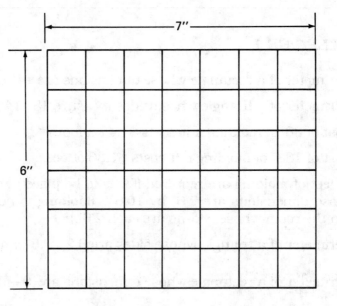

is found by counting the number of square inches in the surface of the rectangle. Thus its area is 42 sq. in.

The area can also be found by multiplying the number of inches in the width of the rectangle by the number of inches in its length:

$$\text{Area} = \text{length} \times \text{width}$$
$$A = l \times w$$
$$A = 6'' \times 7''$$
$$A = 42 \text{ sq. in.}$$

FOR PAINTING

Mr. Ramirez has to estimate the cost of painting the walls and ceiling of the Wheels-and-Deals Automobile Showroom. Its dimensions are 30′ × 50′ × 10′, and a gallon of paint covers 250 sq. ft. and costs $17.00. Can you help?

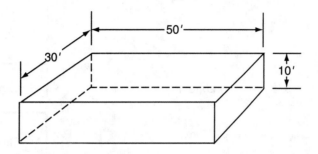

The dimensions of the ceiling are 30′ × 50′. Its area is 30 × 50 or 1500 sq. ft. The dimensions of two walls are 50′ × 10′. The area of the two walls is 1000 sq. ft.

The dimensions of the other two walls are 30′ × 10′. The area of one wall is 300 sq. ft. The area of the two walls is 600 sq. ft.

The combined areas are

$$
\begin{array}{r}
1500 \text{ sq. ft.} \\
1000 \\
+\ \ 600 \\
\hline
3100 \text{ sq. ft.}
\end{array}
$$

If 1 gallon of paint covers 250 sq. ft., then

$$
\begin{array}{r}
12.4 \\
250\overline{)3100.0} \\
\underline{250} \\
600 \\
\underline{500} \\
100\ 0 \\
\underline{100\ 0}
\end{array}
$$

Mr. Ramirez needs at least 12.4 gallons of paint and must purchase 13 gallons. Each gallon costs $17.00. Thus:

$$
\begin{array}{r}
\$17.00 \\
\times\ \ \ \ 13 \\
\hline
51\ 00 \\
170\ 0 \\
\hline
\$221.00
\end{array}
$$

Therefore, $221.00 will cover the cost of the materials needed to paint the showroom. Labor and profit would still have to be considered, however, before Mr. Ramirez could estimate the job.

PRACTICE EXERCISE 2

1. The floor plan of the Wheels-and-Deals Automobile Sales Company is shown below.

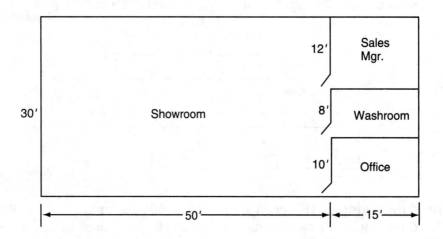

   a. Find the area of the ceiling surface for each room.
   b. Find the total ceiling surface.
   c. If the ceiling height throughout the agency is 10′, find the total surface area excluding the floor.
   d. If 1 gallon of paint covers 250 sq. ft. and costs $17.00, what is the total cost of material needed to paint the ceilings and walls of the agency?
   e. If labor and profit are figured as $0.55 a square foot, what estimate should Mr. Ramirez quote for the job?

2. The floor plan of a house is drawn below.

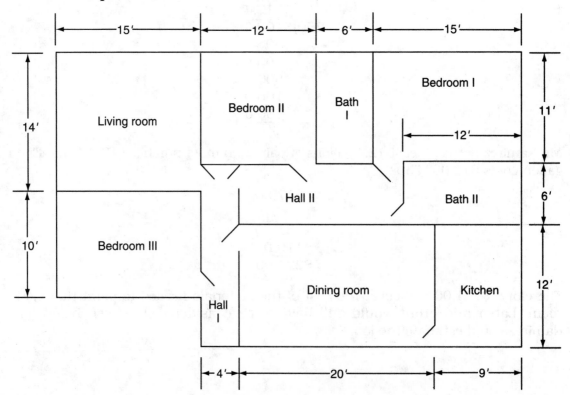

  a. Find the area of the floor surface of each room.
  b. Find the total floor surface.
  c. Find the total ceiling surface.
  d. If the ceiling height throughout the house is 8′, find the surface area of all the walls. (Include the doorways.)
  e. Find the total surface area of the ceilings and the walls.
  f. If 1 gallon of paint covers 200 sq. ft. and costs $12.49, then:
    (1) how many gallons of paint are needed to paint the ceiling and the walls?
    (2) how much money will the paint cost?

3. The plan of the exterior of a house is drawn below.

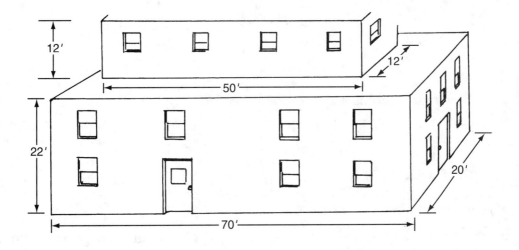

  a. Find the total surface area of the bottom portion of the house, including the windows and doorways. (*Hint:* There are two other sides of the house that cannot be seen.)
  b. Find the area of the top part of the house.
  c. Find the total area of the exterior surface of the house, excluding the roof.
  d. If the paint needed to cover the exterior of the house costs $13.95 a gallon and covers only 150 sq. ft. per gallon, then:
    (1) how many gallons will be required for one coat of paint? for two coats?
    (2) what will be the cost for one coat? for two coats?

FOR WALLPAPERING

Wallpaper can be purchased in single rolls, double rolls, and sometimes triple rolls. A single roll of wallpaper covers approximately 28 sq. ft. How many square feet does a double roll cover? A triple roll?

EXAMPLE: Mr. Ramirez must wallpaper the office space of the Wheels-and-Deals Automobile Salesroom, which measures 10′ by 15′ with a 10′ ceiling height. How many single rolls of wallpaper are needed?

SOLUTION: The dimensions of one pair of walls are 10′ × 10′. The area of one wall is 100 sq. ft., and the area for this pair of walls is 200 sq. ft.

The dimensions of the other pair of walls are 15′ × 10′ or 150 sq. ft. For this pair of walls the area is 300 sq. ft.

The total surface area of the four walls is 200 + 300 = 500 sq. ft.

If one roll of wallpaper covers 28 sq. ft., then

$$\begin{array}{r} 17 \text{ R}24 \\ 28\overline{)500} \\ 28\text{x} \\ \hline 220 \\ 196 \\ \hline 24 \end{array}$$ or 18 single rolls are needed for the job.

How many double rolls would be needed? Triple rolls?

PRACTICE EXERCISE 3

1. A room has dimensions of 12′ × 14′ × 8′.

    a. How many single rolls of wallpaper will Mr. Ramirez need to wallpaper the room? Double rolls?

    b. If the price of wallpaper is $5.25 per single roll, what will be the cost?

2. The washroom is to be wallpapered and has total dimensions of 17′6″ × 11′3″ × 8′.

    a. How many single rolls of wallpaper will Mr. Ramirez need to wallpaper the washroom? Double rolls?

    b. If wallpaper costs $4.98 a single roll, what will be the cost?

3. If a double roll of wallpaper covers 56 sq. ft. and costs $9.75, find the cost of wallpapering two walls each 30′ × 10′.

4. What will it cost to cover 1500 sq. ft. of surface area with wallpaper if one roll costs $6.25 and covers 25 sq. ft.?

5. If a roll of wallpaper costs $7.49 and covers 30 sq. ft., find the cost of wallpapering a ceiling that is 14′ × 15′.

6. The dimensions of a room are 14′ × 20′ × 8′. If a roll of wallpaper covers 32 sq. ft. and the total cost of wallpapering the 4 walls is $99.62, what is the cost of a single roll of wallpaper?

WORKING ON THE OUTSIDE

The landlord of the building housing the Wheels-and-Deals Automobile Sales Company agrees to renovate the exterior of the building, and he also requests estimates from a number of painters and contractors. When Marty, a contractor, arrives, he sees that the outside walls of the building are quite high. He must determine their height if he is to give a reliable estimate. If his estimate is too low, he may lose money on the job; if the estimate is too high, he may not be awarded the job. He needs a ladder to measure the height of the wall, but since he didn't bring a ladder with him, he proceeds to find the height *indirectly.* This is what he does.

He places a mirror on the street between himself and the building wall:

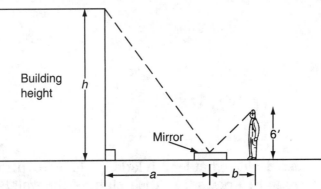

He looks into the mirror until he sights the reflection of the top of the wall. The dotted lines in the diagram indicate the reflection and his line of sight. He marks the spot on the street where he is standing and performs two measurements, the distances a and b. Since he knows his own height, he is able to calculate the height of the wall without directly measuring it.

He uses "similar triangles" since the two "right triangles" formed are similar to each other. Similar figures have the same shape but not the same size. When triangles are similar to each other, their corresponding sides form equal ratios to each other. A *ratio* is the quotient of two numbers in the same unit of measurement. The ratio $\frac{6}{3}$ may also be written as $6 : 3$, which is read as "6 is to 3." The numbers 6 and 3 are called the *terms* of the ratio. In the case we are considering, all the dimensions are measured in feet. When two ratios are equal to each other, the four terms form a proportion. Since the ratio $\frac{2}{4}$ is equal to the ratio $\frac{1}{2}$, we may write $\frac{2}{4} = \frac{1}{2}$. These equal ratios or this equation is called a *proportion.* The 4 and the 1 are called the *extremes,* and the 2 and the 2 are the *means.*

In the diagram, the two corresponding sides are Marty's height and the height of the wall. The ratio of the height of the wall to Marty's height is:

$$\frac{\text{Height of wall}}{\text{Marty's height}} = \frac{h}{6 \text{ ft.}}$$

Marty finds the length a is 9 feet and b is 3 feet.
The ratio of the measured distances a and b is:

$$\frac{\text{Measured distance } a}{\text{Measured distance } b} = \frac{9 \text{ ft.}}{3 \text{ ft.}}$$

These two ratios form a proportion because corresponding sides of similar triangles form a proportion. Thus:

$$\frac{\text{Height of the wall}}{\text{Marty's height}} = \frac{\text{measured distance } a}{\text{measured distance } b}$$

$$\frac{h}{6 \text{ ft.}} = \frac{9 \text{ ft.}}{3 \text{ ft.}}$$

Since the product of the means is equal to the product of the extremes, we obtain:

$$3 \cdot h = 6 \cdot 9$$
$$3h = 54$$

Dividing both sides of the equation by 3 gives

$$\frac{3h}{3} = \frac{54}{3}$$

$$h = 18 \text{ ft.}$$

Marty must also decide what length ladder will be required to reach the top of the wall. Even though he knows that the height of the wall is 18 ft., the ladder cannot be the same length because the ladder must lean against the building at a safe angle. Marty figures that the ladder must reach a height of 18 ft. and for a *safe angle* must be placed about 8 ft. from the foot of the wall. This is drawn in the diagram below.

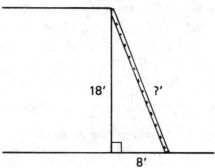

What size ladder is required?

To give a good estimate for a job like this one, you must know some mathematics. Marty knows his mathematics well and from a few calculations decides that he will need a 20-ft. ladder for this job. How did he determine this?

In ancient Greece there was a mathematician by the name of Pythagoras. He found that in any right triangle the *area of the square on the hypotenuse* (the name of the side opposite the right angle) *is equal to the sum of the areas of the squares on the legs.* This statement can be written as a formula:

$$(\text{Hypotenuse})^2 = (\text{leg})^2 + (\text{leg})^2$$

EXAMPLE: What is the length of the leg of a right triangle whose legs are 6 and 8?

SOLUTION: Using the formula

$$(\text{Hypotenuse})^2 = (\text{leg})^2 + (\text{leg})^2$$

gives

$$h^2 = 6^2 + 8^2$$
$$h^2 = 36 + 64$$
$$h^2 = 100$$

Since

$$h \cdot h = 100$$

therefore,

$$h = 10$$

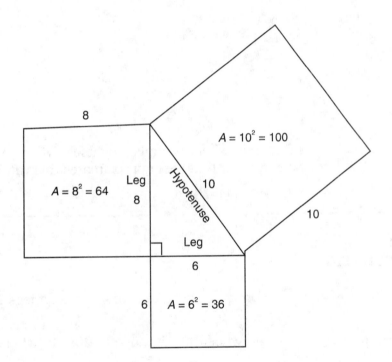

As you see, this right triangle, whose legs are 6 and 8, has a hypotenuse equal to 10.

The length of the ladder can be found by using the same formula:

$$(\text{Hypotenuse})^2 = 18^2 + 8^2$$
$$h^2 = 324 + 64$$
$$h^2 = 388$$
$$h \cdot h = 388$$

Since there is no number that, when multiplied by itself, will result in 388, we select the whole number that best approximates this value:

$$h = 20 \text{ since } 20 \cdot 20 = 400$$

EXAMPLE: What is the length of the leg of a right triangle if the hypotenuse is 24 and the other leg is 18?

SOLUTION:

$$(\text{Hypotenuse})^2 = (\text{leg})^2 + (\text{leg})^2$$
$$24^2 = l^2 + 18^2$$
$$576 = l^2 + 324$$

Subtracting 324 from both sides of the equation gives

$$\begin{array}{r} 576 = l^2 + 324 \\ -\ 324 = \quad -324 \\ \hline 252 = l^2 \end{array}$$

or

$$l^2 = 252$$
$$l \cdot l = 252$$

Since there is no number that, when multiplied by itself, will result in 252, we select the whole number that best approximates this value:

$$l = 16 \text{ since } 16 \cdot 16 = 256$$

PRACTICE EXERCISE 4

1. Find the length of the hypotenuse of a right triangle whose legs measure 5 and 12.

2. Find the leg of a right triangle whose hypotenuse is 20 and whose other leg is 16.

3. If two right triangles are similar to each other, find the height of the smaller triangle if the height of the larger triangle is 14 and its corresponding legs are 3 and 7.

4. In the diagram, $\triangle ABC$ and $\triangle DEF$ are similar to each other. Find length a.

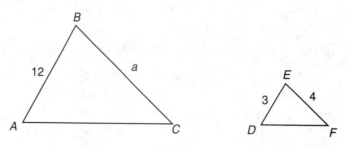

5. A tree casts a shadow 28 ft. long at the same time that a 3-ft. vertical stick casts a 4-ft. shadow. How tall is the tree? (*Hint:* The two triangles formed are similar to each other.)

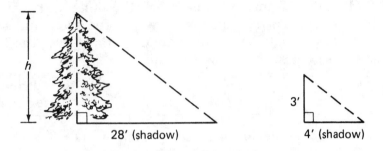

28' (shadow)   4' (shadow)

6. **a.** If the legs of a right triangle are represented by a and b and the hypotenuse by c, find c if $a = 15$ and $b = 20$.
   b Find a, to the nearest whole number, if $b = 6$ and $c = 9$.
   c. Find b if $c = 25$ and $a = 24$.
   d. A 45-ft. flagpole has a wire attached to its top and to a point in the ground 10 ft. from the foot of the flagpole. How long, to the nearest foot, is the wire?

7. What is the ratio of the measurement of side $\overline{AB}$ to side $\overline{BD}$ in triangle *ABD* below?

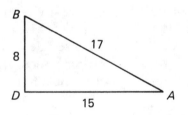

8. What is the ratio of the measurement of the width to the length in rectangle *RSTV* below?

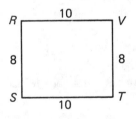

9. If there are 4 boys who wear eyeglasses and 12 boys who do not, what is the ratio of eyeglass wearers to nonwearers?

10. The owners of some lawnmowers are required to put oil in the gas tank when they add gasoline to the lawnmower. If they put in 1 quart of oil to 2 gallons of gas, what is the ratio of the oil to the gas? (*Hint:* 1 gallon = 4 quarts.)

BUYING OFFICE EQUIPMENT

Florence, the office manager, must stock the office with new equipment: desks, chairs, file cabinets, copying machines, word processors and so on.

She checks various catalogs from companies that specialize in office supplies and decides to purchase the following items.

Quantity	Item	Price for each	Cost
1	Workstation—Left-Return Desk with Main Desk and Left Return, Tropic Sand		
1	Workstation—Right-Return Desk with Main Desk and Right Return, Tropic Sand		
1	Single Pedestal Desk, 48 × 30, Black		
4	Double Pedestal Desks, 60 × 30, Tropic Sand		
2	Storage Credenzas, Black		
1	High Back Swivel/Tilt Chair, Gray, Model HON-4831-JJ16		
4	Armless Swivel/Tilt Chairs, Gray, Model HON-4833-JJ16		
8	Guest Arm Chairs, Burgundy, Model HON-4302-GG62		
3	Secretarial Posture Chairs, Forest Green, Model HON-4304-GG98		
4	2-drawer Fire King 25 Legal-Size Files, Black		
4	4-drawer Fire King 25 Legal-Size Files, Black		
2	4-drawer Fire King 31½ Letter-Size Files, Black		
1	Brother WP-1350DS Word Processor		
1	Brother WP-5750DS Word Processor		
1	Octagonal Wall Electric Clock, 12″ Diameter		
1	Granite Wall Electric Clock, 11″ Diameter		
1	Insulated Safe, Model 6530		
1	Canon Desktop Copier, Model PC330		
1	PC Mini-cartridge for copier		
1	Sharp Fax Machine, Model UX 184		
1 carton	Thermal paper for fax machine		
4	Casio Printing Calculators, Model CSO-FR5100L		
1	Casio Printing Calculator, Model CSO-FR520		
1	Sharp Desktop Display Calculator, Model SHR-QS-2122H		

Total cost

The items, specifications, colors, and prices are given in the reproductions of the catalog that follow.

HON

4830 SHELL SERIES

Shell-shaped seat conforms to body, provides excellent lumbar support. Attractive and durable nylon fabric. Gently sloping arms allow chair to be pulled closer to work surface. Black frame. Molded backs.

A **High Back Swivel/Tilt Chair**—Seat: $21\frac{3}{4}$w x $17\frac{3}{4}$d. Back: $21\frac{3}{4}$w x $21\frac{5}{8}$h. Overall height $36\frac{15}{16}$" to $39\frac{11}{16}$". Shpg. wt. 51 lbs.

HON-4831-JJ16	Gray
HON-4831-JJ21	Brown
HON-4831-JJ62	Burgundy

Each $429.00

B **Swivel/Tilt Chair**—Seat: $21\frac{3}{4}$w x $17\frac{3}{4}$d. Back: $21\frac{3}{4}$w x $17\frac{5}{16}$h. Overall height $32\frac{5}{8}$" to $35\frac{5}{8}$". Shpg. wt. 48 lbs.

HON-4832-JJ16	Gray
HON-4832-JJ21	Brown
HON-4832-JJ62	Burgundy

Each $337.00

C **Armless Swivel/Tilt Chair**—Seat: $21\frac{3}{4}$w x $17\frac{3}{4}$d. Back: $21\frac{3}{4}$w x $17\frac{5}{16}$h. Overall height $32\frac{5}{8}$" to $35\frac{3}{8}$". Shpg. wt. 44 lbs.

HON-4833-JJ16	Gray
HON-4833-JJ21	Brown
HON-4833-JJ62	Burgundy

Each $304.00

D **Swivel Arm Chair**—Seat: $21\frac{3}{4}$w x $17\frac{3}{4}$d. Back: $21\frac{3}{4}$w x $17\frac{5}{16}$h. Overall height $34\frac{1}{16}$" to $36\frac{11}{16}$". Shpg. wt. 46 lbs.

HON-4834-JJ16	Gray
HON-4834-JJ21	Brown
HON-4834-JJ62	Burgundy

Each $318.00

B

C

D

Acknowledgment: Reprinted from the 1994 Office Products Catalog, copyright 1993, by permission of United Stationers Supply Co.

Metal Desks

HON

METRO STANDARD SERIES

Woodgrain laminate tops resist marring, staining and scratching. Polished chrome legs and drawer pulls. Full-suspension file drawers. Center drawer lock secures all drawers.

A Storage Credenza—62 x 18. Two box and two file drawers plus dual sliding-door storage. Shpg. wt. 130 lbs.

HON-31881-ML	Putty/Medium Oak
HON-31881-WK	Tropic Sand/Walnut
HON-31881-WP	Black/Walnut

Each $622.00

Double Pedestal Desks—Three box drawers in left pedestal. One box and one file drawer in right pedestal. Central locking center drawer.

B 72 x 36—Recessed modesty panel. Shpg. wt. 208 lbs.

HON-32281-ML	Putty/Medium Oak
HON-32281-WK	Tropic Sand/Walnut
HON-32281-WP	Black/Walnut

Each $645.00

C 60 x 30—Flush modesty panel. Shpg. wt. 176 lbs.

HON-32261-ML	Putty/Medium Oak
HON-32261-WK	Tropic Sand/Walnut
HON-32261-WP	Black/Walnut

Each $517.00

55 x 24—Flush modesty panel. Shpg. wt. 138 lbs.

HON-32444-ML	Putty/Medium Oak
HON-32444-WK	Tropic Sand/Walnut
HON-32444-WP	Black/Walnut

Each $463.00

METRO SPECIAL SERIES

D Workstation—60 x 30 desk with 40 x 18½ keyboard height return. Complete shpg. wt. 219 lbs. Desk and return shipped separately—ORDER BOTH.

Desk	Return	Return Side
PUTTY/MEDIUM OAK		
HON-32271R-ML	HON-31831L-ML	Left
HON-32272L-ML	HON-31830R-ML	Right
TROPIC SAND/WALNUT		
HON-32271R-WK	HON-31831L-WK	Left
HON-32272L-WK	HON-31830R-WK	Right
BLACK/WALNUT		
HON-32271R-WP	HON-31831L-WP	Left
HON-32272L-WP	HON-31830R-WP	Right

Complete...$775.00

E Single Pedestal Desks—Locking pedestal with file drawer and one box drawer. Non-locking center drawer.

48 x 30—Shpg. wt. 115 lbs.

HON-31151-ML	Putty/Medium Oak
HON-31151-WK	Tropic Sand/Walnut
HON-31151-WP	Black/Walnut

Each $373.00

40 x 24—Shpg. wt. 82 lbs.

HON-31001-ML	Putty/Medium Oak
HON-31001-WK	Tropic Sand/Walnut
HON-31001-WP	Black/Walnut

Each $294.00

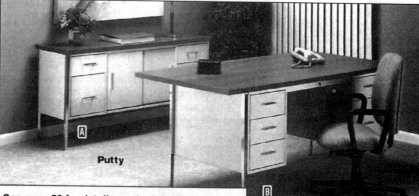

Putty

See page 50 for details on the HON chair shown.

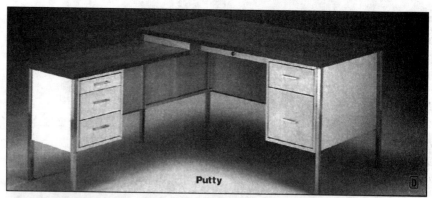

Black

Putty

Tropical Sand

Acknowledgment: Reprinted from the 1994 Office Products Catalog, copyright 1993, by permission of United Stationers Supply Co.

Metal Chairs

OVERTURE SERIES

Backrest provides full-lumbar support. Black oval tube frames. 100% nylon upholstery. Flame-retardant foam and upholstery. Padded arms.

A High Back Swivel/Tilt Chair—Seat height adjusts 17" to 20". Seat: 20¼w x 19d. Back: 20w x 21h. Overall height 36" to 39". Shpg. wt. 48 lbs.

SUP-16503521   Gray
SUP-16503535   Navy
SUP-16503536   Burgundy
Each $261.00

B Swivel/Tilt Chair—Seat height adjusts 17" to 20". Seat: 20¼w x 19d. Back: 20w x 17h. Overall height 32" to 35". Shpg. wt. 36 lbs.

SUP-16103521   Gray
SUP-16103535   Navy
SUP-16103536   Burgundy
Each $219.00

C Guest Chair—Sled Base. Padded arms. Seat: 20¼w x 19d. Back: 20w x 17h. Overall height 34". Shpg. wt. 27 lbs.

SUP-16323521   Gray
SUP-16323535   Navy
SUP-16323536   Burgundy
Each $202.00

HON

4300 SERIES

Molded armrests. Black monochromatic base and frame. Woven nylon upholstery. Thick comfortable cushions.

D Swivel/Tilt Chair—Seat: 20w x 18½d. Back: 20w x 13h. Overall height 30⅜" to 34⅝". Shpg. wt. 35 lbs.

HON-4301-GG12   Dark Gray
HON-4301-GG62   Burgundy
HON-4301-GG98   Forest Green
Each $218.00

E Secretarial Posture Chair—Seat: 17w x 16½d. Back: 14½w x 11½h. Overall height 28⅛" to 35¼". Shpg. wt. 29 lbs.

HON-4304-GG12   Dark Gray
HON-4304-GG62   Burgundy
HON-4304-GG98   Forest Green
Each $182.00

F Guest Arm Chair—Sled Base. Seat: 20w x 18½d. Back: 20w x 13h. Overall height 32½". Shpg. wt. 26 lbs.

HON-4302-GG12   Dark Gray
HON-4302-GG62   Burgundy
HON-4302-GG98   Forest Green
Each $156.00

Acknowledgment: Reprinted from the 1994 Office Products Catalog, copyright 1993, by permission of United Stationers Supply Co.

Converts your file into a data safe.

Insulated File Cabinets

FireKing 🔥

INSULATED FILE CABINETS

Insulation between drawers makes each drawer a separate, fireproof unit. Drill- and pick-resistant high security key lock (standard key lock on Turtle® files). Manufacturer's lifetime warranty on all parts; lifetime replacement of any file damaged beyond repair in a fire.

A FireKing 25® Vertical Files—25" deep—19¾" filing capacity. 20¾" wide.
Legal Size/2-Drawer—27⅞"h. Shpg. wt. 354 lbs.
FIR-22125CBL   Black
FIR-22125CPA   Parchment
Each $870.00
Legal Size/4-Drawer—52¹³⁄₁₆"h. Shpg. wt. 601 lbs.
FIR-42125CBL   Black
FIR-42125CPA   Parchment
Each $1330.00

B FireKing 31½" Deep Vertical Files—26" filing capacity.
Letter Size/4-Drawer—17¹¹⁄₁₆W x 52¹³⁄₁₆h. Shpg. wt. 620 lbs.
FIR-418CBL   Black
FIR-418CPA   Parchment
Each $1595.00
Legal Size/2-Drawer—20¾w x 27⅞h. Shpg. wt. 401 lbs.
FIR-221CBL   Black
FIR-221CPA   Parchment
Each $1050.00
Legal Size/4-Drawer—20¾w x 52¹³⁄₁₆h. Shpg. wt. 691 lbs.
FIR-421CBL   Black
FIR-421CPA   Parchment
Each $1645.00

C Compact Turtle® Files—File letter size documents, front-to-back; legal size, side-to-side. Only 22⅛d x 17¹¹⁄₁₆w. Parchment finish.
2-Drawer—27⅞"h. Shpg. wt. 243 lbs.
FIR-21822CPA...................Each $595.00
4-Drawer—52¹³⁄₁₆"h. Shpg. wt. 405 lbs.
FIR-41822CPA................Each $1060.00

D Computer Media Transformer—Converts insulated files into data safes. Keeps interior temperature below the critical 125 F diskette temperature limit for at least 1 hour. Holds 80 3½" or 40 5¼" diskettes. Key locking. 10" high.
For Letter Size and Turtle Files—10¼w x 11⅞d. Wt. 25 lbs.
FIR-TF125TCM ....Each $295.00
For Legal Size Files—16⅜w x 13⅛d. Wt. 35 lbs.
FIR-TF125HCM.................Each $395.00

E Lateral Files—Hang letter or legal size folders. 22⅛d x 37²⁷⁄₆₄w. Parchment.
2-Drawer—27⅞"h. Shpg. wt. 541 lbs.
FIR-238CPA...................Each $1670.00
4-Drawer—52¹³⁄₁₆"h. Shpg. wt. 801 lbs.
FIR-438CPA....................Each $2420.00

Look at these tested facts:
Paper contents withstood heat to 1700°F for 1 hour...even when placed in the midst of a fiery explosion. Files survived drops from a height of 30 feet. UL Class "C" rated.

Acknowledgment: Reprinted from the 1994 Office Products Catalog, copyright 1993, by permission of United Stationers Supply Co.

Safes & Vaults

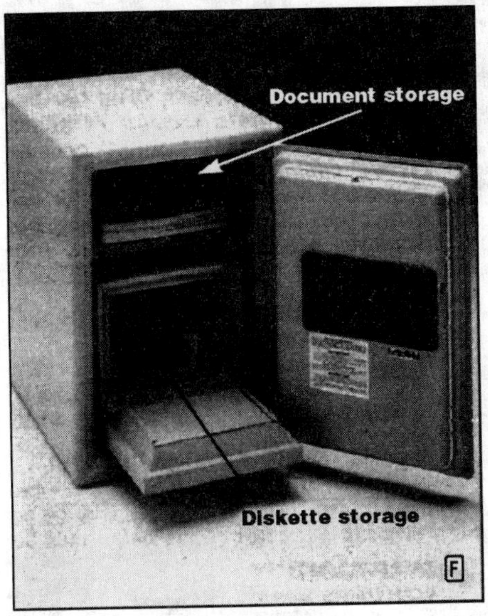

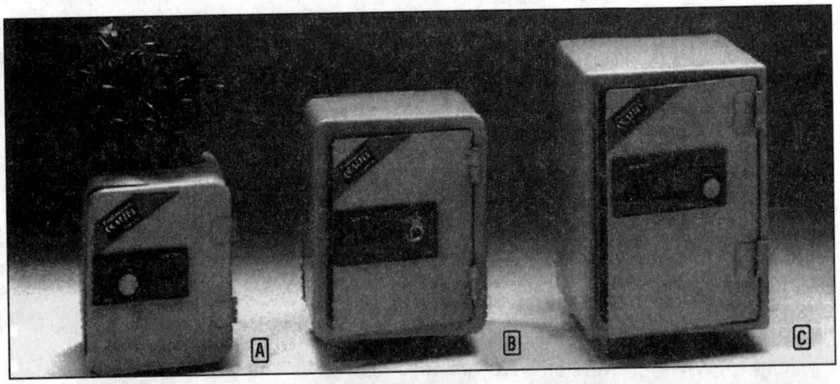

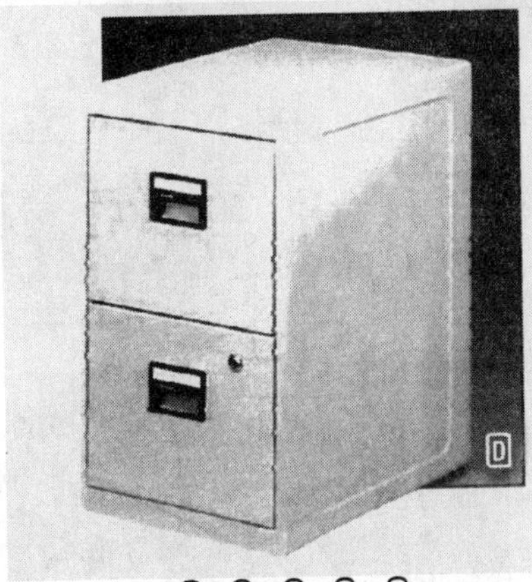

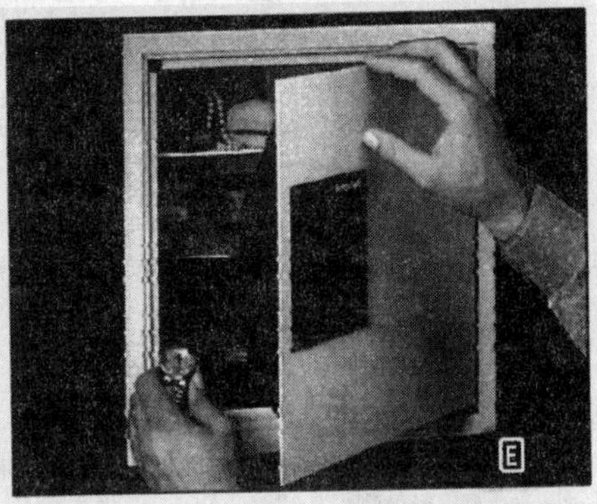

SENTRY® FIRE-SAFE® SAFES

A Model 6250—Less than 18" high, for closet or other small areas. Locking inner drawer. Bolt down capability. Shpg. wt. 81 lbs.
SEN-6250■ ....Each $212.00

B Model 6310—Only 15" deep, yet holds letter size papers. Removable shelf. Bolt down capability. Shpg. wt. 119 lbs.
SEN-6310■ ....Each $294.00

Model 6330—Same as SEN-6310, except 18½" deep. Locking valuables drawer. Shpg. wt. 137 lbs.
SEN-6330■ ....Each $318.00

Model 6380—Same as SEN-6310, except 24" deep. Locking interior drawer. Shpg. wt. 168 lbs.
SEN-6380■ ....Each $407.00

C Model 6530—Large storage area for valuable records. Four live-locking bolts. One adjustable shelf. Bolt down capability. Shpg. wt. 213 lbs.
SEN-6530■ ....Each $642.00

DIMENSIONS

Model No.	Outside H x W x D	Inside H x W x D
SEN-6250	17⅛ x 14¾ x 17¼	13¼ x 10⅞ x 12⅛
SEN-6310	22 x 17¼ x 15	15 3/16 x 12⅛ x 10
SEN-6330	22 x 17¼ x 18½	15 3/16 x 12⅛ x 13½
SEN-6380	22 x 17¼ x 24	15 3/16 x 12⅛ x 19
SEN-6530	26 x 18⅛ x 25½	20 x 14¼ x 18

FEATURES

Model	SEN-6250	SEN-6310	SEN-6330	SEN-6380	SEN-6530
Dove Gray Finish	•	•	•	•	•
Lifetime after-fire replacement	•	•	•	•	•
5-year limited warranty	•	•	•	•	•
Key lock on inner drawer	•		•	•	
UL fire-tested one hour at 1700° F	•	•	•	•	•
Three-number changeable combination lock	•	•	•	•	•
Four live-locking bolts	•				•
Bolt-down system for permanent installation	•	•	•	•	•
Shrouded dial	•	•	•	•	•
Shelf		•			•
Punch-resistant dial					•

FIRE-SAFE® OFFICE FILE

Protects valuable files. UL Classified (up to 1700°F for 1 hr.). Dual latching system keeps file closed in event of fire. Key lock.

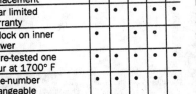

D Model 6000—Two-drawer file accommodates letter, legal, and A4 size files. Exterior 17¼w x 23¾d x 28h. Each drawer 12w x 18d x 9½h. Sabre beige color. Shpg. wt. 180 lbs.
SEN-6000■ ....Each $499.00

KEEP-SAFE® WALL SAFE

Mounts flush to wall—conceal behind a picture. Store valuables without using valuable floor space. Fits any wall with studs on 16" centers.

E Model 7100—Expandable depth allows full use of space inside stud walls. Easy installation with common tools. Double plate steel door with 2 live-locking bolts and 16 ga. steel cabinet, 3-number recessed combination lock. Outside: 15⅞h x 15⅞w x 3¾ to 5 7/16d. Inside: 13⅝h x 13⅝w x 3 7/16 to 4¾d. Shpg. wt. 28.5 lbs.
SEN-7100......................Each $169.99

INSULATED MEDIA SAFE

Fortified 5½" thick walls—twice as thick as most safes. Exterior door is dead-bolted into steel frame—removal of hinge pins won't budge it. Manufacturer's 5-year limited warranty and lifetime after-fire replacement guarantee. Three-number combination lock. Shrouded dial for security. For diskettes, microfiche, photographs, x-rays. UL tested to 1700° F for one hour.

F Model 6760—Dove gray. Holds up to 200 3½", 140 5¼", or 60 8" diskettes. Document area 1537 cu. in. cap. Inside: diskette area, 6½h x 9⅝w x 12d; document area, 6⅞h x 14w x 16d. Overall: 26h x 18⅛w x 25½d. Shpg. wt. 249 lbs.
SEN-6760■ ....Each $1070.00

Acknowledgment: Reprinted from the 1994 Office Products Catalog, copyright 1993, by permission of United Stationers Supply Co.

Word Processors

brother®

PORTABLE WORD PROCESSORS

With PC↔WP conversion software that allows you to convert and exchange files between your Brother Word Processor and your PC. Supports WordPerfect®, Wordstar, pfs: Write™, Microsoft Word, Display Writer. Supports LOTUS® 1-2-3® WK 1 spreadsheet files. Transfer ASCII files to and from PCs.

A Complete Range of Options

All models on this page include:

- **GrammarCheck:** Word Spell 70,000-word Dictionary catches errors, suggests alternatives. Also Punctuation Alert, Redundancy Check, and Word Count.
- **3.5" 720KB Floppy Disk:** MS DOS file compatibility. Disk Copy copies text from one disk to another.
- **Spreadsheet Software:** Balance your checkbook, maintain a telephone/address directory, perform sales analysis, monthly budgeting and financial calculations, or create spreadsheets.
- **Quick Correction System:** Remove a single word or complete line. Automatic relocation after correction.
- **Advanced Word Processing:** Type an abbreviated phrase and the entire phrase will print out. Plus double column printing, block move/copy/delete and search and replace.
- **Also:** Triple pitch typing (10, 12, 15).

A WP-1350DS Writer Series
Portable Daisywheel Word Processor

Includes Random House Electronic Encyclopedia on insertable disk. A comprehensive, easily accessible and usable reference tool for students, writers. Subjects include *Geography, History, Religion, Philosophy, Mythology, Social Science, the Arts, Science, Sports, Law, Government, Leisure.*

- 7 lines by 80 character LCD display
- 13 cps typing speed
- Help key
- Interchangeable daisywheels with various type styles available
- Built-in tutorial
- Page layout view permits you to see how text for three pages will lay out before printing
- Random House® Electronic Encyclopedia on disk (incl.)
- Auto Pagination
- Weight 14.5 lbs.
- Built-in carrying handle and lid cover

BRT-WP-1350DS ....Each $549.95

B WP-5750DS
Write better with 45,000-word Thesaurus

- Easy-to-read 80 character by 20 line 14" Super Flat amber CRT display
- 198 pre-formatted business letters for customizing your correspondence
- 13 cps typing speed
- Dual Screen capability
- Built-in tutorial
- TETRIS arcade game for hours of fun
- Auto Pagination
- Weight: WP 13.2 lbs, CRT 18.1 lbs.

BRT-WP-5750DS ....Each $699.95

C WP-2600Q "Whisper Print"
Daisywheel Word Processor

- 5 x 9 CRT with contrast adjustment
- 15 cps typing speed
- Ultra quiet daisywheel print system. You can talk on phone while printing
- TETRIS arcade game for hours of fun
- Weight 25.6 lbs.
- Built-in carrying handle

BRT-WP-2600Q ....Each $699.95

Acknowledgment: Reprinted from the 1994 Office Products Catalog, copyright 1993, by permission of United Stationers Supply Co.

Clocks

SQ SEIKO QUARTZ CLOCKS

Quartz movements use one AA battery, included (except **A**). One-year manufacturer's warranty. In business over 100 years.

WOOD WALL CLOCKS

A **Pendulum/Chime**—Solid oak "schoolhouse" wall clock features dual Westminster/Whittington quarter hour chime, volume control and nighttime silencer. Uses two C batteries (incl.). Dial face 9" dia., case 13w x 4d x 21h.
SKO-QXH102BC ....Each $175.00

B **Octagonal Wall Clock**—Light stained solid oak frame, white dial, goldtone second hand and bezel. 12 dia. x 1d.
SKO-QXA102BC ....Each $59.50

C **Round**—Solid oak case with goldtone accents. Easy-to-read black Arabic numerals, second hand. Glass lens. 8½" dia. dial, case 11¾ dia. x 1⅛d.
SKO-QXA101DC ....Each $59.50

CONTEMPORARY WALL CLOCKS

D **Granite**—9½" faux granite dial. 11 dia. x 1½d black and goldtone case.
SKO-QXA042KR ....Each $39.50

E **Round**—9" black dial. 10¼ dia. x 1½d case.
SKO-QAGO19K   Black Case
SKO-QAGO19W   White Case
Each $19.95

F **Square**—8½h x 9w dial. 9¾h x 10⅛w x 1½d black case. Gray dial.
SKO-QAGO18H ....Each $19.95

G **Square**—Black dial with raised gold arabic numerals, gold accents, second hand. 8¼" square dial. 9½w x 9½h x 1½d black case.
SKO-QXA030KR ....Each $29.50

H **Square**—8½h x 9w dial. 9¹³⁄₁₆h x 10¼w x 1⁹⁄₁₆d goldtone case with black easy-to-read arabic numerals.
SKO-QXA016GR ....Each $29.50

TRADITIONAL OFFICE CLOCKS

Large, easy-to-read dials.

I **Calendar**—Day/date window, white 11" dial. 12⅕ dia. x 1⅘d case.
SKO-QXL001GR   Goldtone
SKO-QXL001SR   Silvertone
Each $55.00

J **Manager**—11½" white dial with red sweep second hand. 14 dia. x 1d case.
SKO-QXA105BC   Brown
SKO-QXA105KC   Black
Each $39.95

K **24-Hour Manager**—Same as **J** above, but with red 24-hour numerals. Black case.
SKO-QXA111KC ....Each $39.50

Acknowledgment: Reprinted from the 1994 Office Products Catalog, copyright 1993, by permission of United Stationers Supply Co.

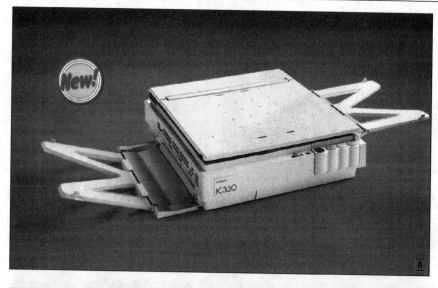

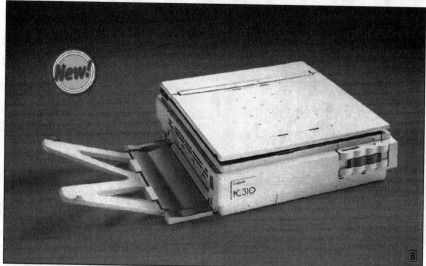

Copiers

Canon® COPIERS

READY TO WORK WHEN YOU ARE!

No Warm-Up—RAPID Fusing System™ for instant start-up.

Space Saving—Compact design—fits practically anywhere.

Versatile—Makes copies of sheets, books and three-dimensional objects onto stationery, postcards, transparencies, labels and more.

Saves Power—Shuts off automatically when not in use for 5 minutes.

Virtually Maintenance Free—Easy-to-change PC Mini-Cartridge includes toner, drum and developer all-in-one for service free performance.

🅰 PC 330 Letter Stack Feed Desktop/Portable Copier

- Automatically feeds up to 50 sheets
- Six copies per minute
- Multiple copy runs, from one to nine copies, or F mode for continuous copying of all 50 sheets
- Accepts paper sizes from business card (2 x 3½) to letter (8½ x 11)
- Automatic and manual exposure control
- Moving copyboard
- Pop-up handle for transport

Delivers multiple copies instantly and automatically. PC Mini-Cartridge sold separately (see box). 14⅛ x 15⅝ x 4¼. Wt. 16.7 lbs.
CAN-PC330......................Each $895.00

PC 330L Letter/Legal Stack Feed Desktop Copier

- Same great features as 🅰 plus copies 8½ x 14 legal size sheets.
- Multiple copies up to legal size

Does not include pop-up handle. 16⅜ x 15⅝ x 4¼. Wt. 17.8 lbs.
CAN-PC330L..................Each $1095.00

🅱 PC 310 Manual Feed Desktop/Portable Copier

- Six copies per minute
- Manual paper feed
- Accepts paper sizes from business card (2 x 3½) to letter (8½ x 11)
- Manual exposure control
- Moving copyboard
- Pop-up handle for transport

Big performance in a compact design. Ideal for the small business or home office. PC Mini-Cartridge sold separately (see box). 14⅛ x 15⅝ x 4¼. Wt. 16.5 lbs.
CAN-PC310......................Each $795.00

PC Mini-Cartridge—All-in-one toner, drum and development unit required for all Canon copiers on this page. Yields approximately 1600 copies. Black.
CAN-E16-BLK ....Each $94.95

Manufacturer's Warranty—Canon copiers are UL listed, include a 90-day manufacturer warranty on machine parts and labor and a 15-day manufacturer warranty on cartridge. All are backed by Canon's nationwide service.

Acknowledgment: Reprinted from the 1994 Office Products Catalog, copyright 1993, by permission of United Stationers Supply Co.

Fax Machines

SHARP.

A ECONOMICAL FAX UX-114
Ideal for home office or small business.

FAX FEATURES
• G3 compatibility • Transmits an 8½ x 11 page in 15 seconds • 10-page automatic document feeder • 16-level halftone control, automatic contrast control • Thermal copier function • Answering machine hook-up • Fax/Telephone automatic changeover • Paper saver • Transaction reports

TELEPHONE FEATURES
• One-touch 15-number autodialer plus 35-number speed dialer • LCD display • Telephone number listing • Redial, On-Hook Dialing, Hold

Plug into any single-line phone jack and regular electric outlet—eliminates the need for separate fax and phone lines. Integrate your answering machine into the same line to expand your communications options.
SHR-UX114 ....Each $695.00

B FAX UX-184 WITH AUTOMATIC PAPER CUTTER
Upgrade of the UX-114 with the convenience of a paper cutter.
SHR-UX184 ....Each $695.00

> **Thermal Paper for UX114 & UX184**—8½" x 98 ft. long. Also for Sharp fax machines 101, 103, 171, 172 and 195. Six rolls per carton.
> SHR-FO-20PR6W.........Carton $39.99

C PLAIN-PAPER FAX UX-1500
Incoming faxes look as good as the original!

FAX FEATURES
• G3 compatibility • Transmits an 8½ x 11 page in 15 seconds • 10-page automatic document feeder • 300 DPI resolution with smoothing and 64-level halftone control • 100-sheet paper tray • Letter and legal size paper handling • Plain paper copier function—up to 30 copies • TAD interface for answering machine hook-up • Fax/Telephone auto-changeover • Out-of-Paper Reception, Memory Transmission • Networking capabilities—Polling, Broadcasting (max. 20 stations) • 28-page memory

TELEPHONE FEATURES
• One-touch 15-number autodialer plus 35-number speed dialer • LCD display • Chain-dialing

Ink jet printed faxes on plain paper can be marked, filed or retransmitted without the thermal paper problems of discoloration, fading or jamming. Powerful 512KB memory provides an extended range of communication capabilities.
SHR-UX1500 ....Each $1495.00

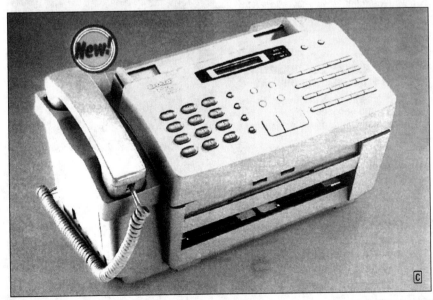

Acknowledgment: Reprinted from the 1994 Office Products Catalog, copyright 1993, by permission of United Stationers Supply Co.

Calculators

Read printouts at a glance without squinting.

CASIO

DESKTOP PRINTING CALCULATORS

Smart-looking "radius" case. Subtotal, total and grand total computations. Large fluorescent numbers punctuated by commas for easy reading. AC only.

A 12-Digit Readout, 2-Color Extra-Large Printing
- Prints extra big type in red and black
- 1.8 lines per second print speed
- Time calculations

Full decimal selection, four-key memory, right shift key, item counter. 7⅞w x 9¾d x 2⅞h. Wt. 2 lbs.
CSO-FR2650......................Each $84.95

B 12-Digit Readout, 2-Color Printing
- Prints in red and black
- 1.8 lines per second print speed

Decimal selection, four-key memory, item counter. Right shift key lets you erase entries. Percent key. 6⅞w x 8¾d x 2⅜h. Wt. 1.5 lbs.
CSO-FR520......................Each $69.95

C 10-Digit Readout, Extra-Fast 1-Color Printing
- Accelerated 2.6 lines per second print speed
- Enlarged digit display

Full decimal selection, independent 4-key memory. Right shift key for error correction, item counter. 8¼w x 10¼d x 3h. Wt. 2.2 lbs.
CSO-FR5100L....................Each $84.95

SHARP

DESKTOP PRINTING CALCULATORS WITH ONE-TOUCH GRAND TOTAL KEY

- Easy-to-read blue fluorescent display with comma punctuation • Well-spaced sculpted keys • 4-key memory • Item counter • Floating and fixed decimal
- Subtotal/total/grand total functions
- Mark-up/profit margin key • Oversize add/equals key

D For General Office Use
- Fast 3 lines per second printing
- Grand total key, right shift key
- Big 12mm digit display

High speed drum printing, black print. Floating point or fixed decimal, three-way rounding. Subtotal key also prints code number, date. 8⅝w x 10¾d x 2⅝h.
SHR-EL-2197GII 12-digit ..Each $109.99
SHR-EL-1197GII 10-digit ..Each 89.99

E For Light Office Use
- Prints 2 lines per second
- Grand total key, right shift key

Prints on standard size paper roll. Mark-up/profit margin key. Choice of floating point or fixed decimal, and add mode, two-way rounding. 7⅝w x 9¹⁄₁₆d x 2⅜h.
12-Digit, 2-Color
SHR-EL-2192G...................Each $74.99
10-Digit, 1-Color
SHR-EL-1192G...................Each 69.99

DESKTOP DISPLAY CALCULATOR 12-DIGIT FLUORESCENT DISPLAY

- Time calculation key calculates hours, minutes and seconds
- Low-profile keyboard with sculpted keys

F Four-key memory, 3-digit punctuation.
Round-up/off/down, adding, fixed/floating decimal, constant and sigma mode selectors. Multiple use, right shift and grand total keys. 8⁹⁄₃₂w x 8³¹⁄₃₂d x 2²⁹⁄₃₂h. Wt. 2.2 lbs.
SHR-QS-2122H ....Each $94.00

Acknowledgment: Reprinted from the 1994 Office Products Catalog, copyright 1993, by permission of United Stationers Supply Co.

PRACTICE EXERCISE 5

1. Find in the catalog each item that Florence ordered and its price, and enter the price in the third column of the table on pp. 56 and 57. Then compute the cost for the number ordered of each item, and write your answers in the last column.

2. a. Figure the total cost of all the office equipment ordered.
   b. If the sales tax is 8%, what is the total tax on the entire cost?
   c. What will be the total cost for all equipment plus the sales tax?
   d. By how much money does the cost of the office equipment exceed $25,000?

3. How much money would Florence have saved by purchasing a 40 × 24 single pedestal desk instead of the 48 × 30 single pedestal desk?

4. How much less would it have cost to order four Compact Turtle files (4 drawer) instead of the four 4-drawer Fire King 25 Legal-size files?

5. a. How much more would it cost to purchase a four-drawer lateral file, legal size, rather than a two-drawer lateral file, letter size?
   b. How many pounds would 8 Fire King 25 legal-size 4-drawer files weigh?
   c. What would it cost for delivery if the trucking company charged $0.15 a pound?

6. What would the savings be if Florence purchased 6 Manager wall electric clocks rather than 6 calendar clocks?

7. a. In a subsequent order, Florence purchased a fire-safe office file. What was the cost?
   b. If the sales tax is 8%, what was the sales tax on this new order?
   c. What was the total cost including the sales tax?

8. a. Extra paper tapes and ink rollers for the Sharp and Casio calculators were ordered to replenish the supply. Find the cost for each item:

5 boxes	Extra Paper Tapes at $8.99 per box
3	Extra Ink Rollers at $3.79 each

   b. Find the total cost for both items.
   c. Find the sales tax at 7%.
   d. Find the total cost of both items, including the tax.

BUYING OFFICE SUPPLIES

No office can function without supplies. Some of the items needed to run an office efficiently are shown in the catalog reproductions that follow.

Wastebaskets

Rubbermaid

RIGID SEAMLESS PLASTIC

E Form 1000®—Attractive seamless design. Made of recycled materials. 6½ gal. 10 sq. x 14h.

RUB-R2103-301	Ebony
RUB-R2103-317	Burgundy
RUB-R2103-332	Graphite Gray
Each	........................$19.99

F High-Gloss Rectangular—6½ gal. 16w x 8¾d x 13⅝h.

RUB-R2505-1	Black
RUB-R2505-21	Putty
Each	........................$12.99

G High-Gloss Round—Black 5 gal. 10⅞ dia. x 15⅜h.

RUB-R2504-1 ....Each $16.99

H Walnut Trim—Rectangular basket, black with walnut-grain corners. 6½ gal. 16½w x 8⅞d x 14h.

RUB-R2503-86 ....Each $16.99

SOFT MOLDED

I Medium Rectangular—7 gal. 14¼w x 10¼d x 15h.

RUB-R2956-0	Beige
RUB-R2956-1P	Black
RUB-R2956-23	Brown
RUB-R2956-82P	Gray
Each	........................$4.15

J Large Rectangular—10¼ gal. 15¼w x 11d x 20h.

RUB-R2957-0	Beige
RUB-R2957-1	Black
RUB-R2957-82	Gray
Each	........................$7.25

FIRE-RESISTANT

Plastic wastebaskets will not burn, melt, warp or contribute fuel to container contents. For hospital, industrial, dormitory or institutional use. HEW approved. UL listed. Polyliners not recommended.

K Rectangular—7 gal. 16⅛w x 8½d x 14h.

RUB-R2824-1	Black
RUB-R2824-34	Sand
Each	........................$30.20

L Round—4½ gal. 10½ dia. x 14h.

RUB-1537-1	Ebony
RUB-1537-17	Burgundy
RUB-1537-32	Graphite Gray
Each	........................$36.49

Universal office products

Tenex®

A Small—4½ gal. 11½w x 8d x 12h.

TEN-16001	Black
TEN-16003	Putty
TEN-16004	Gray
Each	........................$3.95

B Medium—7 gal. 14¼w x 10½d x 15h.

TEN-16021	Black
TEN-16023	Putty
TEN-16024	Gray
Each	........................$5.15

C Large—10¼ gal. 15½w x 11d x 20h.

TEN-16041	Black
TEN-16043	Putty
TEN-16044	Gray
Each	........................$8.15

FIRE-RESISTANT WASTEBASKET

D UL Listed and DHHR Approved—Fire-resistant plastic retains shape, contains fire, prevents spreading. 6¾ gal. 16w x 8d x 14h.

TEN-16101	Black
TEN-16102	Putty
Each	........................$32.95

M Plastic Wastebaskets—Polyethylene resists puncture, cold cracking, chemicals. Easy to clean. 14½w x 10½d x 15h. 7 gallon capacity.

UNV-29901	Black
UNV-29902	Putty
UNV-29903	Gray
Each	........................$5.05

Acknowledgment: Reprinted from the 1994 Office Products Catalog, copyright 1993, by permission of United Stationers Supply Co.

Staplers

STANLEY BOSTITCH

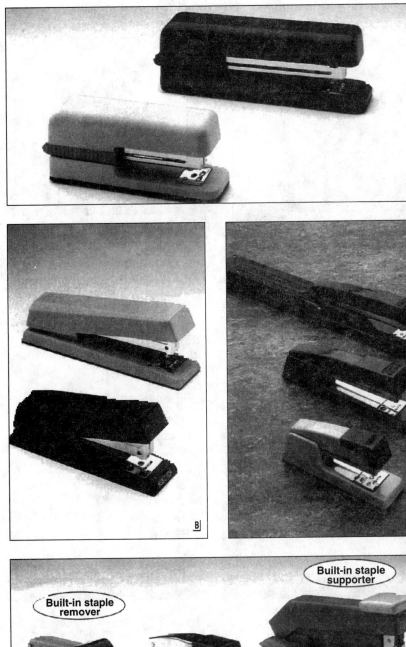

A **Executive Stapler**—Sleek, rounded Comfortwrap™ base pad for hand-held comfort. Staple reload indicator.

No.	Color	Each
FULL STRIP, LOADS 210 STAPLES		
BOS-B2000-BK	Black	$24.95
BOS-B2000-BY	Burgundy	24.95
BOS-B2000-PY	Putty	24.95
HALF STRIP, LOADS 105 STAPLES		
BOS-B2100-BK	Black	13.95
BOS-B2100-BY	Burgundy	13.95
BOS-B2100-PY	Putty	13.95

B **Anti-Jam**—A movable core and floating drive virtually eliminate jams. Use as plier, stapler, tacker or for pinning. Padded base.

No.	Color	Each
FULL STRIP, LOADS 210 STAPLES		
BOS-B660-BK	Black	$16.95
BOS-B660-BGE	Beige	16.95
HALF STRIP, LOADS 105 STAPLES		
BOS-B600-BK	Black	10.95
BOS-B600-BGE	Beige	10.95

C **Deluxe**—Staples, pins and tacks. Staple supply indicator, padded base.

No.	Color	Each
FULL STRIP, LOADS 210 STAPLES		
BOS-B440-BK	Black	$18.95
BOS-B440-BGE	Beige	18.95
BOS-B440-BY	Burgundy	18.95
HALF STRIP, LOADS 105 STAPLES		
BOS-B400-BK	Black	12.95
BOS-B400-BGE	Beige	12.95

D **All-Steel Long Reach**—Full 12" throat depth. Adjustable paper stop has English/metric rulers. Anti-jam floating core to prevent annoying jam-ups. Rubber base anchors to desk. Loads 210 standard staples.
BOS-B440LR   Black..........Each $39.95

E **Premium**—Stapler, tacker, plier, staple remover and letter opener in one unit. Half strip, loads 105 B8 staples (order from staple chart).

No.	Color	Each
BOS-B8RC	Black	$13.45
BOS-B8RCB	Beige	13.45
BOS-B8RCM	Burgundy	13.45

Same as above, without staple remover and letter opener.

No.	Color	Each
BOS-B8C	Black	$12.95
BOS-B8CM	Burgundy	12.95

F **Bright Chrome**—Can be used as a plier. Plastic cap and base, rubber bottom. Half strip, loads 105 staples.
BOS-B9-BK   Black.............Each $13.50

G **Heavy-Duty Front Loader**—Features staple supporter for precise penetrating power and repetitive stapling. Opens for tacking. Full strip, loads 210 standard staples.

No.	Color	Each
BOS-B5	Black	$59.95
BOS-B5J	Gray	59.95

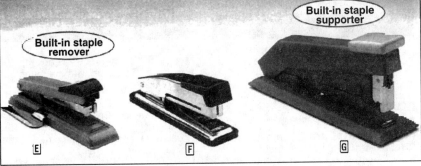

Built-in staple supporter

Built-in staple remover

BOSTITCH STAPLE CHART

Staples	Leg Length	Staples Per Strip	For Model	Price Per Box
STANDARD STAPLES, BOX OF 5000				
BOS-SBS19-1/4CP	¼"	210	All Staplers Using Standard Staples	$1.95
B8 STAPLES, BOX OF 1000				
BOS-SB8-10M	¼"	105	All B8 Prefixed Staplers	1.65
B8 STAPLES, BOX OF 5000				
BOS-STCRP-2115-1/4	¼"	210	All B8 Prefixed Staplers	2.75
BOS-STCR2115-3/8	⅜"	210	B8P Plier	4.95
P3 STAPLES, BOX OF 5000				
BOS-SP19-1/4	¼"	210	P3 Plier	4.15

Acknowledgment: Reprinted from the 1994 Office Products Catalog, copyright 1993, by permission of United Stationers Supply Co.

Desk Calendars

EverReady®

CALENDAR BASE & ORGANIZER
- Compartments for memo pads, paper clips, pencils

Deluxe organizer base for all calendar refills on this page.

Angled base shown with EverReady® two-color calendar and tabbed refill.

A Angled for easier reading and writing. Holds standard #17-style, 3½ x 6 loose-leaf calendar refill (not included). Shown with tabbed index (not incl.), sold below. Rubber feet protect desk surface. Base 8 x 10½.

KEI-J17-00   Black Plastic ....Each $14.40
KEI-J17-19   Putty Plastic .....Each   14.40

3½ x 6 CALENDAR REFILLS

Half-Hour Appointments, 7AM to 5PM— Two-page spread includes daily appointment schedule and current, past and future month calendars for reference and planning. Ample space for notes. Fits #17 bases, sold separately.

B **Two-Color Deluxe**—Two pages for each day, including Saturday and Sunday. Printed in red and black. Shown with angled base **A** sold above.
KEI-E017-50   Refill ....Each $4.25

C **One-Color**—Printed in black only. Two pages for each weekday; one page each for Saturday and Sunday. Shown with base sold on facing page.
KEI-E717-50   Refill ....Each $3.50

Recycled—Same as **C**, but printed on recycled paper.
KEI-E717R-50   Refill ....Each $3.85

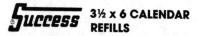

Success®   3½ x 6 CALENDAR REFILLS

D **Half-Hour Appointments, 7AM to 5PM**—Two-page spread: left page ruled for appointments, with past and future calendar month calendars; right page with current month calendar and memo space. Shown with #17 base sold separately.

Two-Color Deluxe—Two pages for each day, including Saturday and Sunday. Printed red and black. Perpetucal® and Calcumetric® converter. Boxed.
SUC-017-00   Refill ....Each $5.00

E **International Version**—Printed in English, French and Spanish. Includes Mexican and Canadian holidays. Two pages for weekdays, one each for Saturday and Sunday. Similar format as **D** above, printed in black.
SUC-707-09   Refill ....Each $3.85

Timepeace®
Calendars with a Difference

ILLUSTRATED DESK CALENDAR REFILLS
Original art and current month calendar. Right page ruled for hourly appointments from 8AM to 5PM, with past and future month calendars. Printed in color on quality recycled paper. Size 3½ x 6, shown with #17 base sold separately.

F **Sports Timepeace®**—Featuring the art of Gary Patterson with his humorous look at sports.
KEI-TM1117-50   Refill.........Each $9.90

G **Timepeace®**—Original verse and dramatic graphics. Meant to make a difference in your day! Includes facts from Women's Hall of Fame.
KEI-E1017-50   Refill ....Each $9.90

Acknowledgment: Reprinted from the 1994 Office Products Catalog, copyright 1993, by permission of United Stationers Supply Co.

Hanging File Folders

Pendaflex ®

NEW IMPROVED HANGING FOLDERS

New Improved Pendaflex hanging folders are 10x more durable than traditional folders because they are laminated at stress points. Pendaflex brand resists abrasion and outlasts all other folders on tear and fold strength. Offers easier set-up and faster tabbing with reinforced teardrop slots.

Improved! All Pendaflex hanging folders now...

Have coated rod tips for smoother sliding.

Laminated at stress points.

Won't fray or wear out.

10x stronger—tabs and rods don't tear out. Hold more weight.

Scored for expansion.

A **Standard Green**—Deep green exterior with lighter green interior helps eliminate between folder misfiles.

No.	Size	Tabs	Box of 25
ESS-4152-1/5	Letter	⅕ cut	$16.85
ESS-4152-1/3	Letter	⅓ cut	18.65
ESS-4152	Letter	None	14.45
ESS-4153-1/5	Legal	⅕ cut	19.75
ESS-4153-1/3	Legal	⅓ cut	21.69
ESS-4153	Legal	None	17.50
ESS-4158	14 x 18*	None	56.19

*X-ray size

Data Size—12 x 15¼ standard green made of stronger, wear-resistant stock, to withstand the weight and bulk of computer printouts. No index tabs.

No.	Capacity	Box of 25
ESS-4159	1"	$135.55

B **Colored**—Categorize subjects with color. With matching ⅕ cut index tabs.

Color	Letter Size	Legal Size
BRIGHT COLORS		
Blue	ESS-415215BLU	ESS-415315BLU
Green	ESS-415215BGR	ESS-415315BGR
Orange	ESS-415215ORA	ESS-415315ORA
Red	ESS-415215RED	ESS-415315RED
Yellow	ESS-415215YEL	ESS-415315YEL
Box/25	$18.20	$21.15
ASSORTED BRIGHTS—Five each of above colors		
Assorted	ESS-415215AST	ESS-415315AST
Box/25	$20.39	$23.69
PASTELS		
Violet	ESS-415215VIO	ESS-415315VIO
Pink	ESS-415215PIN	ESS-415315PIN
Gray	ESS-415215GRA	ESS-415315GRA
Aqua	ESS-415215AQU	ESS-415315AQU
Black	ESS-415215BLA	ESS-415315BLA
Box/25	$18.20	$21.15
ASSORTED PASTELS—Five each of above colors		
Assorted	ESS-415215AS2	—
Box/25	$20.39	

Lighter color interior helps prevent misfiling.

C **Interior Folders**—Useful for subdividing records within a hanging folder or carrying papers from file to desk. Cut to fit inside hanging folders without obscuring tabs. ⅓ cut index tabs. 100 per box, 500 per carton.

Color	Letter Size	Legal Size
BRIGHT COLORS		
Blue	ESS-421013BLU	ESS-435013BLU
Green	ESS-421013BGR	ESS-435013BGR
Orange	ESS-421013ORA	ESS-435013ORA
Red	ESS-421013RED	ESS-435013RED
Yellow	ESS-421013YEL	ESS-435013YEL
Box/100	$16.70	$23.35

C **Interior Folders, Cont.**

Color	Letter Size	Legal Size
PASTELS		
Violet	ESS-421013VIO	ESS-435013VIO
Pink	ESS-421013PIN	ESS-435013PIN
Gray	ESS-421013GRA	ESS-435013GRA
Aqua	ESS-421013AQU	ESS-435013AQU
Black	ESS-421013BLA	ESS-435013BLA
Box/100	$16.70	$23.35
MANILA	ESS-4210-1/3	ESS-4350-1/3
Box/100	$13.79	$19.60

Acknowledgment: Reprinted from the 1994 Office Products Catalog, copyright 1993, by permission of United Stationers Supply Co.

Tapes

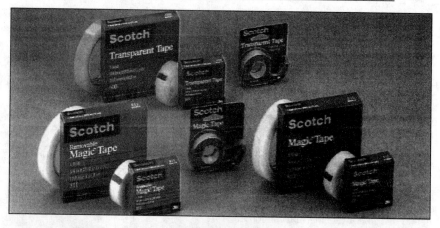

MAGIC™ TAPE

Permanent—Lasts as long as the paper it's used on!
Invisible—Disappears on the job—won't discolor or curl!
Write On It—With pen, pencil or typewriter! 12 rolls per box.

No.	Size	Roll
WITH 1" CORE		
MMM-810-12-12	½" x 1296"	$2.28
MMM-810-34-12	¾" x 1296"	3.01
WITH 3" CORE		
MMM-810-12-25	½" x 2592"	3.80
MMM-810-34-25	¾" x 2592"	5.24
MMM-810-1-25	1" x 2592"	6.76

Self-Dispensers—Refillable.

No.	Size	Roll
MMM-104A	½" x 450"	$1.25
MMM-119A	½" x 800"	1.84
MMM-105A	¾" x 300"	1.25
MMM-122A	¾" x 650"	2.16

REMOVABLE MAGIC TAPE

Just like Magic Tape, but you can remove it leaving no mark. For jobs where temporary attachment is necessary.

No.	Size	Roll
WITH 1" CORE		
MMM-811-12-12	½" x 1296"	$2.45
MMM-811-34-12	¾" x 1296"	3.25
WITH 3" CORE		
MMM-811-12-25	½" x 2592"	4.11
MMM-811-34-25	¾" x 2592"	5.65
MMM-811-1-25	1" x 2592"	7.27

TRANSPARENT TAPE—THE ORIGINAL GENERAL PURPOSE TAPE

Lower priced than Magic Tape, yet ideal for most office needs. Easy to dispense. Clear, long-aging, won't crack or curl.

No.	Width	Rolls/ Pack	Roll
WITH 1" CORE—1296" LENGTH			
MMM-600121296	½"	1	$1.65
MMM-600341296	¾"	1	2.26
WITH 3" CORE—2592" LENGTH			
MMM-600382592	⅜"	3	2.08
MMM-600122592	½"	1	2.84
MMM-600342592	¾"	1	4.15
MMM-60012592	1"	1	5.44

Self-Dispensers—Refillable.

No.	Size	Roll
MMM-144A	½" x 500"	$1.11
MMM-174A	½" x 1100"	2.02

PLASTIC TAPE DISPENSERS

1" Core Dispensers—For tapes up to ¾"w x 1296" long.

A Refillable Hand Dispenser

MMM-H126	For ½" tape	$0.72
MMM-H127	For ¾" tape	.72

B Space Saving Dispenser

MMM-C38-BK	Black	$5.49
MMM-C38-BY	Burgundy	5.49
MMM-C38-PY	Putty	5.49

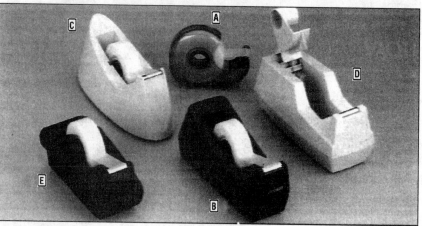

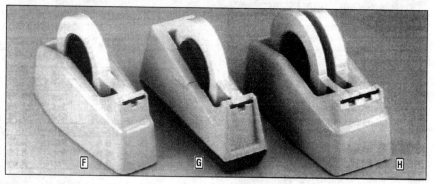

C Decor Dispenser

MMM-C15-BK	Black	$7.67
MMM-C15-PY	Putty	7.67
Extra Core		
MMM-C15CORE		$0.15

D Deluxe Dispenser with Attached Core

MMM-C40-BK	Black	$10.49
MMM-C40-PY	Putty	10.49
MMM-C40-GY	Gray	10.49

E Contemporary Dispenser with Attached Core

MMM-C-4210	Black	$12.95

3" Core Dispensers—For tapes up to 1" wide x 2592" long.

F Heavy-Duty Dispenser

MMM-C23	Beige	$35.98
Extra Core		
MMM-C23CORE		$2.39

G Dispenser with Attached Core

MMM-C25	Putty/Brown	$23.73

H Two Roll Dispenser—Two cores turn independently. Holds a single roll up to 2" x 2592" or two 1" rolls. Weighted body.

MMM-C22	Beige	$54.71
Extra Core		
MMM-C22CORE		$3.24

Acknowledgment: Reprinted from the 1994 Office Products Catalog, copyright 1993, by permission of United Stationers Supply Co.

Ballpoint Pens

PAPER⊗MATE.

A Flexgrip Ball Pen—Rubberized barrel with no-slip grip for added writing comfort. Chrome "arrowhead" tip, protective cap. Refillable.

No.	Ink	Point	Each
PAP-96101	Blue	.7mm (med.)	$1.09
PAP-96201	Red	.7mm (med.)	1.09
PAP-96301	Black	.7mm (med.)	1.09
PAP-96501	Purple	.7mm (med.)	1.09
PAP-96901	Pink	.7mm (med.)	1.09
PAP-96601	Blue	.5mm (fine)	1.09
PAP-96701	Red	.5mm (fine)	1.09
PAP-96801	Black	.5mm (fine)	1.09

B Economy Ballpoint—Long-lasting stick pen at a throw-away price.

No.	Ink	Point	Dozen
PAP-33111	Blue	1.0mm (med.)	$3.60
PAP-33211	Red	1.0mm (med.)	3.60
PAP-33311	Black	1.0mm (med.)	3.60
PAP-33411	Green	1.0mm (med.)	3.60
PAP-33611	Blue	.8mm (fine)	3.60
PAP-33711	Red	.8mm (fine)	3.60
PA-P33811	Black	.8mm (fine)	3.60

C Rubberstik—Rubberized barrel for added writing comfort. Dependable, economical.

Ink Color	Med. Point	Fine Point	Each
Blue	PAP-61101	PAP-61601	$0.59
Red	PAP-61201	PAP-61701	.59
Black	PAP-61301	PAP-61801	.59
Green	PAP-61401	—	.59

WRITE BROS.®

D Stick Pen—Papermate's economy ballpoints. Metal pocket clip. Disposable.

No.	Ink	Point	Price
PAP-93101	Blue	1.0mm (med.)	$5.40/Dz
PAP-93201	Red	1.0mm (med.)	5.40/Dz
PAP-93301	Black	1.0mm (med.)	5.40/Dz
PAP-93601	Blue	.8mm (fine)	5.88/Dz
PAP-93701	Red	.8mm (fine)	5.88/Dz
PAP-93801	Black	.8mm (fine)	5.88/Dz
PAP-91101	Blue	.7mm (prof. fine)	.95/Ea
PAP-91201	Red	.7mm (prof. fine)	.95/Ea
PAP-91301	Black	.7mm (prof. fine)	.95/Ea

E Refillable Erasermate—Click-seal cap with metal clip. Assorted barrel colors.

No.	Ink Color	Point Dia.	Each
PAP-380-64	Blue	Medium	$2.69

F Disposable Erasermate 2—Erases as easily as pencil! Use any standard eraser (ink is permanent after approximately 30 days.) Writes at any angle.

Ink Color	Med. Point	Fine Point
Blue	PAP-39101	PAP-39601
Red	PAP-39201	PAP-39701
Black	PAP-39301	PAP-39801
Each	$1.05	$1.05

ITOYA

G Gripper Ballpoint Pen—Rubber cushion grip for fingertip comfort and nonslip control. See-through barrel for ink supply. Refillable.

Gripper Ballpoint Pen, Cont.

Ink Color	Med. Point	Fine Point	Each
Blue	ITY-OH-20-BE	ITY-OH-10-BE	$0.99
Black	ITY-OH-20-BK	ITY-OH-10-BK	.99
Red	ITY-OH-20-RD	ITY-OH-10-RD	.99

H Gripper Fluorescent Colors Ballpoint Pen—In four dazzling colors. Medium point; Nonrefillable.

No.	Color	Each
ITY-OH-30-PK	Hot Pink	$0.99
ITY-OH-30-LV	Lavender	.99
ITY-OH-30-TQ	Turquoise	.99
ITY-OH-30-OE	Orange	.99

fisher

I Bullet Pocket Pens—Fits in a small pocket when closed, yet opens to a full-sized pen. Solid brass. Medium ballpoint refill, black ink. Refillable.

No.	Finish	Each
FIS-400	Chrome	$15.00
FIS-400B	Black Matte	15.00

A — No-slip grip.

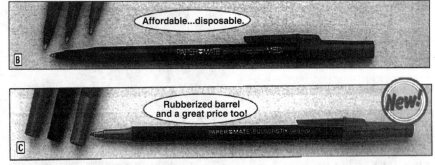

B — Affordable...disposable.

C — Rubberized barrel and a great price too! New!

D — Economy stick pen. PAPER⊗MATE WRITE BROS. MED. PT. USA

E — Refillable erasable ink.

F — Erasable ink. Eraser⊗Mate 2 ERASABLE INK MED PT

G — Cushioned Grip.

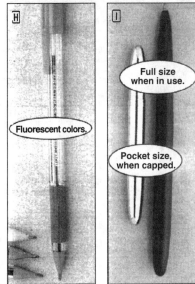

H — Fluorescent colors.

I — Full size when in use. Pocket size, when capped.

Acknowledgment: Reprinted from the 1994 Office Products Catalog, copyright 1993, by permission of United Stationers Supply Co.

Office Papers

A Premium White Xerographic & Laser Printer Paper

- Brightness rating 86
- Controlled moisture content prevents curling
- Special wrap preserves freshness

For all plain paper copiers, duplicators and laser printers, including high speed copiers like Xerox 9500 and 9700. 20-lb. sub. sulfite. Ten reams per carton (except 11 x 17 size).

No.	Size	Ream
UNV-11200	8½ x 11	$ 7.80
UNV-14200	8½ x 14	9.90
UNV-17110*	11 x 17	15.70
*Packed five reams per carton		
3-HOLE PUNCHED		
UNV-11230	8½ x 11	8.25

B Bulk Copy Paper

- Brightness rating 83
- For high volume use
- Our most economically priced paper

White, 20-lb. sub. paper suitable for all plain paper copiers. Sold in carton of ten reams only (except 11 x 17 size).

No.	Size	Carton
UNV-21200	8½ x 11	$75.50
UNV-24200	8½ x 14	95.90
UNV-28110	11 x 17	76.05
*Sold in carton of five reams		
3-HOLE PUNCHED		
UNV-28230	8½ x 11	79.89

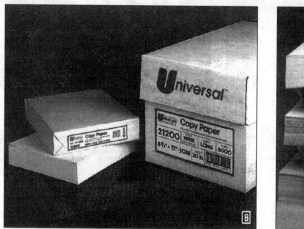

C Premium Colored Xerographic & Laser Printer Paper

- Five colors—ideal for special projects
- Attracts attention to important documents

For positive performance in all plain paper copiers and printers. Also for off-set and spirit duplicators. 20-lb. sub. sulfite, 8½ x 11. Ten reams per carton.

No.	Color	Ream
UNV-11201	Canary	$8.75
UNV-11202	Blue	8.75
UNV-11203	Green	8.75
UNV-11204	Pink	8.75
UNV-11205	Goldenrod	8.75
3-HOLE PUNCHED		
UNV-11231	Canary	9.25

D Five-Ream Convenience Pack

- 20-lb. sub. paper, 8½ x 11
- Ream-wrapped to preserve freshness
- Convenient and reusable carry-and-store box

Multipurpose copier and printer paper provides sharp, high-contrast images. Sold in carton of five reams only.

No.	Color	Carton
UNV-11289	White (86 Brightness)	$39.95
UNV-11292	Assorted Colors*	44.95
*One ream each Canary, Blue, Green, Pink, Goldenrod.		

Reusable carton with hand holes for easy carrying.

E Typewriter Paper

- Brightness rating 86
- White bond, 20-lb. sub.
- Uniform finish for a professional look

High bulk and opacity—delivers clear mimeos, too. Ten reams per carton.

No.	Size	Ream
UNV-81200	8½ x 11	$7.65
UNV-84200	8½ x 14	9.75

Acknowledgment: Reprinted from the 1994 Office Products Catalog, copyright 1993, by permission of United Stationers Supply Co.

Card Files

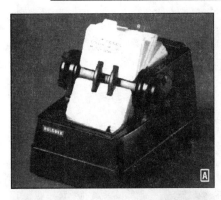

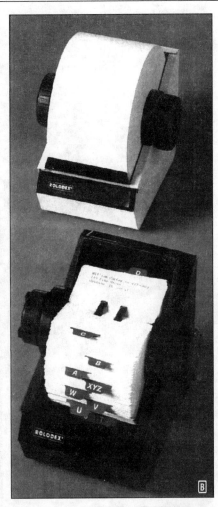

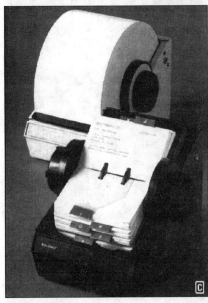

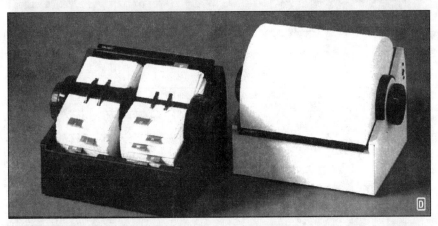

ROLODEX®
BRAND

ROTARY CARD FILES
- Includes full capacity of removable cards and guides, roll-top dustcover, marproof rubber feet.

A Economy Card File—Black poly-styrene file includes 500 1½ x 2¾ cards and 24 index guides with non-insertable A to Z tabs.
ROL-R202-BK......................Each $20.00

B Steel Rotary File—Includes 500 1¾ x 3¼ cards and 24 index guides with non-insertable A to Z tabs.

No.	Color	Each
ROL-1753-BK	Black	$33.50
ROL-1753-PT	Putty	33.50

C Steel File with Locking Cover—Index guides with insertable A to Z tabs. Two keys.

No.	Color	Cap.	Guides	Each
2¼ x 4 CARD SIZE				
ROL-2254D-BK	Black	500	25	$ 40.00
ROL-2254D-PT	Putty	500	25	40.00
ROL-2254-BK	Black	1000	40	61.75
ROL-2254-PT	Putty	1000	40	61.75
ROL-2400S-BK	Black	2000	40	108.00
3 x 5 CARD SIZE				
ROL-5350-BK	Black	500	25	50.50
ROL-3500S-BK	Black	1000	40	101.25
ROL-3502S-BK	Black	2000	60	145.00

D High-Capacity Steel File with Locking Cover—Includes 2000 or 4000 cards. Two wheels in single unit turn separately. Index guides with insertable A to Z tabs. Two keys.

No.	Color	Cap.	Guides	Each
2¼ x 4 CARD SIZE				
ROL-2400-BK	Black	2000	40	$110.25
ROL-2400-PT	Putty	2000	40	110.25
ROL-2400T-BK	Black	4000	60	170.50
3 x 5 CARD SIZE				
ROL-3500T-BK	Black	2000	60	151.75
ROL-3504T-BK	Black	4000	60	195.75

Bates

E Steel Rotary File—Manufacturer's lifetime guarantee! Steel hood protects cards. Includes 500 cards and 26 index guides with insertable A to Z tabs.

No.	Card Size	Color	Each
BAT-SR24C-BK	2¼ x 4	Black	$40.00
BAT-SR35C-BK	3 x 5	Black	50.50

Acknowledgment: Reprinted from the 1994 Office Products Catalog, copyright 1993, by permission of United Stationers Supply Co.

Weekly Appointment Books

Shirt pocket size.

Two sizes!

REFILLS FOR 70-544 C
Appointment Section
KEI-70-905-20 ....Each $6.00
Memo Pad
KEI-80-905-10 ....Pack of 3 $3.40
REFILLS FOR 70-038 C
Appointment Section
KEI-70-907-20.................Each $5.10

Page Format	Page Size	Appt. Ruling	Special Features	Refillable	Color	No.	Price
AT·A·GLANCE® by Keith Clark							
A Ruled, one week per spread.	8¾ x 6⅞	None	Two-color printing, 12-month calendar on each spread. Simulated leather cover.	No	Black	KEI-70-875-05	$10.20
B Ruled, one week per spread.	4 x 2½	None	Shirt pocket size. Tel./add. & expense pages. Simulated leather cover.	No	Assorted	KEI-70-037-00	5.80
C Unruled blue bond, one week per spread.	6¼ x 3¼	None	Tabbed tel./add. & memo pad sections. Genuine leather, wallet style w/3 pockets.	Yes	Black	KEI-70-544-05	41.70
			Without tel./add. section. Simulated leather cover.	No	Black	KEI-70-045-05	6.30
			12-month Academic/Fiscal, July-June. Simulated leather cover.	No	Black	KEI-70-044-05	6.00
	4½ x 2½	None	Shirt pocket size. Simulated leather cover.	No	Assorted	KEI-70-035-00	5.50
					Black	KEI-70-035-05	5.50
			Tabbed tel./add. section. Simulated leather cover.	Yes	Black	KEI-70-038-05	11.20
D Ruled, one week & memo page per spread.	8 x 4⅞	None	Simulated leather cover.	No	Black	KEI-70-150-05	6.95
Timepeace® by Keith Clark Calendars with a Difference							
E Ruled, no appt. times. Weekly spreads have illustration or facts page.	8 x 4⅞	None	One week an office cartoon, the next related facts, tips. Burgundy & black printing on gray paper. Deluxe linen weave cover.	No	Burgundy	KEI-TM225-14	9.30

Acknowledgment: Reprinted from the 1994 Office Products Catalog, copyright 1993, by permission of United Stationers Supply Co.

Telephone Message Forms

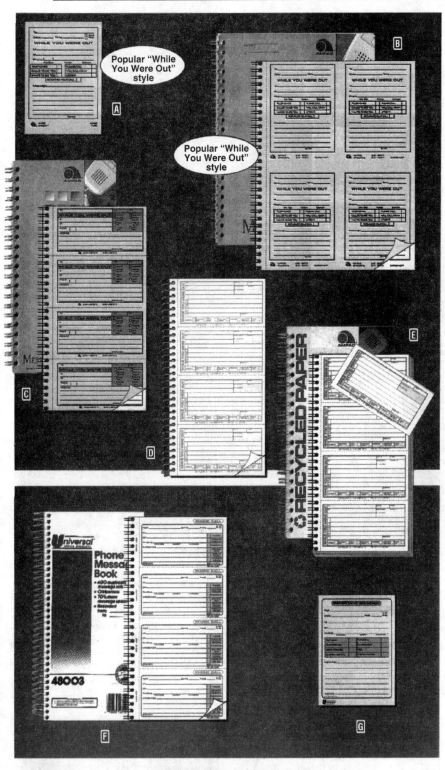

Popular "While You Were Out" style

Popular "While You Were Out" style

AMPAD®

"WHILE YOU WERE OUT" PADS

• 50 sheets per pad, 4 x 5¼

A Standard Pink—High quality paper.
AMP-23-000 ....Pack of 12 $3.00
Embassy Colors—Three pads each of Blue, Orchid, Ivory and Gray.
AMP-23-002 ....Pack of 12 $4.56

Green Cycle—Made from 50% recycled paper containing at least 10% post-consumer waste. White with green printing.
AMP-23-700 ....Pack of 12 $3.36
Stick-On—Standard pink form with self-stick backing attaches anywhere. Removable, restickable.
AMP-44020...............Pack of 12 $11.76

WIREBOUND MESSAGE BOOKS

• Wire binding opens completely flat Black-print carbonless duplicates, bound-in "write prevent" card, date register on cover. Box pack five books.

No.	Book Size	Sets/Book	Each Book
5½ x 4 FORMS, FOUR PER PAGE			
B AMP-23-021	11 x 8½	200	$6.60
AMP-23-421	11 x 8½	400	7.86
5½ x 4 FORMS, TWO PER PAGE			
AMP-23-023	11 x 4¾	200	7.13
AMP-23-022*	5½ x 8½	100	5.15
*Box pack ten books			
2¾ x 5 FORMS, FOUR PER PAGE			
C AMP-23-027	11 x 6	200	5.65
AMP-23-024	11 x 6	400	7.55
D AMP-23-176	11 x 6	400	5.89
E AMP-23-776**	11 x 6	400	7.00
**Recycled paper			

Universal office products **MESSAGE FORMS**

• Quality features, economical prices

F Wirebound Message Book—Black-print carbonless duplicates, with stop card to prevent write-through. Reverse side is ruled for more note space. Wire binding lets book open flat. Form 2¾ x 5, four forms per page. Book 11 x 5½, 400 sets per book. Box pack five books.
UNV-48003 ....Each book $5.90

G Telephone Message Pad—Economical version of the standard pink 4 x 5¼ forms. Ruled note section on back. 50 sheets per pad.
UNV-48023.................Pack of 12 $2.45

3M Post-it™

SELF STICK MESSAGE FORMS

• Adhesive backed to attach anywhere

H Stick-On Message Forms—Eye-catching yellow message form sticks anywhere, peels off, resticks again. 4⅛ x 2⅞. 50 sheets per pad.
MMM-7660 ....Pack of 12 $10.44

Acknowledgment: Reprinted from the 1994 Office Products Catalog, copyright 1993, by permission of United Stationers Supply Co.

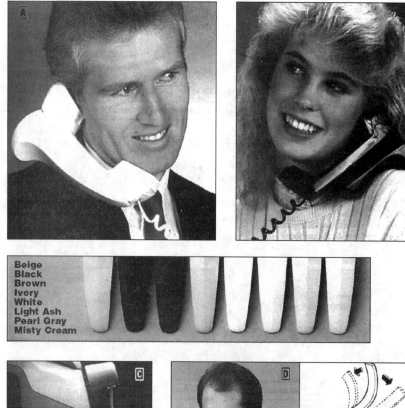

Telephone Accessories

Beige
Black
Brown
Ivory
White
Light Ash
Pearl Gray
Misty Cream

Without   With

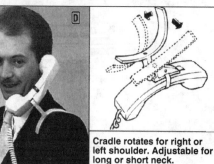

Cradle rotates for right or left shoulder. Adjustable for long or short neck.

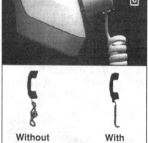

SOFTALK TWISSTOP™

C **Ends Tangled Phone Cords!**—Automatically rotates with phone. Easy to install, requires no tools.

No.	Color	Each
SOF-1500	Clear	$5.99
SOF-1505	Ivory	5.99
SOF-1508	White	5.99

REST-A-PHONE.

SHOULDER RESTS

D Ergonomically designed for handsfree conversations. Holds phone at correct height and angle. Instant switchover for left or right shoulder. Comfortable rubber lined shoulder cradle. Lightweight plastic, attaches with metal band to all handsets.

No.	Color	Each
RES-4001	Beige	$8.49
RES-4002	Black	8.49
RES-4006	Ivory	8.49
RES-4014	Ash/Putty	8.49

FLEXIDUCT.

ADAPTERS AND CORDS

E **In-Line Coupler**—Back-to-back modular jacks connect two base cords for longer reach. Ivory.
WIN-8090Z ....Each $2.19

F **Modular Duplex Jack**—For phone & answering machines. Combine two communication vehicles into a single jack.
WIN-8070Z ....Each $3.29

G **Two-Plug 25-Ft. Extension Cord Adapter**
WIN-8150Z ....Each $5.29

H **Base Cords**—Use to connect modular phone to modular jack. Silver.

No.	Size	Each
WIN-8161Z	7 ft.	$2.49
WIN-8141Z	25 ft.	4.59

I **Handset Coil Cord**—Snaps into all modular handsets.

No.	Length	Color	Each
WIN-8122Z	12 Ft.	Ivory	$3.59
WIN-8120Z	12 Ft.	Beige	3.59
WIN-8124Z	12 Ft.	Almond	3.59
WIN-8132Z	25 Ft.	Ivory	4.89
WIN-8133Z	25 Ft.	Black	4.89
WIN-8131Z	25 Ft.	White	4.89
WIN-8130Z	25 Ft.	Beige	4.89
WIN-8134Z	25 Ft.	Almond	4.89

Softalk®

SHOULDER PHONE RESTS

- Fits either shoulder, keeps hands free
- Self-adhesive backing
- Comfortable! Cushiony, nonslip plastic

A **Softalk**—For regular, Trimline, or Panasonic phones. Plastic. 7⅝ x 2½.

No.	Color	Each
SOF-100	Beige	$8.99
SOF-101	Black	8.99
SOF-103	Brown	8.99
SOF-105	Ivory	8.99
SOF-108	White	8.99
SOF-115	Light Ash	8.99
SOF-133	Pearl Gray	8.99
SOF-135	Misty Cream	8.99

B **Mini Softalk**—For cordless or one-piece phones with short, flat or angular backs. 4¼ x 1¾.

No.	Color	Each
SOF-300	Beige	$5.99
SOF-301	Black	5.99
SOF-305	Ivory	5.99
SOF-308	White	5.99
SOF-315	Light Ash	5.99
SOF-333	Pearl Gray	5.99
SOF-335	Misty Cream	5.99

Acknowledgment: Reprinted from the 1994 Office Products Catalog, copyright 1993, by permission of United Stationers Supply Co.

Quantity	Item	Price for each	Cost
1	Rubbermaid Large Rectangular Wastebasket, RUB-R2957-82		
1	High Gloss Round Wastebasket		
4	High Gloss Rectangular Wastebaskets, Black		
3	Fire-resistant Wastebaskets, Rectangular, Black		
8	Bostitch Deluxe Staplers, BOS-B440BK		
24 boxes	Standard Staples		
8	Desk Calendar Base, Black/EverReady		
8	Desk Calendar Refills, Two-Color Deluxe/EverReady		
12 boxes	Pendaflex Hanging Folders, Standard Green, Letter Size, ESS-4152		
12 boxes	Pendaflex Hanging Folders, Standard Green, Legal Size, ESS-4153		
20 boxes	Interior Folders, Letter Size, ESS-421013 BGR		
20 boxes	Interior Folders, Legal Size, ESS-435013 BGR		
6	Rolodex Rotary Card Files, 3 × 5 Cards, ROL5350-BK		
12 rolls	"Scotch" Brand Magic Tape, Self-Dispensers, MMM119A		
3 cartons	Universal Bulk Copy Paper, 8½ × 11		
4 cartons	Universal Bulk Copy Paper, 8½ × 14		
36 reams	Premium White Xerographic Paper, 8½ × 11		
36 reams	Premium White Xerographic Paper, 8½ × 14		
3 dozen	Papermate Economy Ballpoint Pen, Medium Point, Black		
1 dozen	Papermate Economy Ballpoint Pen, Medium Point, Red		
1 pack	3M "Post-it" Message Forms		
5	Weekly Appointment Books, KEI-70-875-05		
2 packs	Telephone Message Pads, "While You Were Out," Standard Pink		
3	Rest-a Phone Shoulder Rests, Beige		

Total cost

PRACTICE EXERCISE 6

1. Find the price for each item ordered from the catalog shown and enter it in the third column of the table. Then find the cost for the number of items ordered and enter it in the last column.

2. **a.** Find the total cost for all the items ordered.
   b. If the sales tax is computed at 8%, find the amount of tax.
   c. Find the total amount spent for office supplies, including the tax.
   d. Find the total amount spent for office equipment and supplies.
   e. By how much money (if any) does the amount spent for office equipment exceed the amount spent for office supplies?

3. How much extra was spent by purchasing 4 rectangular high gloss wastebaskets rather than 4 soft-molded large rectangular baskets?

4. How much money did Florence save by purchasing 8 Deluxe Staplers instead of an equal number of Executive Staplers?

5. **a.** Subsequently, Pendaflex Desk Drawer Files, calendar pads, and vertical file folders were ordered. Find the cost of each item and the total cost.

5	EverReady Desk Calendars (Refill only), one color
5 boxes	Pendaflex Hanging Folders, data size
2 cartons	Five-Ream Convenience Pack Paper, assorted colors
2 cartons	Typewriter Paper, 8½ × 11

   b. If the sales tax is $7\frac{1}{2}\%$ of the cost, find the sales tax.

   c. What is the total cost, including the tax?

6. Check your answers to Problems 2 and 5 using your calculator.

THE NEW CAR

Optional Equipment

The four salespersons in the Wheels-and-Deals Automobile Salesroom understand that customers seldom purchase an automobile without buying extra (optional) equipment for the car. The Wheels-and-Deals Automobile Sales Company displays the advertising for the extra equipment the customer can purchase in the magazine and literature wall unit.

The Jones family, after deciding on the automobile color and model, decided to look over the catalogs and choose some extra pieces of equipment. These are listed below:

1	AM-FM Multiplex Stereo Radio @ $176.00
1	AM/FM/CB Antenna @ $59.95
1 set	Wheel Covers @ $118.00 a set
1 set	Rubber Floor Mats with Carpet Insert
	Front Seats @ $25.60
	Rear Seats @ $17.50
2	License Plate Frames @ $8.10 each
2 pairs	4 Wheel Splash Guards—Rubber/Stainless Steel @ $16.70 a pair
1	Fire Extinguisher @ $28.90
1	Nameplate @ $5.95
1	Eveready Portable Light @ $19.95
1	Compact Roof Luggage Rack @ $113.00
1	Rear View Mirror @ $15.00
1	Cigarette Lighter @ $10.00
1	Car/Home Vacuum Cleaner @ $39.00

PRACTICE EXERCISE 7

1. Find the cost for the number of each item that the Jones family ordered from the optional equipment catalog.

2. **a.** Find the total cost for all the items ordered.
   b. If the sales tax is computed at 7% of the cost, find the amount of sales tax.
   c. Find the total amount spent for optional equipment, including the tax.

Fuel Costs

Fred, one of the automobile salespersons, realizes that many potential automobile buyers are concerned over the cost of operating an automobile. He has, available to all customers, the "Gas Mileage Guide" published by the U.S. Department of Energy. All types of vehicles (sedans, station wagons, and trucks) and the types of test (city and highway driving) that each undergoes are listed in this pamphlet. Each make of automobile is supplied by the manufacturer, and a city and highway driving test is conducted for each car. If a car got 10 miles per gallon of gasoline in city traffic and 14 miles per gallon on the highway, then the average, a combined miles per gallon, is computed as

$$\frac{10+14}{2} = 12 \, \text{mpg}$$

These fuel costs (in dollars) per 15,000 miles are shown in the table below.

Fuel Costs, in Dollars, per 15,000 Miles

Example: If you pay an average of $1.35 (135 cents) per gallon and your car gets 12 mpg, your fuel cost for 15,000 miles of driving is $1688. If you own a car that gets 20 mpg, your annual fuel cost for 15,000 miles at $1.40 (140 cents) per gallon is $1050.

	Cost (cents per gallon)						
Combined MPG	**150**	**145**	**140**	**135**	**130**	**125**	**120**
50	$450	435	420	405	390	375	360
48	468	453	438	422	406	391	376
46	490	473	456	440	424	408	392
44	512	495	478	461	444	427	410
42	536	518	500	482	464	446	428
40	562	544	526	507	488	469	450
38	592	572	552	533	514	494	474
36	624	604	584	563	542	521	500
34	662	640	618	596	574	552	530
32	704	680	656	633	610	586	562
30	750	725	700	675	650	625	600
28	804	777	750	723	696	699	642
26	866	837	808	779	750	721	692
24	938	907	876	844	812	781	750
22	1022	988	954	920	886	852	818
20	1126	1088	1050	1013	976	938	900
18	1250	1208	1166	1125	1084	1042	1000
16	1406	1359	1312	1265	1218	1172	1126
14	1608	1554	1500	1421	1392	1339	1286
12	1876	1813	1750	1688	1626	1563	1500

PRACTICE EXERCISE 8

1. a. If you pay $1.40 (140 cents) per gallon of gasoline, what is your fuel cost for 15,000 miles if your car gets 16 mpg?
    b. Approximately how much money do you spend for gasoline per month if you drive 15,000 miles in a year?
    c. Approximately how much money do you spend for gasoline per month if you drive 30,000 miles in a year?

2. What is your combined miles per gallon if the cost for 15,000 miles is $574 at $1.30 (130 cents) per gallon?

3. How much money do you pay for a gallon of gasoline if your car gets 28 mpg and your cost for driving 15,000 miles is $804?

4. The Canerossi family owns 3 automobiles. Car A gets 22 mpg, Car B gets 18 mpg, and Car C gets 42 mpg. What is the total cost of driving each car 15,000 miles if gasoline costs $1.35 (135 cents) a gallon?

5. If your car gets 35 mpg and you pay $1.20 (120 cents) per gallon of gasoline, approximately what will be your cost for 15,000 miles of driving?

6. The Jones family travels little during the year. What is the cost of driving 5000 miles if the Joneses' car gets 24 mpg and gasoline costs $1.40 (140 cents) a gallon? (*Hint:* 5000 = $\frac{1}{3}$ of 15,000.)

7. Mary Lou's car gets 40 mpg and gasoline costs $1.50 (150 cents) a gallon. How much will it cost to drive:

    a. 15,000 miles?   c. 22,500 miles?
    b. 30,000 miles?   d. 7500 miles?

8. John travels 30,000 miles each year. His cost for gasoline at $1.30 a gallon is $1772. What is his combined miles per gallon?

9. Hank travels 7500 miles at a cost of $286. If gasoline costs him $1.45 a gallon, what is his combined miles per gallon?

10. Ricardo's car gets a combined 38 mpg, and his total cost for driving is $1184. He pays $1.50 a gallon for gasoline. How many miles does he drive?

Insurance

REQUIRED LIABILITY INSURANCE

Some states require you to obtain a liability insurance policy before you can register your car and obtain license plates. The policy usually provides the insurance required by law and any additional insurance you may buy to meet your particular needs.

This insurance protects you from being sued for damages in the event you injure another person or damage that person's property. Many states also require "no fault" insurance, which pays medical bills due to an accident, no matter who is at fault.

The minimum limits for protection may be $10,000 for injury to one person and $20,000 for injuries to all persons in an accident. This is stated as $10,000/$20,000. Since this is the minimum, many people purchase higher protection, such as $25,000/$50,000, $100,000/$300,000, or even more.

With a $100,000/$300,000 policy you are protected up to $100,000 for injury to one person and up to $300,000 for injuries to all persons in an accident. The higher values afford you greater protection, but they cost more to include.

Table I shows the annual cost of liability insurance for three insurance companies for a typical large city.

Table I   Minimum Coverage (Adult Male, Age 35)

Company	$10,000/$20,000 Bodily Injury	$50,000 Basic No-Fault	$5000 Property Damage	Total
A	$124	$156	$192	$332
B	169	100	144	
C	165	64	209	

Auto insurance premiums vary widely, depending on the characteristics of the insured person. One of the most important factors is age. Inexperienced drivers have a poor safety record, and insurance costs more for this group. Table II shows the annual cost of liability insurance for unmarried males, age 20.

Table II   Minimum Coverage (Unmarried Male, Age 20)

Company	$10,000/$20,000 Bodily Injury	$50,000 Basic No-Fault	$5000 Property Damage	Total
A	$222	$300	$364	$886
B	372	222	322	
C	369	168	486	

PRACTICE EXERCISE 9

1. Compute the total cost of insurance for each of the three companies in Tables I and II. The first one has been done for you as an example.

2. a. How much more must a 20-year-old unmarried male driver pay Company A for the same coverage than an adult male of 35 years of age?
   b. Company B?
   c. Company C?

3. Which company is the most expensive for an adult male, age 35? By how much does it surpass each of the other two companies?

4. Which company is the least expensive for an unmarried male, age 20? How much less is it than each of the other two companies?

5. How much more does Company C charge for $10,000/$20,000 bodily injury insurance for unmarried males, age 20, than Company A?

OPTIONAL INCREASED LIABILITY INSURANCE

Many serious accidents involve personal injury and property damage losses to others that are higher than the minimum requirements, namely, $10,000/$20,000 for bodily injury and $5,000 for property damage. Additional coverage for no-fault insurance is also available.

Table III   Optional Insurance (Adult Male, Age 35)

Company	Bodily Injury Raised to $100,000/$300,000	No-Fault Raised to $100,000
A	$350	$172
B	274	109
C	270	89

Table IV   Optional Insurance (Unmarried Male, Age 20)

Company	Bodily Injury Raised to $100,000/$300,000	No-Fault Raised to $100,000
A	$662	$316
B	629	231
C	627	183

Tables III and IV indicate the *increased* premiums over those in Tables I and II. (The "premium" is the monthly, semiannual, or annual cost to the person insured.) The new premium for no-fault insurance for an adult male, age 35, raised to $100,000 for Company B is now $100 + $109 or a total of $209. Bodily injury insurance for an unmarried male, age 20, for Company A is now $222 + $662 = $884.

PRACTICE EXERCISE 10

1. Find the new premium rates for Company A for adult males, age 35, for both bodily injury and no-fault insurance. Do the same for Companies B and C.

2. What are the new rates for an unmarried male, age 20, for both bodily injury and no-fault insurance for Companies A, B, and C?

3. How much more does Company B charge than Company C for bodily injury coverage of $100,000/$300,000 for an unmarried male, age 20?

4. What are the rates for no-fault insurance raised to $100,000 for Companies A and C for an unmarried male, age 20? How does each compare with Company B's rate? How much more or less are they?

OPTIONAL COMPREHENSIVE AND COLLISION INSURANCE

Comprehensive insurance pays for loss to a car from all causes, such as fire, theft, glass breakage, flood, windstorm, and vandalism and malicious mischief. If you purchase a new car, this type of insurance should seriously be considered, as well as collision insurance.

Collision insurance insures you for damages to your car, without regard to fault, from collision with another car or any other object.

Both comprehensive and collision insurance can be purchased in a policy with a deductible clause. You collect for losses that are larger than the deductible amount. Thus, if you have a loss of $400 and a deductible of $250, you collect $150.

The cost of physical damage insurance varies widely, depending on the make, model, and age of your car. Many other factors also determine the premium for your car: how safe a driver you are, how far you drive to work, whether you drive for pleasure only, and others.

Tables V and VI show the rates for new "intermediate" cars for Companies A, B, and C. To calculate the premiums charged for economy, intermediate, and standard cars, the factors in Table VII can be used. Multiply the premium shown in Table V or VI by the appropriate factor in Table VII to obtain the premium for your car.

Table V   Comprehensive and Collision Insurance (Adult Male, Age 35)

Company	Comprehensive $200 Deductible	Collision $250 Deductible
A	$272	$570
B	295	628
C	325	640

Table VI   Comprehensive and Collision Insurance (Unmarried Male, Age 20)

Company	Comprehensive $200 Deductible	Collision $250 Deductible
A	$474	$1094
B	504	1485
C	545	1475

Table VII   Physical Damage Factor

Type of Vehicle	Comprehensive	Collision
Economy	0.83	0.91
Intermediate	1.00	1.00
Standard	1.38	1.18
Luxury	Many varied rates apply	

EXAMPLE: Mr. La Bianca, who is 35 years old, purchases an economy car and takes out comprehensive insurance with Company A. What premium must he pay?

SOLUTION: Company A's comprehensive premium for an intermediate car is $272. Therefore the comprehensive premium for an economy car will be $272 × 0.83 or

$$
\begin{array}{r}
\$272 \\
\times\ 0.83 \\
\hline
8\ 16 \\
217\ 6 \\
\hline
\$225.76
\end{array}
$$

As a car gets older, the cost of physical damage insurance decreases. Table VIII will enable you to calculate the premiums for vehicles of different ages. Multiply the premium for a new vehicle by the appropriate factor to obtain the premium for your vehicle.

Table VIII Physical Damage Factor

Age of Vehicle	Comprehensive	Collision
New vehicle	1.00	1.00
1–2 years old	.85	.85
3–4 years old	.75	.70
5 years or older	.55	.55

EXAMPLE: Unmarried, 20-year-old Angelo Martini buys a 3-year-old intermediate-size Pontiac. If he insures the car with Company A, what will the premium be?

SOLUTION: Company A's comprehensive premium for a new intermediate car for an unmarried male, age 20, is $474. The physical damage factor is 1.00. Therefore the premium for a 3-year-old intermediate car will be $474 × 1.00 × 0.75 or

$$
\begin{array}{r}
\$474 \\
\times\ 0.75 \\
\hline
23\ 70 \\
331\ 8 \\
\hline
\$355.50
\end{array}
$$

Similarly, Company C's collision premium for an economy car, driven by an adult male, age 35, would be $640 × 0.91 or

$$
\begin{array}{r}
\$640 \\
\times\ 0.91 \\
\hline
6\ 40 \\
576\ 0 \\
\hline
\$582.40
\end{array}
$$

For a 5-year-old car, the premium is reduced further by multiplying by the factor 0.55. Thus $582.40 × 0.55 is

$$
\begin{array}{r}
\$582.40 \\
\times\ 0.55 \\
\hline
29\ 1200 \\
291\ 200 \\
\hline
\$320.3200
\end{array}
$$

or, rounding off to the nearest cent, we obtain $320.32.

As you see, there is a monetary saving in owning an economy car rather than a larger intermediate or standard car.

PRACTICE EXERCISE 11

1. Find Company C's premium for a standard automobile for an unmarried male, age 20, for collision insurance with a $250 deductible.

2. What is Company A's premium for $200 deductible comprehensive insurance for an economy car for an unmarried male, age 20?

3. Find the premium for an adult male, age 35, for comprehensive insurance, $200 deductible, for an intermediate car 4 years old that is insured with Company B.

4. Juan purchases a new economy car and wishes to buy collision insurance, $250 deductible, with Company A. What is the premium for this coverage if Juan is unmarried and 20 years of age?

5. Find the premium for collision insurance, $250 deductible, for a 2-year-old standard Ford, insured with Company B, for an adult male, 35 years old.

6. What is Company A's rate for comprehensive insurance, $200 deductible, for an adult male, age 35, for a 1-year-old economy car?

7. Joe is a 20-year-old unmarried male who has purchased a new standard Chevrolet. He decides to choose Company A as his insuring company. What will his premium amount to for $250 deductible collision insurance?

8. Sam is an unmarried male, 20 years of age, who owns a 2-year-old economy Volkswagen. He desires to have $200 deductible comprehensive and $250 deductible collision insurance with Company C. What is his premium?

Although the cost of insurance is important to all of us, you should be aware of the following:

- Price is only one of the reasons for choosing an insurance company; service is another.

- Some companies reimburse you by paying dividends back to you.

- Most insurers have "merit" rating plans that offer lower rates for safe drivers.

EXAM TIME

1. Find the perimeter of a rectangle whose dimensions are 13′6″ × 17′2″.

2. Find the cost of 35 ft. of molding if each foot costs $.86.

3. What is the area of a rectangular wall that measures 12′ × 19′?

4. If 1 gallon of paint covers 240 sq. ft., find the number of gallons needed to cover four rectangular walls whose dimensions are 13′ × 12′.

5. If wallpaper costs $8.50 a roll and covers 30 sq. ft., find the cost of wallpapering a ceiling with dimensions of 14′ × 22′.

6. What is the cost of four file cabinets if each cabinet costs $69.65?

7. Find the total cost of four chairs and two desks if each chair costs $84.75 and each desk costs $249.95.

8. If the total cost of three typewriters is $897, find the cost of one typewriter.

9. If the total cost of electronic calculators was $340, how many calculators were purchased if the cost of one calculator is $85?

10. **a.** Using the table on page 82, find the fuel cost for 15,000 miles of driving if your car gets 26 mpg and you pay 125¢ per gallon of gasoline.
  b. If the total cost for fuel for a car traveling 15,000 miles and getting 20 mpg is $976, what is the cost per gallon of gasoline?
  c. If the cost of gasoline is 150¢ per gallon and your total fuel cost for 15,000 miles is $662, what is your combined miles per gallon?

11. Find Company C's premium for comprehensive insurance with a $200 deductible for a new standard Buick driven by an adult male, age 35. Use the appropriate tables on p. 86.

12. Find the cost of $250 deductible collision insurance with Company B for a 5-year-old standard car driven by an unmarried male, age 20. Use the appropriate tables.

Now Check Your Answers

Now that you have completed the exam, check your answers against the correct ones, which follow the answers to the practice exercises below.

ANSWERS TO PRACTICE EXERCISES AND EXAM

PRACTICE EXERCISE 1

1. 66"   2. 54'   3. $53.96   4. $15.64   5. 31'
6. 22.6" 7. $40\frac{1'}{3}$ 8. $60.30   9. 132'

PRACTICE EXERCISE 2

1. **a.** Showroom   1500 sq. ft.   Washroom   120 sq. ft.
      Sales Mgr.   180 sq. ft.   Office   150 sq. ft.
   b. 1950 sq. ft.   **c.** 5050 sq. ft.   **d.** $357.00
   e. $2777.50 + $357.00 = $3134.50

2. **a.** Living room   210 sq. ft.   **d.** 464 sq. ft.
      Dining room   240 sq. ft.   512 sq. ft.
      Bedroom I   165 sq. ft.   416 sq. ft.
      Bedroom II   132 sq. ft.   368 sq. ft.
      Bedroom III   150 sq. ft.   400 sq. ft.
      Bath I   66 sq. ft.   272 sq. ft.
      Bath II   72 sq. ft.   288 sq. ft.
      Kitchen   108 sq. ft.   336 sq. ft.
      Hall I   72 sq. ft.   352 sq. ft.
      Hall II   102 sq. ft.   368 sq. ft.
   b. 1317 sq. ft.   **e.** 5093 sq. ft.
   c. 1317 sq. ft.   **f.** (1) 26 gallons
                                (2) $324.74

3. **a.** 3960 sq. ft.   **b.** 1488 sq. ft.   **c.** 5448 sq. ft.
    d. (1) 37 gals. for one coat, 73 gals. for two coats
       (2) $516.15 for one coat, $1018.35 for two coats

PRACTICE EXERCISE 3

1. **a.** 15 single rolls, 8 double rolls   **b.** $78.75

2. **a.** 17 single rolls, 9 double rolls   **b.** $84.66

3. $107.25   **4.** $375.00   **5.** $52.43   **6.** $5.86

PRACTICE EXERCISE 4

1. 13   **2.** 12   **3.** 6   **4.** 16   **5.** 21 ft.

6. **a.** 25   **b.** 7   **c.** 7   **d.** 46 ft.

7. $\dfrac{17}{8}$ **8.** 8:10 or 4:5   **9.** $\dfrac{1}{3}$ **10.** 1:8

PRACTICE EXERCISE 5

1.	Price for each	Cost
	$775.00	$775.00
	775.00	775.00
	373.00	373.00
	517.00	2068.00
	622.00	1244.00
	429.00	429.00
	304.00	1216.00
	156.00	1248.00
	182.00	546.00
	870.00	3480.00
	1330.00	5320.00
	1595.00	3190.00
	549.95	549.95
	699.95	699.95
	59.50	59.50
	39.50	39.50
	642.00	642.00
	895.00	895.00
	94.95	94.95
	695.00	695.00
	39.99	39.99
	84.95	339.80
	69.95	69.95
	94.00	94.00

2. **a.** $24,883.59
    b. $1990.69
    c. $26,874.28
    d. $1874.2

3. $79.00

4. $1080.00

5. **a.** $750.00
    b. 4808 lb
    c. $721.20

6. $90.30

7. **a.** $499.00
    b. $39.92
    c. $538.92

8. **a.** $44.95, $11.37
    b. $56.32
    c. $3.94
    d. $60.26

PRACTICE EXERCISE 6

1.

Price for each	Cost
$7.25	$7.25
16.99	16.99
12.99	51.96
30.20	90.60
18.95	151.60
1.95	46.80
14.40	115.20
4.25	34.00
14.45	173.40
17.50	210.00
16.70	334.00
23.35	467.00
50.50	303.00
1.84	22.08
75.50	226.50
95.90	383.60
7.80	286.80
9.90	356.40
3.60	10.80
3.60	3.60
10.44	10.44
10.20	51.00
3.00	6.00
8.49	25.47
	3378.49

2. a. $3378.49
b. $270.28
c. $3648.77
d. $30,523.05
e. $23,225.51

3. $22.96

4. 48.00

5. a.

Price for each	Cost
$3.50	$17.50
135.55	677.75
44.95	89.90
7.65	15.30
TOTAL:	$800.45

b. $60.03
c. $860.48

PRACTICE EXERCISE 7

1.

$176.00	$28.90
59.95	5.95
118.00	19.95
43.10	113.00
16.20	15.00
33.40	10.00
	39.00

2. a. $678.45
b. $47.49
c. $725.94

PRACTICE EXERCISE 8

1. a. $1312   b. $109.33   c. $218.66   **2.** 34 mpg
3. $1.50   **4.** Car A: $920.45; Car B: $1125.00; Car C: $482.14
5. $515   **6.** $291.67   **7.** a. $562   b. $1124   c. $843   d. $281
8. 22 mpg   **9.** 38 mpg   **10.** 29,995 mi

PRACTICE EXERCISE 9

1.

	Table I	Table II
A	$332	$886
B	413	916
C	448	1023

2. a. $554
    b. $503
    c. $575

3. Company C; $116 more than A, $35 more than B

4. Company A; $30 less than B and $137 less than C

5. $91

PRACTICE EXERCISE 10

1.

	Bodily Injury	No Fault
A	$474	$328
B	443	209
C	435	163

2.

	Bodily Injury	No Fault
A	$884	$616
B	1001	453
C	996	351

3. $5.00

4. Company A: $616, which is $163 more than B
    Company C: $351, which is $102 less than B

PRACTICE EXERCISE 11

1. $1740.50   2. $393.42   3. $221.25   4. $995.54

5. $629.88   6. $191.90   7. $1290.92   8. $384.50 + $1140.91 for a
                                             total premium of $1525.41

EXAM TIME

1. 61'4"   2. $30.10   3. 228 sq. ft.   4. 2.6 gal.   5. $93.50

6. $278.60   7. $339.00 + $499.90 = $838.90   8. $299   9. 4

10. a. $721   b. $1.30   c. 34 mpg   11. $448.50   12. $963.77

How Well Did You Do?

0–9   **Poor.** Reread the unit, redo all the practice exercises, and retake the exam.

10   **Fair.** Reread the sections dealing with the problems you got wrong. Redo those practice exercises, and retake the exam.

11, 12 **Good.** Review the problems you got wrong. Redo them correctly.

13, 14 **Very good!** Continue on to the next unit. You're on your way.

THE EXTRA-FINE & FUSSY MACHINE SHOP

THE TIME RECORD

Mr. Frank is a machinist. He operates the Extra-Fine & Fussy Machine Shop and employs six machinists to work for him. They usually work 7 hours a day, but sometimes they work more or less than that amount. At times, they must work on the weekend.

Mr. Rodriquez is one of the machinists employed by Mr. Frank. One week Mr. Rodriquez worked from Monday to Friday: 7, 7, 8, 9, and 7 hours, respectively. To find the total number of hours worked, you add: 7 + 7 + 8 + 9 + 7 = 38 hours.

PRACTICE EXERCISE 1

Mr. Frank's bookkeeper, Sally Williams, keeps a weekly record of the number of hours each employee works. She uses a special form like this:

EFF MACHINE SHOP
Time Record: Weekly Hours

Week Ending .... **19...**

Employee	Mon.	Tues.	Wed.	Thurs.	Fri.	Sat.	Sun.	Total
A. Rodriquez	7	7	8	9	7	—	—	38
B. White	7	9	7	8	10	4	—	
C. Charles	8	7	7	9	9	—	3	
A. Roth	9	9	8	7	8	4	—	
J. Garcia	8	7	8	9	7	3	4	
J. Kelly	7	9	7	7	8	—	—	
	46							

1. Complete the form by finding the total number of hours worked by each employee. The line for Mr. Rodriquez has already been completed.

2. On Monday the six machinists worked a total of 7 + 7 + 8 + 9 + 8 + 7 = 46 hours. Compute the total number of hours worked by all employees each of the other days of the week.

3. Find the *sum* of the total daily number of hours worked:

    Mon. + Tues. + Wed. + Thurs. + Fri. + Sat. + Sun. = ?
    46   +   ?   +   ? +   ?   + ? + ?   + ?   = ?

4. Find the sum of the column on the right headed "Total."

5. Is the sum the same for Problems 3 and 4? If the sums are not the same, check all of your answers.

PRACTICE EXERCISE 2

1. How many *more* total hours did Mr. Roth work than Ms. Kelly?

2. How many *fewer* hours did Mr. Charles work than Ms. Garcia?

3. The combined total hours worked by A. Rodriquez and B. White is (less than, equal to, more than) the combined total hours worked by A. Roth and J. Kelly. What is the difference, if any?

4. The combined total hours worked by A. Rodriquez, J. Garcia, and C. Charles is (less than, equal to, more than) the combined total hours worked by B. White, A. Roth, and J. Kelly. What is the difference, if any?

5. The total hours worked on Wednesday is (less than, equal to, more than) the total hours worked on Friday. What is the difference, if any?

6. How many more total hours were worked on Monday, Tuesday, and Wednesday than on Thursday, Friday, Saturday, and Sunday?

THE TIME CARD

All the machinists working at the Extra-Fine and Fussy Machine Shop are required to punch a time clock whenever they start or end work. The time card keeps an accurate record of these times.

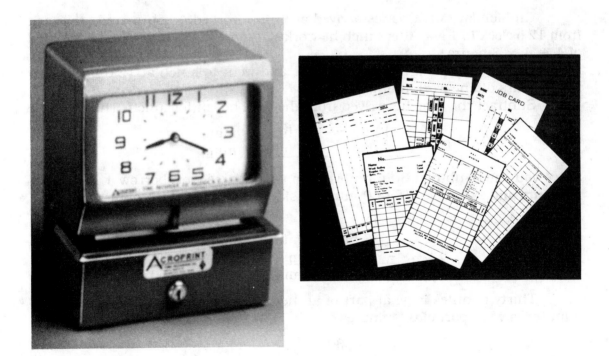

Acknowledgment: Reprinted from the 1981 Office Products Catalog, copyright 1980, by permission of United Stationers Supply Co.

Mr. Charles' time card is shown below.

No. 3 Name: C. Charles Week Ending................ 19......				
	A.M.		**P.M.**		**Hours Worked**
	In	**Out**	**In**	**Out**	
Mon.	8:00	12:00	1:00	5:00	
Tues.	8:30	12:00	12:30	4:00	
Wed.	8:00	12:00	12:30	3:30	
Thurs.	8:00	12:00	12:30	5:30	
Fri.	8:00	11:30	12:00	5:30	
Sat.	—	—	—	—	
Sun.	9:00	12:00			
Total Number of Hours Worked					

On Monday, Mr. Charles arrived at 8 A.M., worked 4 hours, and ate lunch from 12 o'clock to 1 P.M. After lunch he worked 4 more hours until 5 P.M. for a total of 4 + 4 = 8 hours.

On Tuesday, he arrived at 8:30 A.M. and went to lunch at 12 o'clock. How many hours did he work that morning?

Subtract the time he arrived from the time he left for lunch. Thus:

$$12:00 = 12 \text{ hr. } 00 \text{ min.}$$
$$- 8:30 \quad - 8 \text{ hr. } 30 \text{ min.}$$

Since you can't subtract 30 minutes from 00 minutes, you borrow 1 hour from 12 hours and exchange it for 60 minutes since 60 minutes = 1 hour:

$$\begin{array}{cc} 11 & 60 \\ \not{12} \text{ hr.} & \not{00} \text{ min.} \\ - 8 \text{ hr.} & 30 \text{ min.} \\ \hline 3 \text{ hr.} & 30 \text{ min.} \end{array}$$

Thirty minutes is what part of an hour? Since 60 minutes = 1 hour, 30 minutes is what part of 60 minutes?

$$\frac{30}{60} = \frac{1}{2} \text{ hr.}$$

Therefore, Mr. Charles worked 3 hr. 30 min. or $3\frac{1}{2}$ hr. on Tuesday morning.

After starting again at 12:30 P.M., he left at 4 P.M. He worked:

$$\begin{array}{rll} 4\!:\!00 \text{ P.M.} = & 4 \text{ hr.} & 00 \text{ min.} \\ -12\!:\!30 \text{ P.M.} & -12 \text{ hr.} & 30 \text{ min.} \end{array}$$

Since noon or 12 o'clock is the start of P.M. time, you can change 12:30 P.M. to 00:30 when you subtract. Thus:

$$\begin{array}{rll} 3 & 60 & \\ \cancel{4} \text{ hr.} & \cancel{00} \text{ min.} & \\ -0 \text{ hr.} & 30 \text{ min.} & \\ \hline 3 \text{ hr.} & 30 \text{ min.} & = 3\frac{1}{2} \text{hr.} \end{array}$$

Since you can't subtract 30 minutes from 00 minutes, you exchange 1 hour for 60 minutes. Therefore, Mr. Charles worked $3\frac{1}{2} + 3\frac{1}{2}$ hours or 7 hours on Tuesday.

PRACTICE EXERCISE 3

The first one in each problem has been done as an example.

1. Write each time below in terms of *hours* and *minutes*.

    a. 3:10 A.M. = 3 hr. 10 min.   f.   9 P.M.
    b. 5:15 P.M.   g. 11:45 A.M.
    c. 9:40 A.M.   h.   2:15 P.M.
    d. 12 noon   i.   4:25 P.M.
    e. 6:07 A.M.   j.   8:30 P.M.

2. Write each of the following in *time* symbols:

    a.   3 hr. 30 min. = 3:30   f.   7 hr. 12 min.
    b.   2 hr. 10 min.   g.   1 hr. 45 min.
    c. 10 hr. 00 min.   h.   4 hr. 50 min.
    d.   6 hr. 15 min.   i. 12 hr. 06 min.
    e.   9 hr. 40 min.   j. 11 hr. 10 min.

3. Write each of the following, exchanging 1 hr. for 60 min.:

    a. 12 hr. 00 min. = 11 hr. 60 min.
    b.   4 hr. 10 min. = 3 hr. ___ min.
    c.   5 hr. 15 min. = 4 hr.___ min.
    d.   7 hr. 30 min. = 6 hr. ___ min.
    e.   8 hr. 45 min. = 7 hr. ___ min.
    f.   1:10 = 0 hr. ___ min.
    g.   3:20 = 2 hr. ___ min.
    h.   6:05 = 5 hr. ___ min.
    i.   9:50 = 8 hr. ___ min.
    j. 11:48 = 10 hr. ___ min.

4. Change these minutes into parts of an hour:

a. 15 min. $= \dfrac{15 \div 15}{60 \div 15} = \dfrac{1}{4}$ hr.   **f.** 20 min.

b. 45 min.   **g.** 40 min.

c. 10 min.   **h.** 25 min.

d. 30 min.   **i.** 28 min.

e. 50 min.   **j.** 12 min.

5. Change each time below to hours and minutes.

a. $2\dfrac{1}{2}$ hr. $= 2$ hr. 30 min. $\left(\dfrac{1}{\cancel{2}} \times \dfrac{\overset{30}{\cancel{60}}}{1} = \dfrac{30}{1} = 30 \right)$

b. $3\dfrac{1}{4}$ hr.

c. $4\dfrac{3}{4}$ hr.

d. $7\dfrac{2}{3}$ hr.

e. $\dfrac{3}{10}$ hr.

PRACTICE EXERCISE 4

1. Complete the time card for C. Charles on p. 98, figuring the number of hours worked daily.

2. Find the total number of hours he worked during the week.

3. Complete the time card for each of the other five employees.

No. 1 Name: A. Rodriquez Week Ending................. 19....					No. 2 Name: B. White Week Ending................. 19....				
	A.M.		P.M.			A.M.		P.M.	
	In	Out	In	Out	Hours Worked		In	Out	In	Out	Hours Worked

	In	Out	In	Out	Hours Worked		In	Out	In	Out	Hours Worked
Mon.	8:00	12:00	1:00	4:00		Mon.	8:00	12:00	12:30	3:30	
Tues.	8:00	11:30	12:30	4:00		Tues.	8:00	12:00	12:30	5:30	
Wed.	8:00	12:00	12:30	4:30		Wed.	8:00	11:30	12:00	3:30	
Thurs.	8:00	12:00	12:30	5:30		Thurs.	8:00	12:10	1:10	5:00	
Fri.	9:00	12:00	1:00	5:00		Fri.	7:00	11:00	11:30	5:30	
Sat.	—	—	—	—		Sat.	8:00	12:00	—	—	
Sun.	—	—	—	—		Sun.	—	—	—	—	
Total Number of Hours Worked						Total Number of Hours Worked					

No. 4 Name: A. Roth Week Ending 19	A.M.		P.M.		Hours Worked
	In	Out	In	Out	
Mon.	7:30	11:30	12:00	5:00	
Tues.	7:30	11:30	12:00	5:00	
Wed.	8:00	12:00	12:30	4:30	
Thurs.	9:00	12:00	1:00	5:00	
Fri.	8:00	12:00	12:30	4:30	
Sat.	8:00	12:00	—	—	
Sun.	—	—	—	—	
Total Number of Hours Worked					

No. 5 Name: J. Garcia Week Ending 19	A.M.		P.M.		Hours Worked
	In	Out	In	Out	
Mon.	8:00	11:30	12:00	4:30	
Tues.	8:00	12:00	12:30	3:30	
Wed.	7:30	11:30	12:00	4:00	
Thurs.	8:30	12:00	1:00	6:30	
Fri.	8:00	11:30	12:30	4:00	
Sat.	9:00	12:00	—	—	
Sun.	9:00	1:00	—	—	
Total Number of Hours Worked					

No. 6 Name: J. Kelly Week Ending 19	A.M.		P.M.		Hours Worked
	In	Out	In	Out	
Mon.	8:00	12:00	1:00	4:00	
Tues.	8:00	11:30	12:00	5:30	
Wed.	9:00	12:00	12:30	4:30	
Thurs.	8:00	11:30	12:00	3:30	
Fri.	8:00	12:30	1:00	4:30	
Sat.	—	—	—	—	
Sun.	—	—	—	—	
Total Number of Hours Worked					

THE PAYROLL

"Regular hours" are those hours each employee is expected to work. Most employers, including Mr. Frank, consider 35 hours as a regular work week. All hours worked over that are considered as "overtime hours" and eligible for overtime pay. Overtime pay is figured as "time-and-a-half" or "double time." All work done past the regular 35 hours or work done on Saturday is considered as "time-and-a-half." Work done on Sunday is considered as "double time."

Sally Williams, Mr. Frank's bookkeeper, figures the wages the machinists earn each week. She uses a payroll sheet like the one shown below to keep her record.

EFF MACHINE SHOP
Payroll Sheet

Week Ending .... **19...**

Employee	Mon.	Tues.	Wed.	Thurs.	Fri.	Sat.	Sun.	Regular Hours	Overtime Time-and-a-Half	Double Time
A. Rodriquez	7	7	8	9	7	—	—	35	3	—
B. White	7	9	7	8	10	4	—			
C. Charles	8	7	7	9	9	—	3			
A. Roth	9	9	8	7	8	4	—			
J. Garcia	8	7	8	9	7	3	4	35	4 + 3 = 7	4
J. Kelly	7	9	7	7	8	—	—			
S. Williams	7	7	7	7	7	—	—			

Ms. Williams found that A. Rodriquez worked 38 hours last week. This gave him 35 regular hours and 3 overtime hours at time-and-a-half.

J. Garcia worked 39 hours from Monday to Friday. This gave her 35 regular hours and 4 overtime hours. Since she worked 3 hours on Saturday, she earned 4 + 3 = 7 overtime hours at time-and-a-half. She also worked 4 hours on Sunday at double time.

PRACTICE EXERCISE 5

Complete the payroll sheet for Ms. Williams, figuring regular hours, overtime hours at time-and-a-half, and double time for the other five employees.

Regular Pay and Overtime Pay

REGULAR PAY

J. Garcia earned 35 regular hours at $14.40 an hour.

$$
\begin{array}{rl}
\$14.40 & \text{2 decimal places} \\
\times\ 35 & \text{0 decimal place} \\
\hline
72\ 00 & \\
432\ 0 & \\
\hline
\$504.00 & \text{2 decimal places}
\end{array}
$$

TIME-AND-A-HALF

Seven overtime hours at time-and-a-half means that you earn wages for $1\frac{1}{2}$ hours for each hour worked. The amount can be calculated in two ways.

METHOD I

For 7 overtime hours you earn wages for $10\frac{1}{2}$ hours:

$$7 \times 1\frac{1}{2} = \frac{7}{1} \times \frac{3}{2} = \frac{21}{2} = 10\frac{1}{2}\text{ hours}$$

and

$$
\begin{array}{rl}
\$14.40 & \\
\times\ 10\frac{1}{2} & \\
\hline
7\ 20 & \\
144\ 00 & \\
\hline
\$151.20 &
\end{array}
\qquad \left(\frac{1}{2} \times \$14.40 = \$7.20\right)
$$

or

$$
\begin{array}{rl}
\$14.40 & \text{2 decimal places} \\
\times\ 10.5 & \text{1 decimal place} \\
\hline
7\ 200 & \\
144\ 00 & \\
\hline
\$151.20(0) & \text{3 decimal places}
\end{array}
$$

METHOD II

$$
\begin{array}{rl}
\$14.40 & \text{Regular pay} \\
+\ 7.20 & \frac{1}{2}\text{ time} \\
\hline
\$21.60 & \text{Time-and-a-half hourly rate} \\
\times\ 7 & \text{Overtime hours earned} \\
\hline
\$151.20 & \text{Time-and-a-half wages}
\end{array}
$$

DOUBLE TIME

Four hours at double time means that you earn wages for 2 hours for each hour worked. The amount can be calculated in two ways.

METHOD I

$4 \times 2 = 8$ hours at regular pay:

$$
\begin{array}{r}
\$14.40 \\
\times\ 8 \\
\hline
\$115.20
\end{array}
$$

METHOD II

$$
\begin{array}{r}
\$14.40 \\
\times\ 2 \\
\hline
\$28.80 \\
\times\ 4 \\
\hline
\$115.20
\end{array}
$$

Regular pay

Double-time hourly rate

Thus, J. Garcia's wages for the week are:

$$
\begin{array}{r}
\$504.00 \\
151.20 \\
+115.20 \\
\hline
\$770.40
\end{array}
$$

Regular pay
Time-and-a-half
Double time

PRACTICE EXERCISE 6

1. Complete the payroll sheet for Ms. Williams, figuring regular pay and overtime pay for each of the other employees.

Employee	Wages per Hour	Regular		Time-and-a-Half		Double Time		Total Wages
		Hours	Pay	Hours	Pay	Hours	Pay	
A. Rodriquez	$12.40	35		3				
B. White	13.74							
C. Charles	12.80							
A. Roth	15.60							
J. Garcia	14.40	35	$504.00	7	$151.20	4	115.20	$770.40
J. Kelly	13.90							
S. Williams	13.75			—	—	—	—	

2. Find the total weekly payroll for all employees.

3. If Ms. Williams withdrew $5000 from the bank, how much money remained after all the employees had been paid?

4. Using your calculator, check each of the figures in Question 1.

GRAPHING THE DATA

Vertical Bar Graph

In Practice Exercise 1, you completed the table listing the weekly hour record for the six machinists at the Extra-Fine & Fussy Machine Shop. You calculated the total number of hours worked by each employee. The information in this table was later used to compute the weekly earnings for each employee.

Another method of picturing this information is to draw a *graph*. There are various types of graphs, and the one drawn below is a *bar graph*. Since the bars are vertical, this is called a *vertical bar graph*. The bar graph must have a title and a vertical and a horizontal scale.

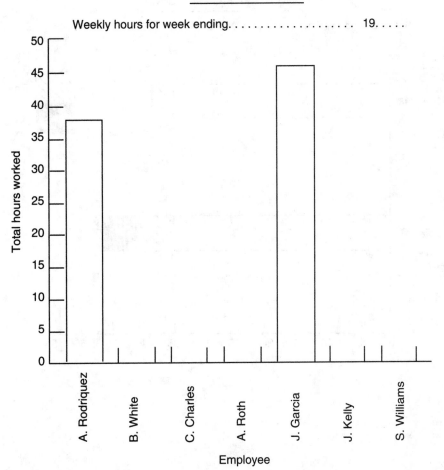

EFF MACHINE SHOP

Weekly hours for week ending. . . . . 19. . . . .

PRACTICE EXERCISE 7

1. Draw the vertical bars for the other employees, using the data from Practice Exercise 1.

Use the graph to answer Problems 2–4.

2. Which employee worked the greatest number of total hours?

3. Which employee worked the least number of total hours?

4. Which employees, if any, worked the same number of total hours?

Horizontal Bar Graph

Sometimes you desire to obtain more than one piece of information from a graph. The vertical bar graph you just completed told you only the total number of hours each employee worked. The number of time-and-a-half or double-time hours was included in the total. Below, regular hours, time-and-a-half, and double-time are pictured on the graph. This is a *horizontal bar graph* since the bars are drawn horizontally. The *legend* is on the right-hand side of the graph.

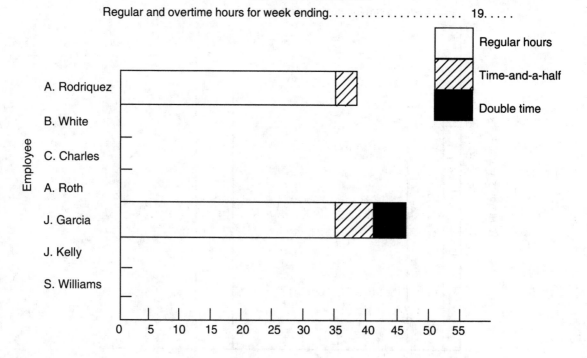

EFF MACHINE SHOP

Regular and overtime hours for week ending. 19. . . . .

PRACTICE EXERCISE 8

1. Draw the horizontal bars for the other employees, using the data from Practice Exercise 5.

Use the graph to answer Problems 2–4.

2. Which employee(s) worked time-and-a-half hours?

3. Which employee(s) earned double-time hours?

4. Which, if any, of the employees worked less than the 35-hour regular work week?

MEASUREMENTS OF LENGTH

Mr. Frank has hired George Green as an apprentice machinist at the Extra-Fine & Fussy Machine Shop. George's first project as a new worker is to take a short training program under Mr. Frank's direction. He must learn how to measure accurately with a ruler.

Mr. Frank begins with linear measurement from "olden days." In those days a "foot" was the actual length of a man's foot, and a "yard" was the length of the distance from a man's nose to the thumb of his outstretched arm.

These units of measurement, as you see, could vary from one person to another. In other words, they were not "standard" units of measure.

PRACTICE EXERCISE 9

Choose three other people. Have each person remove his or her right shoe. Trace the outline of each right foot on a sheet of paper.

1. Compare the tracings of the three right feet.

2. Do the three feet measure the same length?

3. Suppose that each of two people with different-sized feet wanted to purchase 3 feet of linen. Who would receive the smaller quantity, the person with the smaller or the one with the larger feet?

Standard Unit of Measure

The varying sizes of people's feet led to a need for a "standard" unit of measure. The *yard* became the standard. The yard is divided into 3 equal parts, and each part is called a *foot*. Each foot is divided into 12 equal parts. Each of the 12 equal parts is called an *inch*. Thus:

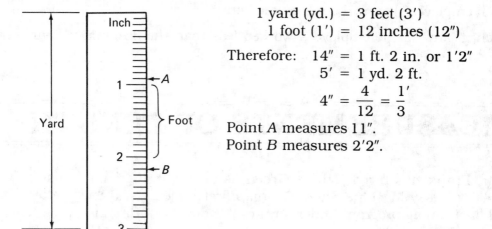

$$1 \text{ yard (yd.)} = 3 \text{ feet (3')}$$
$$1 \text{ foot (1')} = 12 \text{ inches (12")}$$

Therefore: $14" = 1$ ft. 2 in. or $1'2"$
$$5' = 1 \text{ yd. 2 ft.}$$
$$4" = \frac{4}{12} = \frac{1'}{3}$$

Point *A* measures 11".
Point *B* measures 2'2".

PRACTICE EXERCISE 10

1. Change each of these measurements to the indicated units:

   a. $18" = $ _____' _____"   e. $60" = $ _____yd. _____ft.
   b. $42" = $ _____' _____"   f. $10" = $ _____ft.
   c. $16' = $ _____yd. _____ft.   g. $3'4" = $ _____"
   d. 4 yd. = _____ft.   h. $195" = $ _____yd. _____ft. _____in.

2. Using the pictured rulers below, read each measurement and write your answers in the blank spaces at the right:

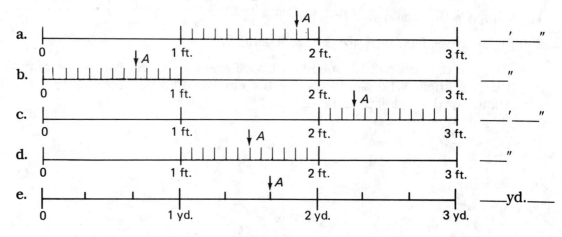

3. Locate each measure on the ruler. Draw an arrow to indicate the point, and label it *A*.

a. 10″

0   1 ft.   2 ft.   3 ft.

b. 1′4″

0   1 ft.   2 ft.   3 ft.

c. 2 yd. 2 ft.

0   1 yd.   2 yd.   3 yd.

d. 5′

0   1 yd.   2 yd.   3 yd.

e. 28″

0   1 ft.   2 ft.   3 ft.

The Inch

The inch is the smallest standard unit in the United States system of linear measure:

On the ruler below the distance *AB* measures 3″.

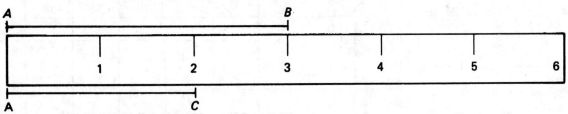

The distance *AC* measures 2″.

PRACTICE EXERCISE 11

1. This is a ruler whose unit of length is 1″. Use this ruler to find the lengths of the line segments shown.

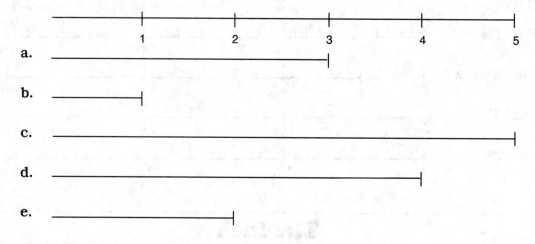

2. A ruler whose unit of length is $\frac{1″}{2}$ is shown below. Find the lengths of line segments *a–j*, and write them in the blanks provided. Then answer Problems 2.a–2.o.

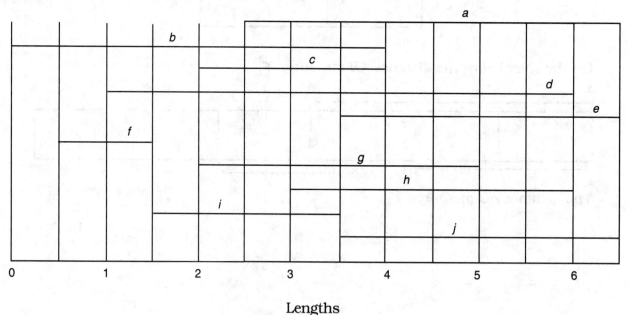

Lengths

a. ____   *c.* ____   *e.* ____   *g.* ____   *i.* ____
b. ____   *d.* ____   *f.* ____   *h.* ____   *j.* ____

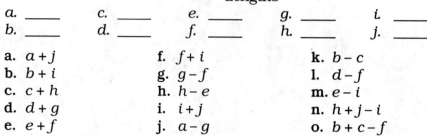

a. $a+j$ **f.** $f+i$ **k.** $b-c$
b. $b+i$ **g.** $g-f$ **l.** $d-f$
c. $c+h$ **h.** $h-e$ **m.** $e-i$
d. $d+g$ **i.** $i+j$ **n.** $h+j-i$
e. $e+f$ **j.** $a-g$ **o.** $b+c-f$

Parts of an Inch

Apprentice George Green has learned the simple rudiments of measurement, but he is a little weak on fractions. Will this prevent him from succeeding as a machinist? Yes, since measurements of less than an inch require him to use fractions.

Some of these are shown below.

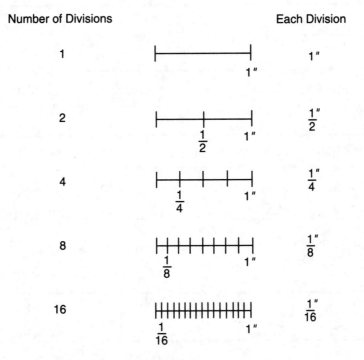

George, when checking the size of this bolt,

discovers that it measures $1\frac{3}{4}''$ excluding the head. If the head measures $\frac{1}{4}''$, the length of the bolt can be determined by adding $\frac{1}{4}''$ to $1\frac{3}{4}''$. Thus:

$$\begin{array}{r} 1\frac{3}{4} \\ + \frac{1}{4} \\ \hline 1\frac{4}{4} = 1 + 1 = 2'' \end{array}$$

PRACTICE EXERCISE 12

1. This is a ruler whose unit of length is $\frac{1''}{2}$. Use the ruler to find the lengths of these line segments:

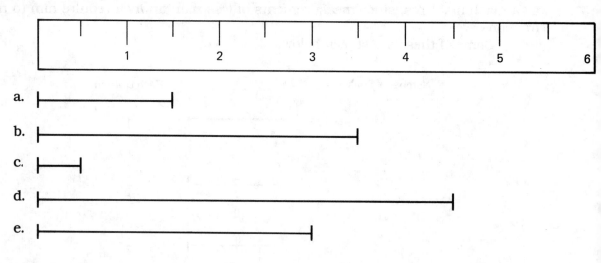

2. A ruler whose unit of length is $\frac{1''}{4}$ is shown. Use the ruler to find the lengths of these line segments:

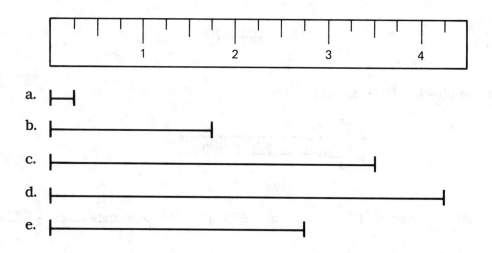

3. A ruler whose unit of length is $\frac{1}{4}$″ is shown below. Find the lengths of line segments *a–j*, and write them in the blanks provided. Then answer Problems 3.a–3.o.

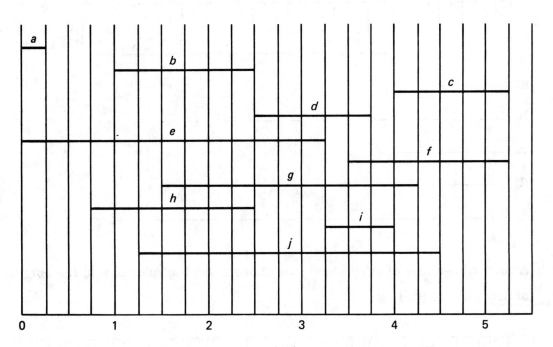

Lengths

a. ____	*c.* ____	*e.* ____	*g.* ____	*i.* ____
b. ____	*d.* ____	*f.* ____	*h.* ____	*j.* ____

a. $a+j$ **f.** $f+i$ **k.** $b-c$

b. $b+i$ **g.** $g-f$ **l.** $f-d$

c. $c+h$ **h.** $e-h$ **m.** $e-i$

d. $d+g$ **i.** $i+j$ **n.** $h+j-i$

e. $e+f$ **j.** $g-a$ **o.** $b+c-f$

4. A ruler whose unit of length is $\frac{1}{8}$" is shown. Use the ruler to find the lengths of these line segments:

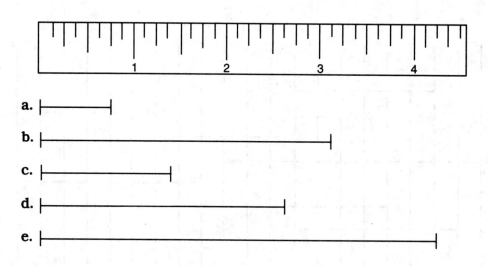

    a.

    b.

    c.

    d.

    e.

5. A ruler whose unit of length is $\frac{1}{16}$" is shown. Use the ruler to find the lengths of these line segments:

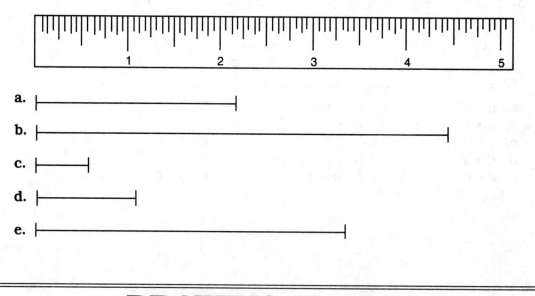

    a.

    b.

    c.

    d.

    e.

DRAWING PLANS

At the Extra-Fine & Fussy Machine Shop, Mr. Green will have to draw plans that agree with the information provided him. He must be able to draw actual size plans like the upper horizontal panel shown on page 115.

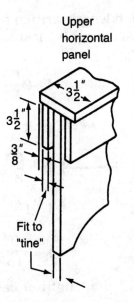

The plan requires the ability to draw lines of different lengths. It is essential that Mr. Green learn to use a ruler correctly.

PRACTICE EXERCISE 13

1. Look at each line segment drawn below and *estimate* its length. Fill in each estimate in the first column of the table provided.

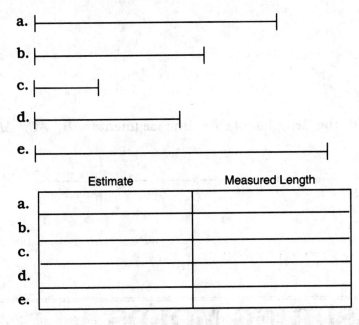

2. Use your ruler to measure the actual length of each line segment shown in Problem 1. Fill in the measured length in the second column of the table.

3. How accurate were your estimates? Would you rely on their accuracy if you were George Green?

4. Draw a line segment for each length written below. Start each line segment at the vertical line provided. Accuracy is most important.

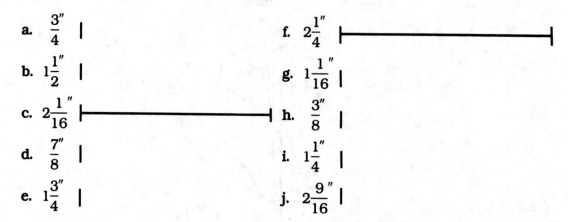

a. $\frac{3''}{4}$ |

b. $1\frac{1}{2}''$ |

c. $2\frac{1}{16}''$ |————————————|

d. $\frac{7''}{8}$ |

e. $1\frac{3}{4}''$ |

f. $2\frac{1}{4}''$ |————————————————|

g. $1\frac{1}{16}''$ |

h. $\frac{3''}{8}$ |

i. $1\frac{1}{4}''$ |

j. $2\frac{9}{16}''$ |

5. Use the ruler shown to find the length of each line segment. Line segment $AB = \frac{9''}{16}$.

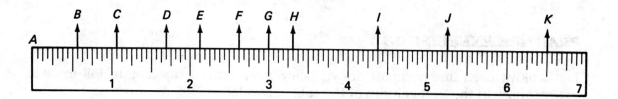

a. $\overline{AB}$ f. $\overline{AG}$
b. $\overline{AC}$ g. $\overline{AH}$
c. $\overline{AD}$ h. $\overline{AI}$
d. $\overline{AE}$ i. $\overline{AJ}$
e. $\overline{AF}$ j. $\overline{AK}$

6. Use your ruler to find the lengths of the line segments $\overline{AB}$, $\overline{AC}$, $\overline{AD}$, $\overline{AE}$, $\overline{AF}$, $\overline{AG}$, and $\overline{AH}$.

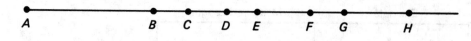

MEASURING MOLDS

A mold is a device used by machinists to form an accurate model or shape. The dimensions of a mold must be found before it can be duplicated.

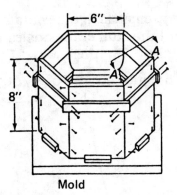

Mold

These dimensions may include the length, width, height, and the angle that is formed when two sides meet. An *angle* is the opening between two intersecting lines, and it is measured by the number of degrees in the opening of the angle. A *protractor* is a measuring instrument used to find the number of degrees in an angle.

This is angle *A*, written as ∠ A:

Different types of angles are listed below.

Angle	Name	Drawing
Greater than 0° but less than 90°	Acute angle	
Equal to 90°	Right angle	
Greater than 90° but less than 180°	Obtuse angle	
Equal to 180°	Straight angle	
Greater than 180° but less than 360°	Reflex angle	

A *triangle* is a closed geometric figure containing three angles. The angles are measured in degrees, and the sum of the angles in a triangle is *always* 180°. In $\triangle EFG$, if $\angle E = 40°$ and $\angle F = 75°$, then

$$\angle G = 65° \text{ since } 40° + 75° + \angle G = 180°:$$
$$115° \quad + \angle G = 180°$$

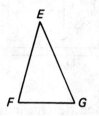

Subtracting 115° from both sides of the equation results in

$$\angle G = 65°$$

PRACTICE EXERCISE 14

1. In the following cross section of a simple mold, find the lengths of the line segments $\overline{AB}, \overline{BC}, \overline{CD}, \overline{DE}, \overline{EF}, \overline{FA}, \overline{GH}, \overline{HI}, \overline{IJ}, \overline{JG}$.

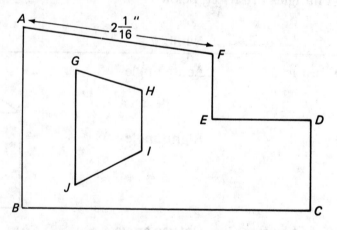

2. A cross section of a simple mold is shown below. Find the lengths of the line segments $\overline{AB}, \overline{BC}, \overline{CD}, \overline{DM}, \overline{MA}, \overline{EF}, \overline{FG}, \overline{GE}$.

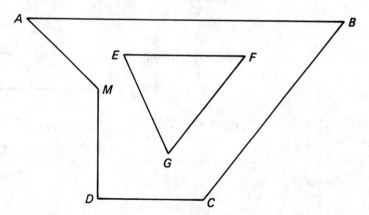

3. In the figure below, classify each of the angles: ∠A, ∠B, ∠C, ∠D, ∠E, ∠F, ∠G, ∠H, ∠I, ∠J. The arc indicates the angle.

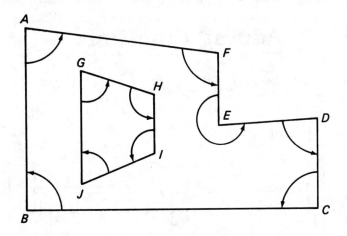

4. If ∠A = 50° and ∠C = 70° in △ABC, find the number of degrees in ∠B.

5. In △XYZ, ∠X = 70° and ∠Y = ∠Z. Find the number of degrees in ∠Y.

6. Find the number of degrees in ∠A of △ABC if ∠B = 90° and ∠C = 60°.

7. In the figure below, classify each of the angles: ∠A, ∠B, ∠C, ∠D, ∠E, ∠F, ∠G, ∠M. The arc indicates the angle.

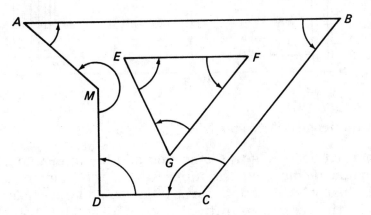

WORKING WITH FRACTIONS

Adding Fractions

Mr. Frank gives George Green a copy of a metal wall shelf that he must build. What is the length of dimension A?

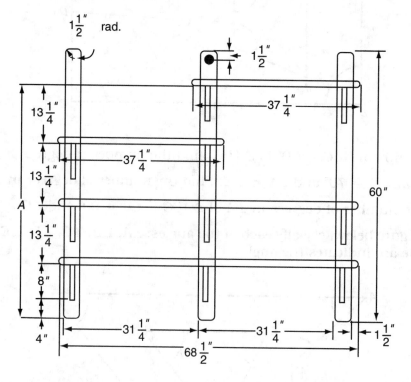

To find the length of A, you add $13\frac{1}{4}'' + 13\frac{1}{4}'' + 13\frac{1}{4}'' + 8'' + 4''$. Since the denominator of each fraction is the same, you add the numerators and keep the same denominator. (*Remember:* The numerator of the fraction is the top part, while the denominator is the bottom part.) Then you add the whole numbers, and combine with the result obtained by adding the fractions. Thus:

$$
\begin{array}{r}
13\frac{1}{4}'' \\
13\frac{1}{4}'' \\
13\frac{1}{4}'' \\
8\ \ '' \\
+\ 4\ \ '' \\
\hline
51\frac{3}{4}''
\end{array}
$$

REMINDER! *The denominators of all the fractions must be alike before you can add.*

In this mold, you see that the dimensions are not the same, so you must change each denominator to a common denominator:

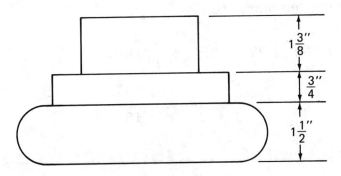

The *lowest* common denominator for these three fractions, $\frac{3}{8}$, $\frac{3}{4}$, and $\frac{1}{2}$, is 8 since the denominators 8, 4, and 2 will all divide into the number 8.

EXAMPLE: What is the common denominator for the fractions $\frac{1}{4}$, $\frac{1}{2}$, $\frac{5}{8}$, and $\frac{11}{16}$?

SOLUTION: What number will 4, 2, 8, and 16 all divide into without leaving any remainder? You see that 16 is the lowest common denominator since

$$4\overline{)16}$$
$$2\overline{)16}$$
$$8\overline{)16}$$
$$\text{and} \quad 16\overline{)16}$$

Once you have found the lowest common denominator, or L.C.D., for the four fractions, you must change each fraction to this new denominator. Therefore:

$$\frac{1}{4} = \frac{?}{16}, \quad \frac{1}{2} = \frac{?}{16}, \quad \text{etc.}$$

As you can see, you must multiply the denominator of the fraction $\frac{1}{4}$ by 4 to get the value 16. But, if you do this, you must multiply the numerator by 4 also so that the value of the fraction remains the same. Multiplying by $\frac{4}{4}$ is the same as multiplying by 1 since $\frac{4}{4} = 1$.

$$\frac{1}{4} \times \frac{4}{4} = \frac{4}{16}$$

EXAMPLE: What should you multiply the fraction $\frac{1}{2}$ by to get the new denominator, 16?

SOLUTION: Let's multiply by $\frac{8}{8}$. Thus:

$$\frac{1}{2} \times \frac{8}{8} = \frac{8}{16}$$

Then, for the height of the mold, we have:

$$1\frac{3}{8} \qquad = \frac{3}{8}$$
$$\frac{3 \times 2}{4 \times 2} = \frac{6}{8}$$
$$+\ 1\frac{1 \times 4}{2 \times 4} = \frac{4}{8}$$
$$\overline{2} \qquad \overline{\frac{13}{8}} = 1\frac{5}{8}'' \quad \text{and} \quad 2 + 1\frac{5}{8} = 3\frac{5}{8}''$$

PRACTICE EXERCISE 15

1. Find the lowest common denominator for each group of fractions.

   a. $\frac{1}{2}, \frac{3}{4}, \frac{3}{8}$ d. $\frac{1}{2}, \frac{1}{4}, \frac{11}{32}$

   b. $\frac{5}{8}, \frac{5}{16}, \frac{1}{2}$ e. $\frac{1}{2}, \frac{11}{16}, \frac{3}{4}$

   c. $\frac{3}{32}, \frac{1}{4}, \frac{3}{8}$ f. $\frac{7}{8}, \frac{21}{32}, \frac{13}{16}, \frac{1}{2}$

2. Add each group of fractions in Problem 1.a–1.f.

3. Find the total width *B* of this room divider:

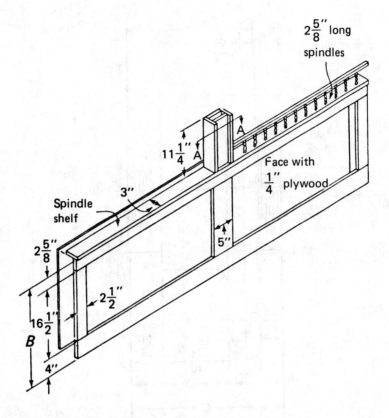

4. Find dimension *A* in the tine detail.

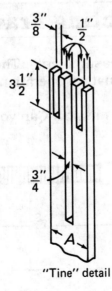

"Tine" detail

5.   Find the missing dimensions *A*, *B*, and *C*.

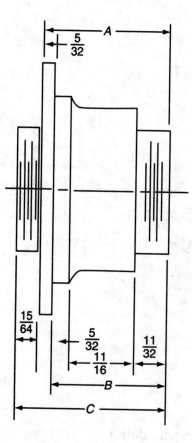

Subtracting Fractions

Many plans omit some measurements. These missing dimensions can be found by using one of the mathematical operations of adding, subtracting, multiplying, and dividing.

In this bolt, dimension *A* is missing. Can you help Mr. Green find it?

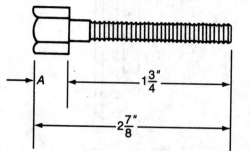

Dimension A can be determined by subtracting $1\frac{3''}{4}$ from $2\frac{7''}{8}$. To subtract fractions you proceed in the same way as to add them. Therefore the first step in subtracting these mixed numbers is to find the lowest common denominator. The L.C.D. for the two fractions $\frac{7}{8}$ and $\frac{3}{4}$ is 8. Thus:

$$2\frac{7}{8} \qquad \text{becomes} \qquad 2\frac{7}{8} \quad = \quad \frac{7}{8}$$
$$-\ 1\frac{3}{4} \qquad\qquad\qquad -\ 1\frac{3}{4} \times \frac{\times\ 2}{\times\ 2} = \frac{6}{8}$$
$$1 \qquad\qquad\qquad \frac{1}{8} = 1\frac{1''}{8} = A$$

What would occur if the longer dimension were $2\frac{1''}{8}$ rather than $2\frac{7''}{8}$? The procedure would be the same as that shown above except for a slight change. Thus:

$$2\frac{1}{8} \qquad = \quad \frac{1}{8}$$
$$-\ 1\frac{3}{4} \times \frac{\times\ 2}{\times\ 2} = \frac{6}{8}$$

But you can't subtract $\frac{6}{8}$ from $\frac{1}{8}$, so you borrow 1 from the 2 and exchange it for $\frac{8}{8}$. Therefore:

$$2\frac{1}{8} = 1 + 1 + \frac{1}{8} \qquad \text{and} \qquad 2\frac{1}{8} = 1\frac{9}{8}$$
$$= 1 + \frac{8}{8} + \frac{1}{8} \qquad\qquad\qquad -\ 1\frac{6}{8} = 1\frac{6}{8}$$
$$2\frac{1}{8} = 1\frac{9}{8} \qquad\qquad\qquad\qquad \frac{3}{8} \quad \text{Thus } A = \frac{3''}{8}.$$

PRACTICE EXERCISE 16

1. Write each of the following mixed numbers in equivalent form:

   a. $4\frac{5}{8} = 3\frac{?}{8}$

   b. $2\frac{1}{2} = 1\frac{?}{2}$

   c. $4\frac{3}{4} = 3\frac{?}{4}$

   d. $6\frac{3}{16} = 5\frac{?}{16}$

   e. $5 = 4\frac{?}{2}$

   f. $28\frac{1}{32} = 27\frac{?}{32}$

   g. $15\frac{7}{8} = 14\frac{?}{8}$

   h. $6\frac{17}{32} = 5\frac{?}{32}$

   i. $5\frac{1}{4} = 4\frac{?}{4}$

   j. $3\frac{5}{8} = 3\frac{?}{16} = 2\frac{?}{16}$

2. Subtract:

   a. $5\frac{5}{8}$
   $-2\frac{1}{8}$

   b. $6\frac{3}{4}$
   $-5\frac{1}{4}$

   c. $9\frac{13}{16}$
   $-1\frac{5}{8}$

   d. $6\frac{7}{8}$
   $-4\frac{1}{2}$

   e. $10\frac{9}{16}$
   $-5\frac{3}{8}$

   f. $7\frac{1}{4}$
   $-3\frac{3}{4}$

   g. $8\frac{3}{16}$
   $-2\frac{1}{2}$

   h. $5\frac{1}{8}$
   $-3\frac{5}{16}$

   i. $2\frac{7}{32}$
   $-\frac{7}{8}$

   j. $17\frac{1}{16}$
   $-13\frac{3}{32}$

3. Find the missing dimensions A and B in the diagram below.

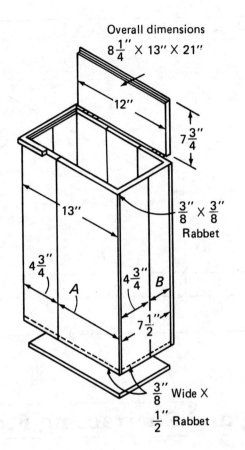

Overall dimensions
$8\frac{1}{4}'' \times 13'' \times 21''$

12"

$7\frac{3}{4}''$

13"

$\frac{3}{8}'' \times \frac{3}{8}''$
Rabbet

$4\frac{3}{4}''$

A

$4\frac{3}{4}''$ B

$7\frac{1}{2}''$

$\frac{3}{8}''$ Wide X

$\frac{1}{2}''$ Rabbet

4. Find the missing dimension A in the end board shown below.

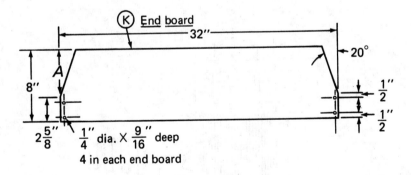

Ⓚ End board
—32"—
20°

8"
A

$\frac{1}{2}''$

$\frac{1}{2}''$

$2\frac{5}{8}''$ $\frac{1}{4}''$ dia. X $\frac{9}{16}''$ deep
4 in each end board

5.   Find the missing dimensions *A* and *B* in this drawing:

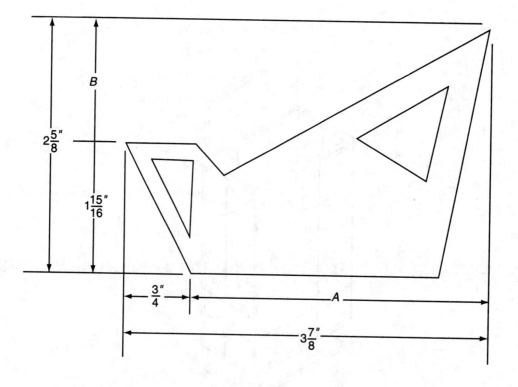

Adding *and* Subtracting Fractions

Sometimes Mr. Green may have to combine adding and subtracting fractions to find a missing dimension. In the diagram below, he must find dimension *A*.

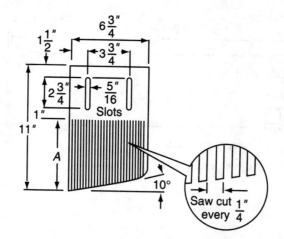

To find dimension *A*, you add the dimensions $1''$, $2\frac{3}{4}''$, and $1\frac{1}{2}''$ together and then subtract their sum from the overall length of $11''$. Thus:

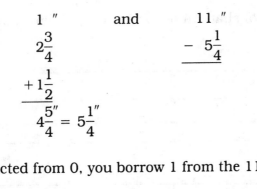

Since $\dfrac{1}{4}$ can't be subtracted from 0, you borrow 1 from the 11 and exchange it for

$\dfrac{4}{4}$. Therefore $11 = 10\dfrac{4}{4}$.

$$11 = 10\dfrac{4}{4}$$
$$- \ 5\dfrac{1}{4} = \ 5\dfrac{1}{4}$$
$$5\dfrac{3}{4} \quad A = 5\dfrac{3}{4}''$$

PRACTICE EXERCISE 17

1. A cabinet of the type shown below has adjustable shelves. If they are inserted as indicated in the drawing, what is dimension A?

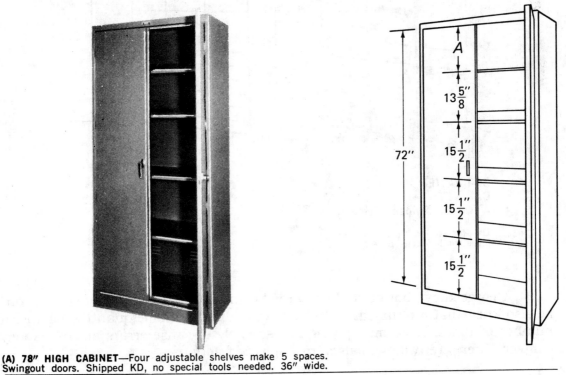

(A) 78″ HIGH CABINET—Four adjustable shelves make 5 spaces. Swingout doors. Shipped KD, no special tools needed. 36″ wide.

Acknowledgment: Reprinted from the 1981 Office Products Catalog, copyright 1980, by permission of United Stationers Supply Co.

2. Find the length of shank *A* of the 1″ drill bit shown below.

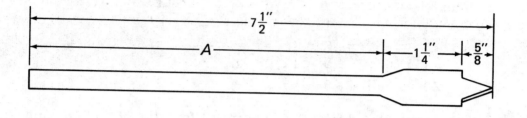

Multiplying Fractions

A metal storage cabinet was received by George with instructions from Mr. Frank. George must figure the amount of material necessary to reproduce the cabinet. The cabinet was of the type shown below but had the dimensions indicated on the drawing. Each of the ten drawers measured $3\frac{1}{8}''$, the bottom was $4\frac{3}{8}''$, and the top was $\frac{7}{8}''$.

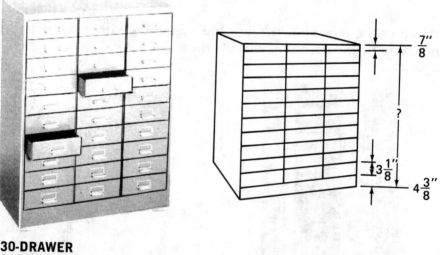

**30-DRAWER
CABINETS**
**— WITH 2 DIVIDERS IN EACH
DRAWER**
Use for storing small parts or
letter and legal size papers.

George knew that he could add the heights of the 10 drawers together, but he also knew that multiplying is faster than adding the same quantity together a number of times. Also, multiplying is less likely to cause errors and is usually simpler. There are two methods of multiplying fractions.

Acknowledgment: Reprinted from the 1981 Office Products Catalog, copyright 1980, by permission of United Stationers Supply Co.

METHOD I

To multiply fractions together, you can multiply the numerators and multiply the denominators. If the problem contains a mixed number, you change it to an improper fraction. Thus:

$$10 \times 3\frac{1}{8} = \frac{10}{1} \times \frac{25}{8} = \frac{250}{8}$$
$$= 31\frac{2}{8}$$
$$= 31\frac{1}{4}{}''$$

METHOD II

A second method, called "cancellation," allows you to work with smaller numbers. To cancel, find a number that divides evenly into one of the denominators *and* one of the numerators. In this example, 2 is divided into 10 and 8.

$$10 \times 3\frac{1}{8} = \frac{\overset{5}{\cancel{10}}}{1} \times \frac{25}{\underset{4}{\cancel{8}}}$$
$$= \frac{125}{4}$$
$$= 31\frac{1}{4}{}''$$

Adding the top and bottom dimensions of the storage cabinet to this sum:

$$
\begin{array}{r|l}
31\frac{1}{4} & \frac{2}{8} \\[4pt]
4\frac{3}{8} & \frac{3}{8} \\[4pt]
+\ \frac{7}{8} & \frac{7}{8} \\[2pt]
\hline
35 & \frac{12}{8} = 1\frac{4}{8} = 1\frac{1}{2}
\end{array}
$$

gives $35 + 1\frac{1}{2}{}'' = 36\frac{1}{2}{}''$, the overall height of the storage cabinet.

You may use either Method I or Method II when you do the practice exercises.

PRACTICE EXERCISE 18

1. Multiply:

    a. $\dfrac{3}{4} \times \dfrac{7}{8}$ f. $\dfrac{15}{16} \times \dfrac{3}{5}$

    b. $\dfrac{1}{2} \times 2\dfrac{1}{4}$ g. $1\dfrac{7}{8} \times 4\dfrac{1}{4}$

    c. $5\dfrac{3}{8} \times 8$ h. $16 \times 5\dfrac{1}{32}$

    d. $28 \times \dfrac{7}{8}$ i. $7\dfrac{1}{2} \times 6\dfrac{11}{16}$

    e. $12\dfrac{3}{4} \times 4\dfrac{1}{4}$ j. $3\dfrac{3}{8} \times 2\dfrac{1}{2}$

2. Find the overall height of a storage cabinet of the type shown below but with the following dimensions: each of the four drawers measures $12\dfrac{5}{8}''$, and the top and the bottom of the cabinet each measures $1\dfrac{1}{8}''$.

Creative use of 4-drawer laterals organizes your office space into efficient, individual work areas.

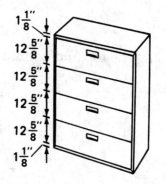

Acknowledgment: Reprinted from the 1981 Office Products Catalog, copyright 1980, by permission of United Stationers Supply Co.

3. If each of the 26 spaces is $\frac{3''}{8}$ and the top margin measures $1\frac{1''}{4}$, how long is the sheet of paper shown below?

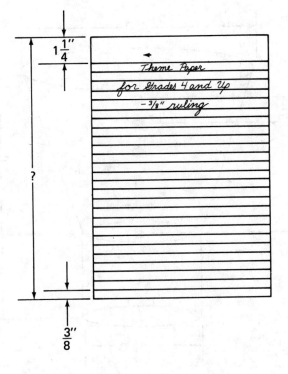

4. This steel bolt is to be produced:

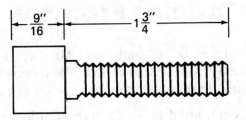

    a. How many inches of material will be needed to produce 24 *dozen* bolts?
    b. How many feet of material will be required?
    c. If steel costs $1.36 a foot, what will the total cost be to produce the necessary number of bolts?

5.   The funnel-shaped adapter, used to connect two aluminum ducts of different sizes, is frequently made by the tinsmith. Find dimension *A* if the holes are equally spaced.

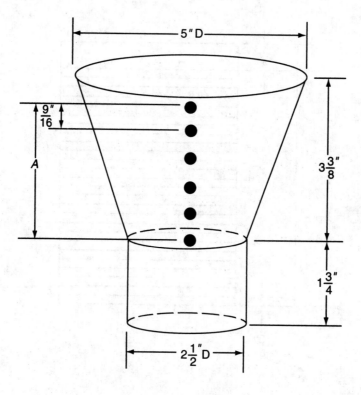

Dividing Fractions

A metal storage cabinet is 78″ high, and Mr. Green is told to locate points inside the cabinet to place 3 shelves so that all 4 spaces will be equally spaced. First, he measures the inside dimension of the cabinet and finds its height to be 72″. If the thickness of each shelf is $\dfrac{3''}{4}$, where should he mark the locations of the shelves?

Since the thickness of each shelf is $\dfrac{3''}{4}$, and there are 3 shelves, he multiplies:

$$3 \times \frac{3''}{4} = \frac{3}{1} \times \frac{3}{4}$$
$$= \frac{9''}{4}$$
$$= 2\frac{1}{4}''$$

Then he subtracts this quantity from the 72″ height:

$$72'' = 71\frac{4}{4}''$$

$$-\ 2\frac{1}{4}'' = \ 2\frac{1}{4}''$$

$$69\frac{3}{4}''$$

The total space, $69\frac{3}{4}''$, must now be divided into 4 equal spaces.

REMINDER! *Dividing a fraction by a whole number is the same as multiplying by the reciprocal of the whole number.* Thus:

The reciprocal of 4 is $\frac{1}{4}$.

The reciprocal of $\frac{3}{4}$ is $\frac{4}{3}$.   Therefore:

$$69\frac{3}{4} \div 4 = 69\frac{3}{4} \times \frac{1}{4}$$

$$= \frac{279}{4} \times \frac{1}{4}$$

$$= \frac{279}{16}$$

$$= 17\frac{7}{16}''$$

Each space between two shelves is found to be $17\frac{7}{16}''$.

PRACTICE EXERCISE 19

1. Write the reciprocal of each of the following fractions or mixed numbers. (*Remember:* Change each mixed number to an improper fraction before finding its reciprocal.)

   a. $\dfrac{7}{8}$ c. $\dfrac{5}{16}$ e. $\dfrac{5}{3}$ g. 8 i. $\dfrac{31}{32}$

   b. $\dfrac{1}{4}$ d. $4\dfrac{1}{4}$ f. $6\dfrac{1}{2}$ h. $2\dfrac{3}{8}$ j. $4\dfrac{5}{16}$

2. Divide:

   a. $\dfrac{1}{4} \div \dfrac{1}{2}$ d. $24\dfrac{1}{2} \div 8$ g. $6 \div \dfrac{11}{16}$ i. $1\dfrac{15}{16} \div \dfrac{3}{4}$

   b. $2\dfrac{1}{4} \div \dfrac{3}{8}$ e. $6\dfrac{1}{8} \div 4\dfrac{3}{4}$ h. $\dfrac{7}{16} \div 7$ j. $12\dfrac{17}{32} \div 6\dfrac{3}{16}$

   c. $\dfrac{31}{32} \div 1\dfrac{7}{7}$ f. $5\dfrac{1}{2} \div 3\dfrac{3}{8}$

3. Find the height of each drawer in a cabinet of the type shown below with the dimensions indicated on the drawing.

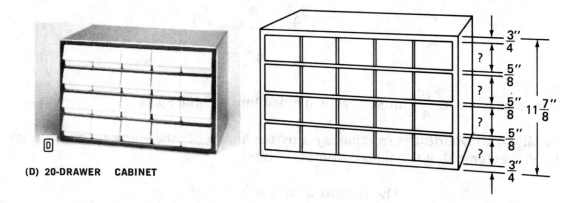

(D) 20-DRAWER CABINET

4. The candelabra shown below has 7 holes equally spaced in the center post, starting $1\frac{3}{8}''$ from the top. It is $6\frac{5}{8}''$ from the top of the center post to the last hole. What is the distance between centers of neighboring holes?

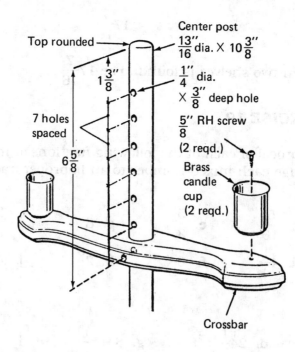

Acknowledgment: Reprinted from the 1981 Office Products Catalog, copyright 1980, by permission of United Stationers Supply Co.

ENLARGING

The drawing on a blueprint is usually smaller than the actual size of the object. It is Mr. Green's job to redraw the part from the blueprint to its full size. The "scale factor" written on the blueprint tells him how he can enlarge the drawing.

EXAMPLE: A steel bolt with a scale factor of $\dfrac{1''}{2} = 1''$ was sent to Mr. Green for enlarging. How should he go about doing this?

SOLUTION: This "scale factor" meant that each $\dfrac{1''}{2}$ on the drawing was equivalent to a real dimension of 1".

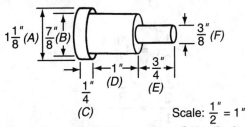

Scale: $\dfrac{1''}{2} = 1''$

Mr. Green measured all the parts of the bolt as illustrated in the diagram above. Since each $\dfrac{1''}{2}$ on the blueprint was equivalent to the real size of 1", he had to discover how many half-inches there were in each dimension. Thus:

$$(A)\quad 1\frac{1''}{8} \div \frac{1}{2} = \frac{9}{\overset{}{\underset{4}{\cancel{8}}}} \times \frac{\overset{1}{\cancel{2}}}{1}$$

$$= \frac{9}{4}$$

$$= 2\frac{1''}{4}$$

and

$$(B)\quad \frac{7''}{8} \div \frac{1}{2} = \frac{7}{\overset{}{\underset{4}{\cancel{8}}}} \times \frac{\overset{1}{\cancel{2}}}{1}$$

$$= \frac{7}{4}$$

$$= 1\frac{3''}{4}$$

Another method of finding the real size is to use a proportion, which you learned earlier:

$$\frac{\frac{1''}{2}}{1''} = \frac{1\frac{1''}{8}}{x''}$$

Since the product of the means is equal to the product of the extremes, you get

$$\frac{1}{2} \cdot x = 1 \cdot 1\frac{1}{8}$$

$$\frac{1}{2}x = 1\frac{1}{8}$$

To undo the multiplication of $\frac{1}{2}$ and x, you divide both sides of the equation by $\frac{1}{2}$. Thus:

$$\frac{1}{2}x = 1\frac{1}{8}$$

$$\frac{\frac{1}{2}x}{\frac{1}{2}} = \frac{\frac{9}{8}}{\frac{1}{2}}$$

$$x = \frac{9}{8} \div \frac{1}{2}$$

$$x = \frac{9}{\overset{}{\cancel{8}}} \times \frac{\overset{1}{\cancel{2}}}{1}$$

$$x = 2\frac{1''}{4}$$

Using a proportion to find *(B)*, you get

$$\frac{\frac{1''}{2}}{1''} = \frac{\frac{7''}{8}}{x''}$$

$$\frac{1}{2} \cdot x = \frac{7}{8} \cdot 1$$

$$\frac{\frac{1}{2}x}{\frac{1}{2}} = \frac{\frac{7}{8}}{\frac{1}{2}}$$

$$x = \frac{7}{8} \div \frac{1}{2}$$

$$x = \frac{7}{\underset{4}{\cancel{8}}} \times \frac{\overset{1}{\cancel{2}}}{1}$$

$$x = \frac{7}{4} = 1\frac{3}{4}''$$

If the scale factor was $\frac{1''}{4} = 1''$, what would Mr. Green do to each dimension to achieve a full-sized drawing? Certainly, it is clear that he would divide by $\frac{1}{4}$, or he could solve his problem by setting up a proportion and solving it. Use either method when solving the practice exercises.

PRACTICE EXERCISE 20

1. Find the actual size of each dimension in the illustrative example of the steel bolt drawn on page 137.

2. Find the actual dimensions of the bolt shown below if the scale factor is $\frac{1''}{4} = 1''$.

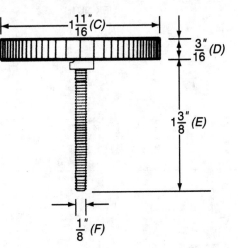

3. Find the actual dimensions of the mold of a tapered block shown below if the scale factor is $\frac{3''}{32} = 1''$.

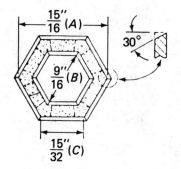

SHRINKING

Very small parts for mechanical objects are usually drawn to a larger scale than their actual sizes. There are times when it is necessary to "shrink" the blueprint drawing to its real size.

EXAMPLE 1: If a hex nut measures $\frac{3''}{8}$ on a blueprint and the scale factor is $1'' = \frac{1''}{4}$, what is the real size of the nut?

SOLUTION: You must multiply the measured dimension, $\frac{3''}{8}$, by $\frac{1}{4}$ to find the real size.

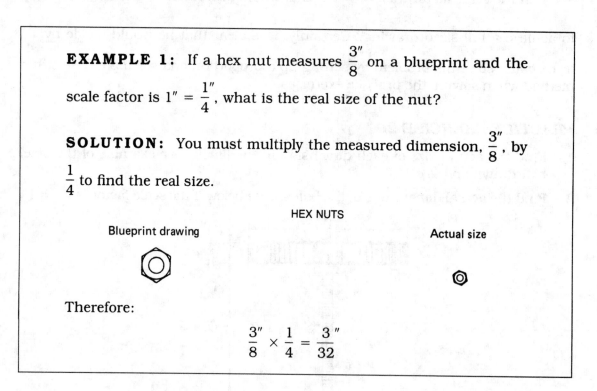

HEX NUTS

Blueprint drawing   Actual size

Therefore:

$$\frac{3''}{8} \times \frac{1}{4} = \frac{3''}{32}$$

EXAMPLE 2: What is the real measurement if the blueprint measurement is $\dfrac{7''}{8}$ and the scale factor is $1'' = \dfrac{1''}{8}$?

SOLUTION:

$$\frac{7''}{8} \times \frac{1}{8} = \frac{7''}{64}$$

If, instead, you use proportions, you obtain the following equations.

Example 1:

$$\frac{\dfrac{1''}{1''}}{\dfrac{1''}{4}} = \frac{\dfrac{3''}{8}}{x}$$

$$1 \cdot x = \frac{1}{4} \cdot \frac{3}{8}$$

$$x = \frac{3''}{32}$$

Example 2:

$$\frac{\dfrac{1''}{1''}}{\dfrac{1''}{8}} = \frac{\dfrac{7''}{8}}{x''}$$

$$1 \cdot x = \frac{1}{8} \cdot \frac{7}{8}$$

$$x = \frac{7''}{64}$$

PRACTICE EXERCISE 21

1. Using a "scale factor" of $1'' = \dfrac{1''}{8}$, find the real length of

   a. $\dfrac{1''}{4}$

   b. $\dfrac{1''}{2}$

   c. $1\dfrac{1''}{2}$

   d. $5\dfrac{7''}{8}$

   e. $1\dfrac{13''}{16}$

2.   Using a "scale factor" of $\frac{1''}{2}$ = 1", find the real length of

   a. $\frac{3''}{8}$ **c.** $2\frac{1''}{4}$ **e.** $\frac{15''}{32}$

   b. $1\frac{1''}{2}$ **d.** $6\frac{11''}{16}$

3.   The length of each screw on a blueprint is written below. If the scale factor is $1'' = \frac{1''}{4}$, find the actual length of each screw.

SCREWS

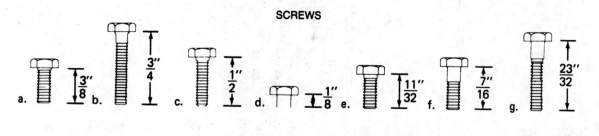

4.   A sleeveless drum sander is shown below. On the blueprint it measures $\frac{3''}{4}$ by $\frac{13''}{16}$. If the scale factor is $1'' = \frac{3''}{4}$, find its real size.

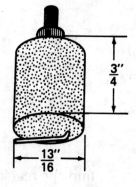

EXAM TIME

1. Mr. Jones worked 7, $7\frac{1}{2}$, $8\frac{1}{4}$, $8\frac{1}{2}$, and 9 hours last week. If he earns $13.84 an hour for 35 hours a week and time-and-a-half for all overtime hours, then:

   a. What was his regular pay?
   b. What was the total number of hours he worked last week?
   c. How many overtime hours did he work?
   d. How much overtime pay did he receive?
   e. What was his gross pay?

2. Joan worked from 8:30 A.M. to 12 o'clock. She then went to lunch, returned at 12:30, and left work at 4:45 P.M. How many hours did she work, excluding lunch?

3. Twenty minutes is what part of an hour?

4. $6\frac{1}{2}$ hours = ___6___ hr. ___?___ min.

5. 16 inches = ___?___ ft. ___?___ in.

6. Find the lowest common denominator for the fractions $\frac{5}{8}$, $\frac{3}{16}$, and $\frac{3}{4}$.

7. What is the length of the machine screw shown below?

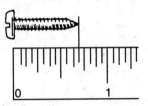

8. What is the length of the *front leg* in this reduced diagram?

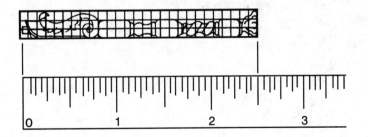

9. Which line segment in the diagram below measures $2\frac{3}{16}''$?

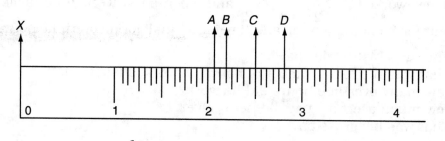

10. If a scale factor is $1'' = \frac{1''}{4}$, what is the real size of a measurement of $\frac{1''}{4}$?

11. $2\frac{7}{8} = 1\frac{?}{8}$

12. Find dimensions A and B in the drawing below.

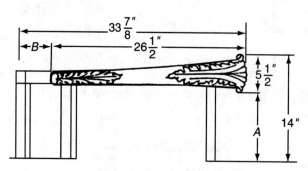

13. Draw a line segment measuring $3\frac{1''}{4}$, starting from point X below.

$\bullet$
X

14. Add: $5\frac{5}{8}$

$+\ 2\frac{15}{16}$

15. Subtract $7\frac{13}{32}$

$-\ 4\frac{5}{8}$

16. Find dimension *A* in this caster:

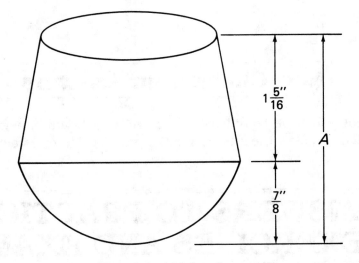

17. If there are seven equally spaced holes in this metal rod, find the distance, *A*, between neighboring holes.

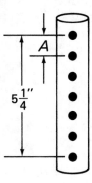

18. **a.** Measure the length of each of the sides, $\overline{AB}$, $\overline{BC}$, $\overline{CD}$, $\overline{DE}$, $\overline{EA}$, in the diagram below.

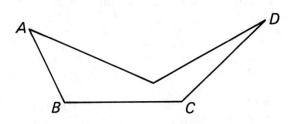

    b. Find the perimeter of the figure in Problem 18.a (*AB* + *BC* + *CD* + *DE* + *EA*).

19. Multiply: $2\dfrac{7}{8} \times 3\dfrac{1}{2}$.

20. Divide: $5\dfrac{3}{4} \div 1\dfrac{1}{8}$.

Now Check Your Answers

Now that you have completed the exam, check your answers against the correct ones, which follow the answers to the practice exercises below.

ANSWERS TO PRACTICE EXERCISES AND EXAM

PRACTICE EXERCISE 1

1. Mr. White   45 hr.   Ms. Garcia   46 hr.
    Mr. Charles   43 hr.   Ms. Kelly   38 hr.
    Mr. Roth   45 hr.

2. Tuesday   48 hr.
    Wednesday   45 hr.
    Thursday   49 hr.
    Friday   49 hr.
    Saturday   11 hr.
    Sunday   7 hr.

3. 255   4. 255   5. Yes

PRACTICE EXERCISE 2

1. 7   2. 3   3. Equal to   4. Less than, 1 hr.
5. Less than, 4 hr.   6. 23 hrs.

PRACTICE EXERCISE 3

1. a. 3 hr. 10 min.   b. 5 hr. 15 min.   c. 9 hr. 40 min.   d. 12 hr. 00 min.
    e. 6 hr. 07 min.   f. 9 hr. 00 min.   g. 11 hr. 45 min.   or
    h. 2 hr. 15 min.   i. 4 hr. 25 min.   j. 8 hr. 30 min.   00 hr. 00 min.

2. a. 3:30   b. 2:10   c. 10:00   d. 6:15   e. 9:40
    f. 7:12   g. 1:45   h. 4:50   i. 12:06   j. 11:10

3. **a.** 11 hr. 60 min. **b.** 3 hr. 70 min. **c.** 4 hr. 75 min. **d.** 6 hr. 90 min.
   e. 7 hr. 105 min. **f.** 0 hr. 70 min. **g.** 2 hr. 80 min. **h.** 5 hr. 65 min.
   i. 8 hr. 110 min. **j.** 10 hr. 108 min.

4. **a.** $\frac{1}{4}$ hr.   **f.** $\frac{1}{3}$ hr.

   b. $\frac{3}{4}$ hr.   **g.** $\frac{2}{3}$ hr.

   c. $\frac{1}{6}$ hr.   **h.** $\frac{5}{12}$ hr.

   d. $\frac{1}{2}$ hr.   **i.** $\frac{7}{15}$ hr.

   e. $\frac{5}{6}$ hr.   **j.** $\frac{1}{5}$ hr.

5. **a.** 2 hr. 30 min. **b.** 3 hr. 15 min. **c.** 4 hr. 45 min.
   d. 7 hr. 40 min. **e.** 18 min.

PRACTICE EXERCISE 4

1. | | | | | | |
   |---|---|---|---|---|---|
   | Mon. | 8 | Thurs. | 9 | Sat. | 0 |
   | Tues. | 7 | Fri. | 9 | Sun. | 3 |
   | Wed. | 7 | | | | |

2. 43 hr.

3.

	Rodriquez	White	Roth	Garcia	Kelly
Mon.	7	7	9	8	7
Tues.	7	9	9	7	9
Wed.	8	7	8	8	7
Thurs.	9	8	7	9	7
Fri.	7	10	8	7	8
Sat.	—	4	4	3	—
Sun.	—	—	—	4	—

PRACTICE EXERCISE 5

Employee	Regular Hours	Time-and-a-Half	Double Time
Rodriquez	35	3	—
White	35	6 + 4 = 10	—
Charles	35	5	3
Roth	35	6 + 4 = 10	—
Garcia	35	4 + 3 = 7	4
Kelly	35	3	—
Williams	35	—	—

PRACTICE EXERCISE 6

1.

Employee	Wages per Hour	Regular		Time-and-a-Half		Double Time		Total Wages
		Hours	Pay	Hours	Pay	Hours	Pay	
A. Rodriquez	$12.40	35	$434.00	3	$ 55.80	—	—	$489.80
B. White	13.74	35	480.90	10	206.10	—	—	687.00
C. Charles	12.80	35	448.00	5	96.00	3	$76.80	620.80
A. Roth	15.60	35	546.00	10	234.00	—	—	780.00
J. Garcia	14.40	35	504.00	7	151.20	4	115.20	770.40
J. Kelly	13.90	35	486.50	3	62.55	—	—	549.05
S. Williams	13.75	35	481.25	—	—	—	—	481.25

2. $4378.30

3. $621.70

PRACTICE EXERCISE 7

1.

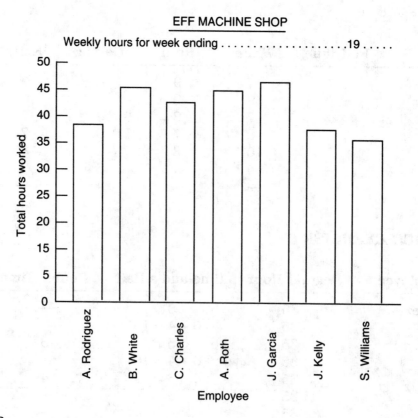

EFF MACHINE SHOP

Weekly hours for week ending . . . . .19 . . . . .

2. J. Garcia

3. Ms. Williams

4. A. Roth and B. White

PRACTICE EXERCISE 8

1.

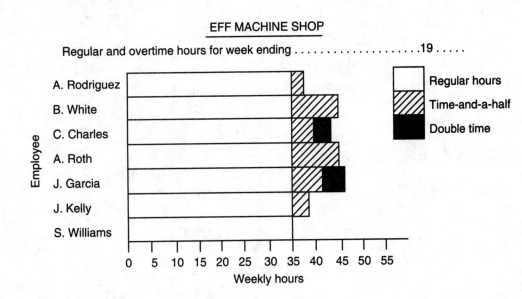

2. All except S. Williams

3. Mr. Charles and Ms. Garcia

4. None

PRACTICE EXERCISE 10

1. a. 1'6"   b. 3'6"   c. 5 yd. 1 ft.   d. 12'
    e. 1 yd. 2 ft.   f. $\frac{5}{6}$ ft.   g. 40"   h. 5 yd. 1 ft. 3 in.

2. a. 1'10"   b. 8"   c. 2'3"   d. 18"   e. 1 yd. 2 ft.

3.

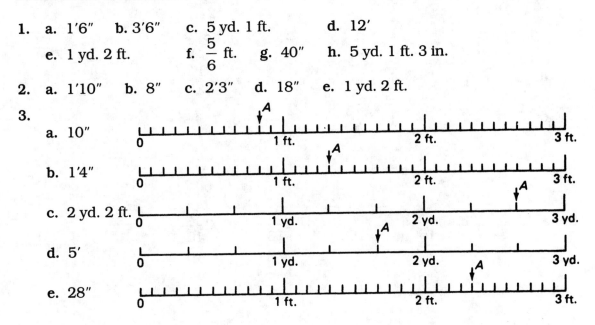

a. 10"

b. 1'4"

c. 2 yd. 2 ft.

d. 5'

e. 28"

PRACTICE EXERCISE 11

1. a. 3″ b. 1″ c. 5″ d. 4″ e. 2″

2.   **Lengths**

a. 4″	*c.* 2″	*e.* 3″	*g.* 3″	*i.* 2″
b. 4″	*d.* 5″	*f.* 1″	*h.* 3″	*j.* 3″

a. 7″	f. 3″	k. 2″
b. 6″	g. 2″	l. 4″
c. 5″	h. 0″	m. 1″
d. 8″	i. 5″	n. 4″
e. 4″	j. 1″	o. 5″

PRACTICE EXERCISE 12

1. a. $1\frac{1}{2}''$ b. $3\frac{1}{2}''$ c. $\frac{1}{2}''$ d. $4\frac{1}{2}''$ e. 3″

2. a. $\frac{1}{4}''$ b. $1\frac{3}{4}''$ c. $3\frac{1}{2}''$ d. $4\frac{1}{4}''$ e. $2\frac{3}{4}''$

3.   **Lengths**

a. $\frac{1}{4}''$	*c.* $1\frac{1}{4}''$	*e.* $3\frac{1}{4}''$	*g.* $2\frac{3}{4}''$	*i.* $\frac{3}{4}''$
b. $1\frac{1}{2}''$	*d.* $1\frac{1}{4}''$	*f.* $1\frac{3}{4}''$	*h.* $1\frac{3}{4}''$	*j.* $3\frac{1}{4}''$

a. $3\frac{1}{2}''$	f. $2\frac{1}{2}''$	k. $\frac{1}{4}''$
b. $2\frac{1}{4}''$	g. 1″	l. $\frac{1}{2}''$
c. 3″	h. $1\frac{1}{2}''$	m. $2\frac{1}{2}''$
d. 4″	i. 4″	n. $4\frac{1}{4}''$
e. 5″	j. $2\frac{1}{2}''$	o. 1″

4. a. $\frac{3}{4}''$ b. $3\frac{1}{8}''$ c. $1\frac{3}{8}''$ d. $2\frac{5}{8}''$ e. $4\frac{1}{4}''$

5. a. $2\frac{3}{16}''$ b. $4\frac{7}{16}''$ c. $\frac{9}{16}''$ d. $1\frac{1}{16}''$ e. $3\frac{5}{16}''$

PRACTICE EXERCISE 13

2. Measured Lengths

   a. $2\dfrac{9}{16}''$ b. $1\dfrac{13''}{16}$ c. $\dfrac{11''}{16}$ d. $1\dfrac{9}{16}''$ e. $3\dfrac{1}{8}''$

5. a. $\dfrac{9}{16}''$ b. $1\dfrac{1}{16}''$ c. $1\dfrac{11''}{16}$ d. $2\dfrac{1}{8}''$ e. $2\dfrac{5}{8}''$

   f. $3''$ g. $3\dfrac{5}{16}''$ h. $4\dfrac{3''}{8}$ i. $5\dfrac{1}{4}''$ j. $6\dfrac{1}{2}''$

PRACTICE EXERCISE 14

1. $AB = 1\dfrac{7''}{8}$ $BC = 3\dfrac{1}{8}''$ $CD = \dfrac{15''}{16}$ $DE = 1\dfrac{1}{16}''$

   $EF = \dfrac{11''}{16}$ $FA = 2\dfrac{1}{16}''$ $GH = \dfrac{3''}{4}$

   $HI = \dfrac{5''}{8}$ $IJ = \dfrac{13''}{16}$ $JG = 1\dfrac{3''}{16}$

2. $AB = 3\dfrac{3''}{8}$ $BC = 2\dfrac{3''}{8}$ $CD = 1\dfrac{1}{8}''$ $DM = 1\dfrac{1''}{8}$

   $MA = 1\dfrac{1''}{16}$ $EF = 1\dfrac{5''}{16}$ $FG = 1\dfrac{5''}{16}$ $GE = 1\dfrac{1}{8}''$

3. $\angle A$ = acute $\angle$ $\angle F$ = obtuse $\angle$
   $\angle B$ = right $\angle$ $\angle G$ = acute $\angle$
   $\angle C$ = right $\angle$ $\angle H$ = obtuse $\angle$
   $\angle D$ = acute $\angle$ $\angle I$ = obtuse $\angle$
   $\angle E$ = reflex $\angle$ $\angle J$ = acute $\angle$

4. $\angle B = 60°$

5. $\angle Y = 55°$

6. $\angle A = 30°$

7. $\angle A$ = acute $\angle$ $\angle E$ = acute $\angle$
   $\angle B$ = acute $\angle$ $\angle F$ = acute $\angle$
   $\angle C$ = obtuse $\angle$ $\angle G$ = acute $\angle$
   $\angle D$ = right $\angle$ $\angle M$ = reflex $\angle$

PRACTICE EXERCISE 15

1. a. 8   b. 16   c. 32   d. 32   e. 16   f. 32

2. a. $1\dfrac{1}{8}$ b. $1\dfrac{7}{16}$ c. $\dfrac{23}{32}$ d. $1\dfrac{3}{32}$ e. $1\dfrac{15}{16}$ f. $2\dfrac{27}{32}$

3. $23\dfrac{1}{8}''$ 4. $3\dfrac{1}{2}''$ 5. $A = 1\dfrac{11}{32}$ $B = 1\dfrac{3}{16}$ $C = 1\dfrac{37}{64}$

PRACTICE EXERCISE 16

1. a. $3\frac{13}{8}$ b. $1\frac{3}{2}$ c. $3\frac{7}{4}$ d. $5\frac{19}{16}$ e. $4\frac{2}{2}$

  f. $27\frac{33}{32}$ g. $14\frac{15}{8}$ h. $5\frac{49}{32}$ i. $4\frac{5}{4}$ j. $3\frac{10}{16}=2\frac{26}{16}$

2. a. $3\frac{1}{2}$ b. $1\frac{1}{2}$ c. $8\frac{3}{16}$ d. $2\frac{3}{8}$ e. $5\frac{3}{16}$

  f. $3\frac{1}{2}$ g. $5\frac{11}{16}$ h. $1\frac{13}{16}$ i. $1\frac{11}{32}$ j. $3\frac{31}{32}$

3. $A = 8\frac{1}{4}''$ $B = 2\frac{3}{4}''$ 4. $A = 5\frac{3}{8}''$ 5. $A = 3\frac{1}{8}''$ $B = \frac{11}{16}''$

PRACTICE EXERCISE 17

1. $11\frac{7}{8}''$ 2. $5\frac{5}{8}''$

PRACTICE EXERCISE 18

1. a. $\frac{21}{32}$ b. $1\frac{1}{8}$ c. 43 d. $24\frac{1}{2}$ e. $54\frac{3}{16}$

  f. $\frac{9}{16}$ g. $7\frac{31}{32}$ h. $80\frac{1}{2}$ i. $50\frac{5}{32}$ j. $8\frac{7}{16}$

2. $52\frac{3}{4}''$ 3. $11''$

4. a. $666''$ b. $55\frac{1}{2}'$ c. $75.48

5. $2\frac{13}{16}''$

PRACTICE EXERCISE 19

1. a. $\frac{8}{7}$ b. $\frac{4}{1}$ c. $\frac{16}{5}$ d. $\frac{4}{17}$ e. $\frac{3}{5}$

  f. $\frac{2}{13}$ g. $\frac{1}{8}$ h. $\frac{8}{19}$ i. $\frac{32}{31}$ j. $\frac{16}{69}$

2. a. $\frac{1}{2}$ b. 6 c. $\frac{31}{64}$ d. $3\frac{1}{16}$ e. $1\frac{11}{38}$

  f. $1\frac{17}{27}$ g. $8\frac{8}{11}$ h. $\frac{1}{16}$ i. $2\frac{7}{11}$ j. $2\frac{5}{198}$

3. $2\frac{1}{8}''$ 4. $\frac{41}{48}''$

PRACTICE EXERCISE 20

1. $(A) = 2\frac{1}{4}''$ $(B) = 1\frac{3}{4}''$ $(C) = \frac{1}{2}''$ $(D) = 2''$ $(E) = 1\frac{1}{2}''$ $(F) = \frac{3}{4}''$

2. $(C) = 6\frac{3}{4}''$ $(D) = \frac{3}{4}''$ $(E) = 5\frac{1}{2}''$ $(F) = \frac{1}{2}''$

3. $(A) = 10''$ $(B) = 6''$ $(C) = 5''$

PRACTICE EXERCISE 21

1. a. $\frac{1}{32}''$ b. $\frac{1}{16}''$ c. $\frac{3}{16}''$ d. $\frac{47}{64}''$ e. $\frac{29}{128}''$

2. a. $\frac{3}{16}''$ b. $\frac{3}{4}''$ c. $1\frac{1}{8}''$ d. $3\frac{11}{32}''$ e. $\frac{15}{64}''$

3. a. $\frac{3}{32}''$ b. $\frac{3}{16}''$ c. $\frac{1}{8}''$ d. $\frac{1}{32}''$ e. $\frac{11}{128}''$ f. $\frac{7}{64}''$ g. $\frac{23}{128}''$

4. $\frac{9}{16}''$ by $\frac{39}{64}''$

EXAM TIME

1. a. $484.40   b. $40\frac{1}{4}$ c. $5\frac{1}{4}$ d. $108.99   e. $593.39

2. $7\frac{3}{4}$ 3. $\frac{1}{3}$ 4. 30 min.   5. 1 ft. 4 in.   6. 16   7. $\frac{11}{16}''$

8. $2\frac{1}{2}''$ 9. XB 10. $\frac{1}{16}''$ 11. $1\frac{15}{8}$ 12. $A = 8\frac{1}{2}''$ $B = 7\frac{3}{8}''$

13. $\overset{\bullet}{X}\rule{6cm}{0.4pt}$ 14. $8\frac{9}{16}$

15. $2\frac{25}{32}$ 16. $2\frac{3}{16}''$ 17. $\frac{7}{8}''$

18. a. $AB = \frac{7}{8}''$ $BC = 1\frac{1}{4}''$ $CD = 1\frac{1}{4}''$ $DE = 1\frac{3}{8}''$ $EA = 1\frac{7}{16}''$ b. $6\frac{3}{16}''$

19. $10\frac{1}{16}$ 20. $5\frac{1}{9}$

How Well Did You Do?

0–19   **Poor.** Reread the unit, redo all the practice exercises, and retake the exam.

20–22   **Fair.** Reread the sections dealing with the problems you got wrong. Redo those practice exercises, and retake the exam.

23–26   **Good.** Review the problems you got wrong. Redo them correctly.

27–30   **Very good!** Continue on to the next unit. You're on your way.

INTERMISSION: FUN TIME

You have been working hard to improve your math skills. Now take a break! The games and puzzles that follow are based on mathematics and are fun to do.

MATHEMATICAL MAZE

This maze contains 25 words used in mathematics. Some are written horizontally, some are written vertically, some are upside down and must be read from bottom to top, and others go diagonally in various directions. When you locate a word in the maze, draw a ring around it. The word PERCENT has already been done for you. The answer to this puzzle and other answers will follow at the end of the Intermission.

```
R E F L E X K D U B R E P O R P
C F A O X B A N U M E R A T O R
Q D I V I S O R (P E R C E N T) F
B I C J O D A C R T N B C A G A E
A F E B G G O R E I L P A O E N A
G F T W Y E T M E H S R T H I M R
E E W D Y L O A L C I O A C O N G
V R B E C O T D G I I A R B N E D
I E R E C F D M X A V I C N R Y T
T N C T D Z N D D M I E D E S H
A C T D O Z B E E U E B A R A G I
G E D O T Z B E R N M B E S Y R A
E L P I B A O B T L E E D T R T
N T M B U O A B R B A C S
E R I B S T L P R A
S R A R M U L T I P L Y B R A S
```

155

PRIME TIME

There are two kinds of numbers, primes and composites. A *prime* number is a number, greater than 1, that has, as its only factors, itself and 1. The smallest prime number is 2 since its factors are 2 and 1. Another prime number is 11 since $11 \times 1 = 11$.

A *composite* number has additional factors other than itself and 1. Six is a composite number since 2 and 3 are factors of 6. What about 15? This is also a composite number since it has factors of 1, 3, 5, and 15.

In the array of numbers below, shade in the spaces containing prime numbers and leave the composite numbers alone.

2	3	4	5	6	7	8	9	10
11	12	13	14	15	16	17	18	19
20	21	22	23	24	25	26	27	28
29	30	31	32	33	34	35	36	37
38	39	40	41	42	43	44	45	46
47	48	49	50	51	52	53	54	55
56	57	58	59	60	61	62	63	64
65	66	67	68	69	70	71	72	73
74	75	76	77	78	79	80	81	82
83	84	85	86	87	88	89	90	91
92	93	94	95	96	97	98	99	100

How many prime numbers were you able to find? If you found more or fewer than 25 prime numbers, review your calculations.

GAMES OF CHANCE

The Happy Days High School conducts an open house each spring for incoming students. Each department prepares a program to introduce the students to topics that they will be taught.

The mathematics department decided to demonstrate games of chance as its contribution. The teachers set up, on their table, two games that illustrated *probability* and were intended to show the foolhardiness of gambling.

One game was the roulette wheel pictured below. It has four numbers—1, 2, 3, and 4—a spinner, and a betting board where the bettors place their wagers. The bettors were given play money to use as they wished.

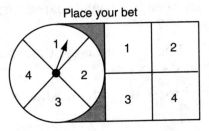

Place your bet

After awhile, all the guests had lost their play money. Although they had won at times, in the end the big winner was the student who ran the game.

He explained as follows why the bettors lost. The probability of their winning each time they made a bet was one chance out of four possibilities. The one chance was the number they chose to bet on, and the four possibilities were the four numbers on the roulette wheel. In other words, their chance, or probability, of winning was 1 out of 4 or $\frac{1}{4}$, and their chance of losing was 3 out of 4, or $\frac{3}{4}$. Their chance of winning could be represented as .25, and their chance of losing as .75. The bettors were able to see that they had a greater chance of losing than of winning. As a matter of fact, their chance of losing was three times that of winning.

They asked whether there was a game they could play that would give them a better chance of winning. They were told that any game that gives the player equal chances of winning and of losing is the best. In one game like this, you try to guess whether a coin will land with the head or the tail up. In this game you make a choice and there are two possibilities, so the probability, or chance, of winning or of losing is 1 out of 2, or $\frac{1}{2}$. This is equivalent to .5, or 50%. You have as good a chance of winning as of losing.

EXAMPLE: What is the probability of winning if you want to select the jack of spades from the spade picture cards? What is the probability of losing?

SOLUTION: There are three possibilities, jack, queen, and king. Therefore, if you select one card, you have one chance out of three, or $\frac{1}{3}$, of winning. You have two chances out of three of losing, so your chances are $\frac{2}{3}$ of losing.

PRACTICE EXERCISE

1. You want to select the ace of hearts from all four aces. What is your chance of doing so?

2. A roulette wheel consists of eight equally spaced numbers. What is the chance of the spinner landing on 5?

3. What is the chance of rolling a 3 with a die that has one of the numbers 1, 2, 3, 4, 5, and 6 on each of its six faces?

4. If a day of the week is chosen at random, what is the probability that it is a Sunday?

5. What is the probability that the card you chose from the 13 clubs in a deck is not the 9?

6. Using the roulette wheel below, answer the following questions:

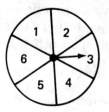

a. What is the chance of the spinner landing on 2?
b. What is the probability of the spinner landing on 5?
c. What is the probability that the spinner will *not* land on 4?
d. What is the chance that the spinner will land on 4?
e. What is the sum of the probabilities found in answer to questions c and d above?
f. What is the chance of the spinner not landing on 2?
g. What is the sum of the probabilities found in answer to questions a and f above?
h. Is the sum found in answer to question e the same as that found in answer to g?
i. If so, what is that sum?
j. Is it true that the chance of winning and the chance of losing will always total 1? Check this result in the two illustrative examples explained earlier. What do you find?

TWO-LETTER CLUE

In each of these ten mathematical words, at least two letters are the same and are given in the column at the right. Can you supply the missing letters in each word?

1. The sum of things.   T __ T __ __
2. The name of all numbers below zero.   __ E __ __ __ __ __ E
3. The top number in a fraction.   __ __ __ __ R __ __ __ R
4. A fraction whose value is 1 or greater than 1.   __ __ P __ __ P __ __
5. An angle whose measure is 180.   __ T __ __ __ __ __ T
6. 1 out of 100 is the same as 1 __ .   __ E __ E __ __
7. A number that is being added.   __ D D __ __ __
8. The bottom number in a fraction.   __ __ N __ __ __ N __ __ __ __
9. The number you are dividing by.   __ I __ I __ __ __
10. The number that is being subtracted from.   __ __ N __ __ N __

THE DI- PUZZLE

In the column on the right, complete the DI- mathematical words to fit the definitions.

1. A six-faced cube with dots on its faces.   D I __
2. Any one of the nine Arabic numbers.   D I __ __ __
3. A number by which another is divided.   D I __ __ __ __ __
4. The amount by which one quantity is greater or less than another.   D I __ __ __ __ __ __ __ __
5. Either the length or the width of a rectangle.   D I __ __ __ __ __ __ __
6. The number that is subjected to the operation of division.   D I __ __ __ __ __ __
7. A straight line through the center of a circle.   D I __ __ __ __ __ __
8. The length of a straight line between two points.   D I __ __ __ __ __ __
9. A straight line from one vertex in a polygon to its opposite vertex.   D I __ __ __ __ __ __
10. A picture of a geometric figure.   D I __ __ __ __ __

BATTLESHIP

This is a game for two players. Before you can start the game, you must learn how to locate, or plot, and to read points on a graph called a *rectangular coordinate system*. The point where the horizontal and vertical lines, or the horizontal or vertical axes, meet is called the "origin."

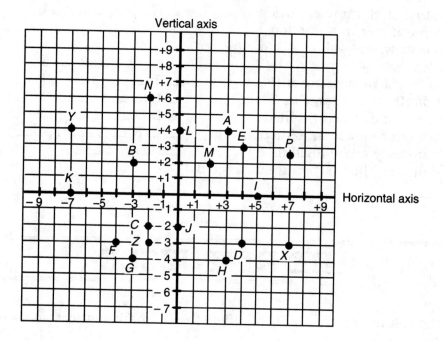

A point in this system is located by a horizontal and a vertical value, called a *pair of coordinates*, or an *ordered pair*. To start, you must begin from the origin and then move first horizontally and then vertically. The directions must always follow that order, hence the name "ordered pair."

You are familiar with the number line with positive numbers to the right and negative numbers to the left. This is the same for the horizontal axis in the graph above. Values on the vertical axis are positive when moving up and negative when moving down.

A pair of coordinates is written as (+7, –3). This pair means that you start at the origin and move 7 spaces to the right in the horizontal direction, and then move 3 spaces down. You will arrive at the point designated by the letter *X*.

EXAMPLE: What is (are) the location (the coordinates) of point *A*?

SOLUTION: Point *A* is 3 spaces to the right and 4 spaces up. Its coordinates are written as (+3, +4).

EXAMPLE: There is a point on the graph designated by the co-ordinates $(-3, +2)$. What point corresponds to these values?

SOLUTION: If you start at the origin and move 3 spaces to the left and then 2 spaces up, you land on point B.

EXAMPLE: What are the coordinates of point I?

SOLUTION: Starting at the origin, you move 5 spaces to the right; therefore the first coordinate is $+5$. Since you now stay where you are, the second coordinate is 0. Thus point I has coordinates $(+5, 0)$.

PRACTICE EXERCISE

Use the graph on p. 160 to answer these questions.

1. Which point corresponds to the coordinates $(+4, -3)$?

2. What are the coordinates of point C?

3. What are the coordinates of the origin?

4. Name the coordinates of each point: D, E, F, G, H, J, K, L, and M.

5. Locate the point whose coordinates are $\left(7, 2\frac{1}{2}\right)$. What letter corresponds to that point?

Now that you can locate points from given coordinates and can read the coordinates of given points, you are ready to play the game "Battleship."

Both you and your opponent have two graphs (see p. 162). On one (graph A), you record your opponent's hits and misses on your fleet; on the other (graph B), your hits and misses on his or her fleet. The first person to knock out the opponent's entire fleet wins the game.

The graphs below will give you the idea of the game and the strategies you can use to record hits and misses on your opponent's fleet. Your opponent should also follow the same instructions.

Each player is allowed three destroyers, which are graphed as pairs of points, like *A*, *B*, and *C*; two battleships, which are graphed as triples, like *D* and *E*; and one aircraft carrier, graphed as four points, like *F*. The points may be placed horizontally as in *C*, vertically as in *A*, or diagonally as in *D*. You should not reveal where you have placed your ships, and your opponent should not let you see the locations of his or her fleet.

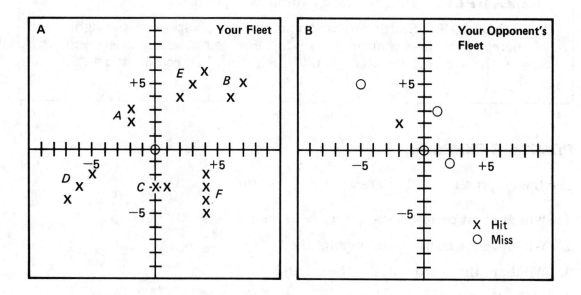

The idea is for you to call out a pair of coordinates and for your opponent to inform you whether you had a "hit" or a "miss." You should record your selections on graph B so that you are aware of the points you have chosen and whether or not you hit one of the enemy ships. You can record your selections with an "X" for a hit and a "O" for a miss. When your opponent calls out a pair of coordinates, record on graph A only the hits. When a ship is knocked out of the fight (that is, when all the coordinates have been correctly named), you must tell your opponent the type of ship that was sunk. You do not notify your opponent that he or she had a hit on, say, your aircraft carrier, but you must reveal when an entire ship has been knocked out of the battle. Good luck!

ANSWERS

MATHEMATICAL MAZE

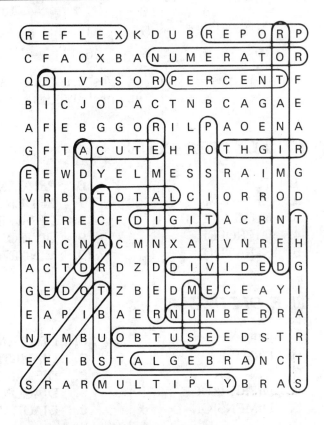

PRIME TIME

		4		6		8	9	10
	12		14	15	16		18	
20	21	22		24	25	26	27	28
	30		32	33	34	35	36	
38	39	40		42		44	45	46
	48	49	50	51	52		54	55
56	57	58		60		62	63	64
65	66		68	69	70		72	
74	75	76	77	78		80	81	82
	84	85	86	87	88		90	91
92	93	94	95	96		98	99	100

GAMES OF CHANCE: PRACTICE EXERCISE

1. $\frac{1}{4}$ 2. $\frac{1}{8}$ 3. $\frac{1}{6}$ 4. $\frac{1}{7}$ 5. $\frac{12}{13}$

6. a. $\frac{1}{6}$ b. $\frac{1}{6}$ c. $\frac{5}{6}$ d. $\frac{1}{6}$ e. 1 f. $\frac{5}{6}$

g. 1 h. yes i. 1 j. Yes. The examples shown confirm the fact that the probabilities of winning and losing always add up to 1.

TWO-LETTER CLUE

1. TOTAL
2. NEGATIVE
3. NUMERATOR
4. IMPROPER
5. STRAIGHT

6. PERCENT
7. ADDEND
8. DENOMINATOR
9. DIVISOR
10. MINUEND

THE DI-PUZZLE

1. DIE
2. DIGIT
3. DIVISOR
4. DIFFERENCE
5. DIMENSION

6. DIVIDEND
7. DIAMETER
8. DISTANCE
9. DIAGONAL
10. DIAGRAM

BATTLESHIP: PRACTICE EXERCISE

1. D   2. (–2, –2)   3. (0, 0)

4. *D*: (+4, –3), *E*: (+4, +3),
   F: (–4, –3), *G*: (–3, –4),
   H: (+3, –4), *J*: (0, –2),
   K: (–7, 0), *L*: (0, +4),
   M: (+2, +2)

5. *P*

THE MORE-FOR-LESS DEPARTMENT STORE

The More-for-Less Department Store is a large store with many employees. A wide assortment of merchandise is sold in its many different departments. To serve customers properly and efficiently, the store's employees must have a variety of mathematical and other skills.

THE SALES SLIP

A sales slip is the written record of a sale. Sales slips are usually issued in duplicate (two copies) or triplicate (three copies) so that the customer can have a copy while the store is able to keep the other copies. The sales slip has many purposes. For the customer, it is a record of what was purchased, the date, and the cost. If the customer wants to return an item for any reason, the sales slip proves that the item was purchased in the store and verifies the price. The store uses its copies to control inventory and to figure commission for its employees.

The men's clothing department is having a sale. Some of the sale items are shown on p. 166.

MEN'S DRESS SHIRTS
19.75 Short sleeve
25.75 Long sleeve

MEN'S LONG SLEEVE SPORT SHIRTS
19.88

MEN'S TURTLENECK KNIT SHIRTS
17.95

SAVE 5.00
MEN'S LONG SLEEVE VELOUR SHIRTS
25.00

MEN'S HOODED SWEATSHIRTS
31.99

MEN'S FASHION SWEATSHIRTS
20.59

SAVE up to 30.00
MEN'S SKI JACKETS
87.50

Acknowledgment: Reprinted by courtesy of AMES Department Stores, Inc.

Mr. Sena wishes to purchase 2 turtleneck knit shirts, 1 short-sleeve dress shirt, 2 sports shirts, 1 hooded sweatshirt, 1 ski jacket, and 1 velour shirt. Ms. Lee, the salesperson, fills out a sales slip and determines the total cost of this merchandise.

The sales slip shows the name of the store and its address. The customer's name, George Sena, and his address follow. The symbol "@" is read as "at," and "@ $2.99" means "at 2.99 each."

	MORE-FOR-LESS DEPT. STORE		
	Main Street		
	Anytown, USA		
	March 15		19 — —
SOLD TO	*Mr. George Sena*		
ADDRESS	*69 Dover Drive*		
CLERK *Nancy*	**DEPT.** *Men's Cloth.*		**AM'T REC'D** $250.00
QUAN.	**DESCRIPTION**		**AMOUNT**
2	*Knit shirts @ 17.95*		35	90
1	*Short-sleeve dress shirt*		19	75
2	*Sport shirts @ 19.88*		39	76
1	*Hooded sweat shirt*		31	99
1	*Ski jacket*		87	50
1	*Velour shirt*		25	00
		TOTAL	239	90

POSITIVELY NO EXCHANGES MADE UNLESS
THIS SLIP IS PRESENTED WITHIN 5 DAYS.

Mr. Sena paid for his purchase with five $50 bills. Ms. Lee indicated that fact in the spaced headed "Amt. Rec'd." After completing the sales slip, which shows that the total cost was $239.90, she had to give her customer the correct change. She subtracted $239.90 from $250 in the following way:

$$\begin{array}{r} \$250.00 \\ -\ 239.90 \\ \hline \$10.10 \end{array}$$
{ *Remember:* Line up the decimal points when subtracting decimal fractions. }

PRACTICE EXERCISE 1

1. a. Using the sales items in the men's department, complete this sales slip and find the total amount of the purchases:

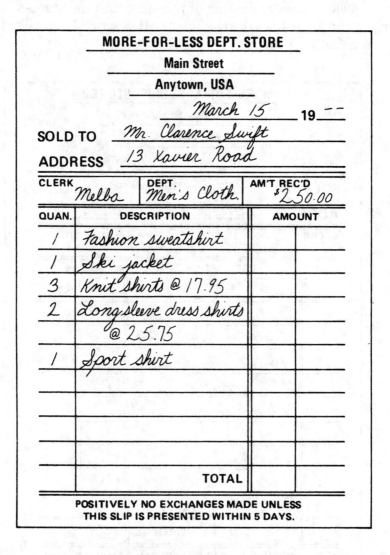

\multicolumn{3}{c}{**MORE-FOR-LESS DEPT. STORE**}
\multicolumn{3}{c}{**Main Street**}
\multicolumn{3}{c}{**Anytown, USA**}
		March 15 19 --
SOLD TO	\multicolumn{2}{l}{*Mr. Clarence Swift*}
ADDRESS	\multicolumn{2}{l}{*13 Xavier Road*}
CLERK *Melba*	**DEPT.** *Men's Cloth.*	**AM'T REC'D** $250.00
QUAN.	**DESCRIPTION**	**AMOUNT**
1	*Fashion sweatshirt*	
1	*Ski jacket*	
3	*Knit shirts @ 17.95*	
2	*Long sleeve dress shirts @ 25.75*	
1	*Sport shirt*	
	TOTAL	

**POSITIVELY NO EXCHANGES MADE UNLESS
THIS SLIP IS PRESENTED WITHIN 5 DAYS.**

   b. How much change would Mr. Swift receive from the five $50 bills offered in payment?

2. If Mr. Thomas bought 2 fashion sweatshirts and paid for them with a $50 bill, how much change would he receive?

3. Ms. Rosario purchased 1 knit shirt, 1 velour shirt, and 3 long-sleeve dress shirts for a friend. What was the total cost of these items of clothing?

4. Ms. Prince saw the following advertisement:

She purchased 1 complete set of thermal underwear and 2 pairs of briefs. Ms. Todd, the salesclerk, had to write out the sales slip, figure the total, and compute the change. Can you help her do it?

a. Prepare the sales slip for Ms. Prince's purchases.

MORE-FOR-LESS DEPT. STORE

Main Street

Anytown, USA

March 15 ___ **19** --

SOLD TO _Ms. M. Prince_

ADDRESS _1 Apache Circle_

CLERK	DEPT.	AM'T REC'D
Karen	_Ladies' Cloth._	$30.00

QUAN.	DESCRIPTION	AMOUNT
	TOTAL	

**POSITIVELY NO EXCHANGES MADE UNLESS
THIS SLIP IS PRESENTED WITHIN 5 DAYS.**

Acknowledgment: Reprinted by courtesy of AMES Department Stores, Inc.

   b. Compute the total cost of the two items.

   c. Will $30 cover the cost of the bill? If so, how much change (if any) will Ms. Prince receive?

5. According to the advertisement, how much would 5 pairs of briefs cost?

THE SALES TAX

As you have already learned, in most cities and states a sales tax is charged on the sale of certain items. The amount of tax varies from place to place, and so do the items (e.g., restaurant bills, food, sporting goods, cosmetics) to which the tax applies.

The sales tax is usually expressed as a percentage of the price of each article being taxed. It may be as low as 3% or as high as 8%, depending on locality. The tax is calculated by multiplying the amount of the bill by the decimal equivalent of the sales tax percent.

Tim Smith saw this advertisement for a sale in the sporting goods department:

He purchased 4 pairs of ice skates, 1 pair for each member of his family. He bought a pair of hockey skates for himself and a pair for his son. He purchased figure skates for his wife and daughter.

Acknowledgment: Reprinted by courtesy of AMES Department Stores, Inc.

His sales slip looked like this:

1 Men's hockey skates	=	$42.99
1 Boy's hockey skates	=	29.99
1 Ladies' figure skates	=	39.99
1 Girl's figure skates	=	24.99
Subtotal		$137.96

The subtotal is the total before the tax is added to it. The final total is the sum of the subtotal and the sales tax.

If the sales tax is 8% of the subtotal, you must first change 8% to its equivalent decimal fraction. Since 8% means 8 hundredths, it is written as $\frac{8}{100}$ or 0.08. The amount of the sale, $137.96, is then multiplied by this decimal equivalent. Thus:

$$
\begin{array}{ll}
\$137.96 & \text{Multiplicand (2-place decimal)} \\
\times\ 0.08 & \text{Multiplier (2-place decimal)} \\
\hline
\$11.0368 & \text{Product (4-place decimal)}
\end{array}
$$

> *Remember:* The number of decimal places in the product is the sum of the decimal places in the multiplier and in the multiplicand.

Rounding off $11.03(6)8 to two decimal places, you get $11.04 since the number in the thousandths place is more than 5. The final total is:

$$
\begin{array}{r}
\$137.96 \\
+\ 11.04 \\
\hline
\$149.00
\end{array}
$$

PRACTICE EXERCISE 2

1. Change each percent to its decimal equivalent.

    a. 6%

    b. 3%

    c. $5\frac{1}{2}\%$

    d. 7.5%

    e. 15%

    f. 4%

    g. $33\frac{1}{3}\%$

    h. 60%

    i. 13%

    j. 4.8%

    k. 6.25%

    l. $\frac{1}{2}\%$

    m. 125%

    n. 1%

    o. 47%

    p. 110%

2. Round off each of the following to two decimal places:

    a. $34.267   f. $432.7519
    b. $18.873   g. $96.062
    c. $436.753   h. $80.286
    d. $29.8963   i. $12.8639
    e. $23.7539   j. $65.8991

3. If the rate is 8%, calculate the sales tax on each of rounded-off amounts 2.a–2.e above.

4. Using your calculator, find the sales tax on items 2.f–2.j above if the rate is 4%.

5. Calculate the sales tax on a total sale of $47.50 if the rate is $7\frac{1}{2}$%.

6. Mr. Roberts is a do-it-yourselfer and an amateur cabinet maker. He went to the More-for-Less Department Store and purchased the three tools shown in this ad:

SAVE 10.00

JIG SAW

29.99   OUR REGULAR 39.99

- Double insulated, single speed; mitre adjustment

SAVE 10.00

FINISHING SANDER

39.99   OUR REGULAR 49.99

- Double insulated, single speed; cut work time down

SAVE 10.00

5/8 H.P. ROUTER

59.99   OUR REGULAR 69.99

- Lubricated ball bearings
- Double insulated feature
- Adjust to vertical depth

    a. Calculate the subtotal for the merchandise.
    b. Find the sales tax if the rate is 8%.
    c. What is the final total?
    d. Will $100 be enough to pay the bill? How much change (if any) will Mr. Roberts receive?

Acknowledgment: Reprinted by courtesy of AMES Department Stores, Inc.

7. Mr. Washington bought the jigsaw and router advertised on p. 172.

    a. What is the subtotal for his purchases?

    b. Find the sales tax if the rate is $7\frac{1}{2}\%$.

    c. What is the final total?

    d. How much change will he get from two $50 bills?

8. Ms. Young fishes and bought the fishing combo and tackle box shown in this advertisement:

    a. What is the subtotal for her purchases?

    b. Find the sales tax if the rate is 5%.

    c. What is the final total?

    d. If she pays for her purchase with one $20 bill, one $10 bill, and two $5 bills, will she receive change? If so, how much?

DISCOUNTS

Most stores, such as the More-for-Less Department Store, have sales on a regular basis. When a store runs a *sale*, it offers some or all of its merchandise at a *discount* from regular prices. Discounts are shown in advertisements in a number of ways, which are described on the following pages. Regardless of the way a discount is expressed, it is designed to save customers money when they make purchases of discounted merchandise.

Acknowledgment: Reprinted by courtesy of AMES Department Stores, Inc.

Discount in Dollars and Cents

The simplest form of discount is shown in terms of dollars and cents. Notice the following advertisement:

CASSETTE TAPE RECORDER
OUR REGULAR 29.95 **24.99**

• Push-button operation; auto recording level control and built-in mike; rugged design
Sold in Jewelry Dept.

Here a cassette tape recorder regularly priced at $29.95 is being offered at a sale price of $24.99. How much money will Mr. Cohen save if he purchases one tape recorder at the sale price? Of course, the discount can be determined by subtracting the sale price from the regular price:

$$\begin{array}{r} \$29.95 \\ -\ 24.99 \\ \hline \$\ \ 4.96 \end{array}$$

PRACTICE EXERCISE 3

1. The drug department of the More-for-Less Department Store is running a sale:

ADHESIVE BANDAGES	SATIN SKIN CREAM	DENTAL FLOSS
99¢	$1.99	$1.19
with this coupon	with this coupon	with this coupon
Regularly $1.79	Regularly $2.79	Regularly $1.79

Acknowledgment (tape recorder): Reprinted by courtesy of AMES Department Stores, Inc.

   a. Calculate the amount saved on each item shown on p. 174.

   b. How much money would be saved if you purchased all three items at the sale price rather than the regular price?

   c. Jill purchased 1 jar of Satin Skin Cream, 2 boxes of bandages, and 1 package of dental floss. What was the total cost?

2. Durant's Rental Center ran a Christmas sale. Its advertisement is shown on p. 176.

   a. Mr. Gorman, the salesperson, sold an 8-hp snow blower, a Sportfire 440 engine sled, and a Liquifire sled. How much did the customer save by purchasing these items at the sale prices?

   b. What is the sales tax on these purchases if the rate is 8%?

   c. How much is the total bill?

3. The jewelry department of the More-for-Less Department Store offered a sale on 14 karat gold-plated neck chains:

14KT. GOLD PLATED NECK CHAINS

SMALL	**6.99**	OUR REGULAR 11.99
MEDIUM	**8.99**	OUR REGULAR 15.99
LARGE	**12.99**	OUR REGULAR 21.99

- Hearts on cable chain
- Choose from 3 size hearts; velvet box
  Sold in Jewelry Dept.

   a. Calculate the amount saved on each size chain.

   b. How much is saved if all the items are purchased?

   c. What is the cost of 2 small chains, 3 large chains, and 1 medium chain? How much would be saved from the regular cost of these items?

Acknowledgment: Reprinted by courtesy of AMES Department Stores, Inc.

MAKE IT A DEERE CHRISTMAS

JOHN DEERE

Buy Him the Best and SAVE BIG!

Liquifire
Liquid-cooled lightning

SAVE $739

Burn your reputation across the snow on this total performance sled. It's powered by an oil-injected 440 Fireburst engine. Then tuned by an exclusive Firecharger chambered exhaust. Runs on straight gasoline. Has a two-way cooling system, wide ski stance and long-travel slide-rail suspension.

reg. $3638

Now $2899

Other great sleds:	Reg.	Xmas
Sportfire 440 engine	$2910	**$2350**
Trailfire 440 engine	$2650	**$2150**
Spitfire 340 engine	$1766	**$1459**

Snow Blowers

8- or 10-hp snow blowers have a two-stage engine five forward speeds and reverse

	reg.	XMAS
8-hp	$894.	**$695**
10-hp	$1014	**$781**

Chained Lightning
from $155⁹⁵

JOHN DEERE

Buy any saw over 50cc and receive $55⁰⁰ wood cutter kit for only $10⁰⁰. It includes 2¹/₂ gal. gas can, qt. of bar oil, 2 8-Oz. cycle oil. 1 pr. cather work gloves, filing wedge & deluxe filling guide. Or a 10% Christmas discount off any saw.

For All Your Rental or Sales Needs
DURANT'S RENTAL CENTER

DANBURY
744-3434
90 FEDERAL ROAD
Formerly United Rent-Alls

JOHN DEERE
Authorized Sales & Service
Your Friendly Neighbor Since 1959

NEW MILFORD
355-3114
150 DANBURY ROAD
Next to Diamond International

Acknowledgment: Reprinted by courtesy of TViews, a Division of CIS, New Milford, CT.

Discounts as Fractions of Original Prices

Sometimes discounts are expressed in terms of fractions of original prices, as in this advertisement:

The advertisement for the entire stock of light fixtures says, "$\frac{1}{3}$ off our regular price." This means that the fixtures are being sold for two thirds *of* the original price. If the regular price of a fixture is $45, the discount can be calculated as shown:

$$\frac{1}{\cancel{3}_{1}} \times \$\cancel{45.00}^{15.00} = \$15.00 \qquad \text{("of" means "to multiply")}$$

Ms. Gonzalez is saving $15.00 by purchasing the fixture on sale. She will have to pay only

$$
\begin{array}{ll}
\$45.00 & \\
\underline{-15.00} & \text{Discount} \\
\$30.00 & \text{For the fixture}
\end{array}
$$

Other advertisements say, "Save 1/2." This means that one half of the original price is to be subtracted from the original price to find the discount or sale price.

Ms. Stein purchases a blouse whose original price was $12. She sees that the sale price is $5.99. Is this equivalent to "1/2 off"? Ms. Rose, the salesclerk, said that it is and showed Ms. Stein in the following way:

$$\frac{1}{\cancel{2}_{1}} \times \$\cancel{12.00}^{6.00} = \$6.00 \text{ savings}$$

Acknowledgment: Reprinted by courtesy of AMES Department Stores, Inc.

The item costs

$$\begin{array}{r} \$12.00 \\ -\ 6.00 \\ \hline \$\ 6.00 \end{array}$$

The store charges $5.99 so that the item appears to be even more of a bargain.

PRACTICE EXERCISE 4

1. If the regular price of a jacket is $120 and you save 1/3, what do you pay?

2. The two items shown below are on sale in the drug department, and the advertisement shows both the sale and the regular prices. Check each item to see whether the advertisement is honest, that is, whether the discount is accurate. (Approximations are permitted.)

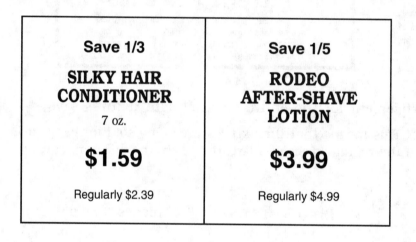

Save 1/3	Save 1/5
SILKY HAIR CONDITIONER	**RODEO AFTER-SHAVE LOTION**
7 oz.	
$1.59	**$3.99**
Regularly $2.39	Regularly $4.99

Discounts as Percentages of Original Prices

Discounts are often expressed in terms of a percent of the original price of an article of merchandise, as in the following advertisement:

30% off

Wide selection of name-brand watches for men and women

$14 to $35

Reg. 19.99 to $50.
Sport and dress styles. Digitals included.

Ms. Marks saw the advertisement and decided to purchase a watch whose regular price was $19.99. Since the advertisement says that she can save 30% *off* this price, she changed 30% to its decimal equivalent, 0.30, and multiplied to find her discount. Thus:

$$\begin{array}{r} \$\ 19.99 \\ \times\ 0.30 \\ \hline \$5.9970 \end{array}$$

Rounding off the discount of $5.99(7)0 results in the value $6.00. Subtracting that from the regular price gave Ms. Marks a sale price of

$$\begin{array}{r} \$19.99 \\ -\ 6.00 \\ \hline \$13.99 \end{array}$$

PRACTICE EXERCISE 5

1. **a.** What discount would be expected if the regular price of a pair of fashion pants is $34.95, and a store has this sign in its window?

25% OFF
OUR REGULAR PRICE

ENTIRE STOCK

**JUNIORS' and MISSES'
DENIM and CORDUROY JEANS**
100% cotton prewashed denim and 100% cotton corduroy. Carpenter jeans, too. Big variety of colors. Sizes 5/6-15/16, 8-18 and 14-20.

FASHION PANTS
Textured or 2 way stretch poly, other fabrics. Trouser looks, waist treatments, belted styles and pockets. Many colors. Sizes 5/6-15/16, 8-18.

    b. What would the sale price be?
    c. What would the pants cost if the sales tax rate is 5%?

2. **a.** From the following advertisement, what discount can be expected if two neck chains regularly priced at $21.95 each are purchased?

30% OFF
OUR REGULAR PRICE

**ENTIRE STOCK
MEN'S and LADIES'
NECK CHAINS**
• Perfect fashion accent
• Choose 14k gold filled or sterling silver in many styles and lengths
Sold in Jewelry Dept.

    b. What is the cost for the two neck chains?

    c. The sales tax rate is $7\frac{1}{2}$%. What is the amount of the tax?

    d. What is the total cost?

Acknowledgment: Reprinted by courtesy of AMES Department Stores, Inc.

3. The More-for-Less Department Store has run the offer shown below. How much will a coat whose regular price is $169.95 cost a senior citizen who shops on Wednesday if the sales tax rate is 4%?

SENIOR CITIZENS
20% OFF   Mon. & Wed. Only

Discounts as Percentages
of Everything Purchased

Sometimes stores offer discounts of a certain percentage off on *everything* purchased. An advertisement may say "10% off on all merchandise in the store—today only" if a store wishes to encourage a large number of sales on a particular day. Suppose that the More-for-Less Department Store has the following sign in its window:

LOW   SENIOR CITIZENS
PRICES   15% OFF ALL   Mon. & Wed. Only
MERCHANDISE

Ms. Smith is a senior citizen and comes into the More-for-Less Department Store on Monday to buy some items for a trip she is planning. The store had run the following advertisement:

GE
AM/FM STEREO
CASSETTE PLAYER

14.99
OUR REGULAR 24.99

• Automatic shut off

VIVITAR
35MM COMPACT CAMERA

27.77
OUR REGULAR 39.99

• Built in flash
• Motorized advance & rewind
• Auto film setting

BONUS PACK
12 EXTRA EXPOSURES
200   3 PACK
Kodak
Gold
200
COLOR PRINT FILM
1 36 EXP ROLL
2 24 EXP ROLLS

KODAK
SUPER 200 GOLD
35MM 24 EXP. 3 PK.

8.66
OUR REGULAR 11.99

• Multi-purpose indoor or outdoor film
• Total 84 exposure

Ms. Smith purchased the following items:

1 35mm compact camera	@ $27.77 =	$27.77
1 3-pack of color film	@ $ 8.66 =	8.66
1 cassette player	@ $14.99 =	14.99
	Subtotal	$51.42

She was entitled to the 15% discount offered all senior citizens on Monday. Thus:

$$15\% = 0.15 \qquad \text{and} \qquad \begin{array}{r} \$ \ 51.42 \\ \times \ \ \ 0.15 \\ \hline 257\ 10 \\ 514\ 1 \\ \hline \$ \ 7.71(3)0 \end{array}$$

Ms. Smith's discount is $7.71, so her bill is $51.42 – $7.71 = $43.71.

If the sales tax rate is 5%, then

$$5\% = 0.05 \qquad \text{and} \qquad \begin{array}{r} \$ \ 43.71 \\ \times \ \ \ 0.05 \\ \hline \$2.18(5)5 \end{array}$$

which is equivalent to $2.19 when rounded off to the nearest cent. Her total bill is

$$\begin{array}{r} \$43.71 \\ + \ 2.19 \\ \hline \$45.90 \end{array}$$

PRACTICE EXERCISE 6

1. A local store, to encourage sales, ran the following advertisement and offered an additional 10% discount on all items purchased:

TOOTHBRUSH Hard, Soft or Medium	JURGENSEN'S CREME FORMULA 2 oz.	TINGLY MOUTHWASH
77¢	**$1.78**	**$1.17**
with this coupon	with this coupon	with this coupon
Regularly $1.29	Regularly $2.19	Regularly $1.49

The Landers and Cullen families purchased the following items:

    4   Tingly mouthwashes
    2   medium toothbrushes
    2   Jurgensen's Creme Formulas

   a. What was the total bill?
   b. How much was the discount?
   c. What was the final cost for the items purchased?

2. Mr. Oak purchased $38.67 worth of merchandise and was allowed a 15% discount. What was his final cost?

3. A sofa cost $449.95 and a chair cost $219.95. Ms. Chu was allowed a 12% discount when she paid for her purchases.

   a. What was the discount?
   b. What was the final cost of the purchases?
   c. If sales tax is computed at $4\frac{1}{2}\%$, find the amount of the tax.
   d. What was Ms. Chu's final bill?

Discounts for Buying More Than One Item

Stores sometimes wish to encourage customers to buy more than one item of a particular kind. They then offer special discounts for such purchases.

This advertisement says, "2 boxes [Kordite Trash Bags] 3.00; our regular 2.59 ea." If a customer buys *one* box of trash bags, she pays $2.59, but if she buys *two* boxes, she pays only $3. This is a large saving because the bags will cost only $3.00 ÷ 2 = $1.50 per box, instead of $2.59. The customer saves

$$\$2.59 - \$1.50 = \$1.09 \text{ per box}$$

and this amounts to a large discount.

Acknowledgment: Reprinted by courtesy of AMES Department Stores, Inc.

PRACTICE EXERCISE 7

1. **a.** Ms. Barnaby noticed these advertisements:

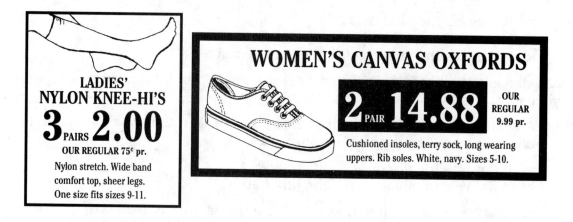

LADIES'
NYLON KNEE-HI'S
3 PAIRS **2.00**
OUR REGULAR 75¢ pr.
Nylon stretch. Wide band comfort top, sheer legs. One size fits sizes 9-11.

WOMEN'S CANVAS OXFORDS
2 PAIR **14.88** OUR REGULAR 9.99 pr.
Cushioned insoles, terry sock, long wearing uppers. Rib soles. White, navy. Sizes 5-10.

She purchased a pair of women's canvas oxfords and 9 pairs of nylon knee-hi's. What was the cost of these 10 items?

b. Calculate her total bill after adding a sales tax charge of 8%.

c. How much did she save by purchasing 9 pairs of nylon knee-hi's at the sale price instead of at the regular price?

d. If Ms. Barnaby paid for her purchase with one $20 bill, how much change did she receive?

2. **a.** Mr. Leary saw the advertisement shown below for a sale in the boy's clothing department. He purchased 1 dickey shirt, 2 flannel shirts, and 3 pairs of crew socks.

BOYS'
MARKSMAN
CREW SOCKS
2.99
OUR REGULAR 3.99
3 pairs per package. Cotton/nylon blend in sizes 6-8½ and 9-11.

JR. BOYS'
KNIT SHIRTS
5.99
Poly/cotton. Permanent press and long sleeves. Screen print on front. In jr. boys' sizes 4-7.

BOYS'
DICKEY SHIRTS
8.00
OUR REGULAR 10.99
Poly LaCoste shirt and acrylic dickey insert. In solids. Sizes 8-18.

BOYS'
FLANNEL SHIRTS
12.99 OUR REGULAR 14.99
100% cotton. Long sleeves. Many plaids in sizes 8-18.
JR. BOYS' SIZES 4-7 9.98

Calculate the amount of money he saved on each item, compared to its regular price.

b. Calculate his total saving.

c. Calculate the total bill for his purchases.

Acknowledgment: Reprinted by courtesy of AMES Department Stores, Inc.

3. Ms. Shaw read that the advertisement shown offered 1/3 off on braided rugs.

ENERGY SAVER

1/3 OFF

OUR REGULAR PRICE

ENTIRE STOCK
BRAIDED RUGS

20"x30" SIZE	65"x100" SIZE
20"x40" SIZE	95"x135" SIZE
30"x50" SIZE	15" CHAIR PADS

- 80% nylon/20% misc. fibers, double core, reversible
- Sturdy tubular construction; green, brown, gold, rust

a. At 1/3 off, what will it cost Ms. Shaw to purchase a rug selling at a regular price of $135?

b. What is the sale price of a braided rug whose regular price is $69.99?

c. Ms. Shaw bought a number of braided rugs of different sizes whose original prices were $44.95, $19.95, and $89.95, respectively.
   (1) What was the discount for each rug if she was allowed 1/3 off?
   (2) Find the sale price of each item.
   (3) Find the total cost for all three items, excluding the sales tax.

Acknowledgment: Reprinted by courtesy of AMES Department Stores, Inc.

4. Ms. Fitz purchased a gold chain, a charm bracelet, and a serpentine bracelet, as advertised below. Their regular prices were $24.95, $39.95, and $47.50, respectively. She received a 40% discount on all of them.

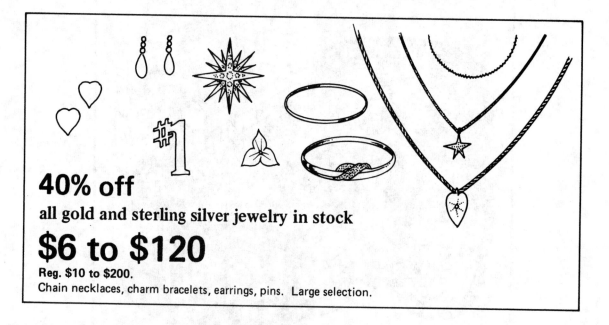

40% off

all gold and sterling silver jewelry in stock

$6 to $120

Reg. $10 to $200.
Chain necklaces, charm bracelets, earrings, pins. Large selection.

    a. Calculate the amount Ms. Fitz saved on each of these items.
    b. Find the actual selling price of each item.
    c. Find the amount of Ms. Fitz's total bill.

5. The More-for-Less Department Store offers a 15% discount on all merchandise for senior citizens on Mondays and Wednesdays. Ms. Johnson, aged 70, saw the advertisement on p. 187 and bought a medium-sized quilted body warmer for herself and a large one as a gift.

QUILTED BODY WARMER

MEDIUM SIZE

14.99

Poly/cotton cover, Fortrel poly fill. Converts to twin, fullsize comforter. Fits up to 5'6". Bright patterns.

LARGE SIZE 5'6" or TALLER **16.99**

a. How much is the discount on each purchase?
b. What will Ms. Johnson's total savings be?
c. What will her total bill be after she receives her discount?

COMMISSIONS

In some stores, salesclerks receive *commissions* on the sales they make. These commissions may be offered in addition to their salaries, or they may be paid in place of regular salaries. They are offered to salespersons to encourage them to make special efforts to complete sales. A commission is usually expressed as a percentage of the amount of a sale completed. The particular percentage offered varies in different departments or businesses.

At the More-for-Less Department Store commissions are paid to sales-clerks in departments selling "large items" (the more expensive merchandise), in addition to their salaries. Some of the departments in which commissions are paid include furniture, major appliances (television sets, washers and dryers, refrigerators, ovens, etc.), and rugs and carpeting.

Acknowledgment: Reprinted by courtesy of AMES Department Stores, Inc.

EXAMPLE: On a busy Saturday, Mr. Silver, a salesclerk in the major appliance department, sold two 13″ color TV sets and three 25″ color TV sets, at the prices shown in this advertisement:

SAVE 40.00

13″ COLOR TV WITH REMOTE

179.99

OUR REGULAR 219.99

• 181 channel cable

• Sleep timer
• On screen display

SAVE 70.00

25″ COLOR TV WITH REMOTE

349.99

OUR REGULAR 419.99

• 181 channel cable compatible tuning

13″ GUSDORF COLOR TV STAND
OUR REGULAR 24.99
16.99

25″ GUSDORF COLOR TV STAND
OUR REGULAR 27.99
19.99

Salespersons in the major appliance department receive commission of 7% on all sales they make, in addition to their salaries. How much did Tim earn this Saturday if his salary is $50 a day plus commissions?

SOLUTION: Tim calculated his earnings for Saturday as follows:

2 13″ color TV sets	@ $179.99	=	$ 359.98
3 25″ color TV sets	@ $349.99	=	+ 1049.97
	Total sales		$ 1409.95

At 7% commission: 7% = 0.07   so

$$\times \quad 0.07$$
$98.69(6)5 $98.70
rounded off
to the
nearest cent

Thus Mr. Silver earned

$ 50.00	Salary
+ 98.69	Commission
$148.69	Total earnings for Saturday

EXAMPLE: Mr. Willis is a salesperson in the rugs and carpeting department and receives a 12% commission on all sales he makes. Mr. and Mrs. Grant came in to purchase wall-to-wall carpet for a room measuring 12 feet wide and 21 feet long. They purchased the carpet shown in this advertisement:

Level-loop easy-care 100%
nylon with foam-rubber backing....*Reg. 21.88* 1/2 price 10.94 sq. yd.
installed wall-to-wall

How much commission did Mr. Willis earn on this sale?

SOLUTION: The carpet is selling at $10.94 per square yard. Mr. Willis first calculated the number of square yards the Grants needed. Since each dimension is measured in feet, it is necessary to change it into yards. A foot is a *smaller* unit than a yard, and to change from a smaller unit to a larger one, you *divide*. Specifically, you divide by the number of units of the smaller measurement that is contained in the larger measurement.

In this case, 3 feet is equal to 1 yard, so Mr. Willis divides by 3. He finds that 12 feet wide ÷ 3 = 4 yards wide. Likewise, 21 feet long = 7 yards long, since 21 ÷ 3 = 7. The room is 4 yards wide and 7 yards long.

The area of the rectangular room is found by using the formula

$A = l \cdot w$, where l = length and w = width
$A = 7 \cdot 4$
$A = 28$ sq. yd.

Since the carpet costs $10.94 per square yard, the Grants will be charged $10.94 × 28 = $306.32. The total sale is $306.32, and Mr. Willis's commission is calculated as follows:

12% = 0.12   so

$$
\begin{array}{r}
\$ \ \ 306.32 \\
\times \ \ \ \ \ 0.12 \\
\hline
6\ 12\ 6\ 4 \\
30\ 63\ 2 \\
\hline
\$36.75(8)4
\end{array}
= \$18.45
$$

rounded off to the nearest cent

PRACTICE EXERCISE 8

1. Mr. Bradford purchased the color TV and stand advertised below.

Calculate the amount of the commission on this sale if salespersons in the TV department earn a commission of 15%.

2. Ms. Carr is a part-time salesclerk in the clock department and receives a 10% commission on all her sales. She sold 3 AM/FM L.E.D. clock radios, 1 AM/FM digital clock radio, and 5 Seville clock radios at the prices shown below.

    a. Calculate the commission on these sales.

    b. If Ms. Carr earns $4.25 an hour plus commission, what will her earnings be if she works 4 hours and sells 2 Seville clock radios?

3. On a very busy day Ms. Evans, a salesperson in the major appliances department, sold two refrigerators and one range at the prices shown in the advertisement on p. 191.

Acknowledgment: Reprinted by courtesy of AMES Department Stores, Inc. *(clock radios)*

Chip & Dent Sale

*All merchandise brand new & fully warranteed.
Only cosmetic damage, not mechanical.*

WASHER
Reg. $399
$339

WASHER
Reg. $309
$259

FROST-FREE REFRIGERATOR
adjustable shelves, high
efficiency model Reg. $619. **$499**

30" CONVECTION RANGE
with infinite burner control,
time bake cycle, clock &
60 min. timer. Reg $409. **$329**

    a. If her rate of commission is 7%, calculate the amount of commission on these sales.

    b. What will Ms. Evans earn for this 8-hour day if her salary is $4.52 an hour plus commissions?

4. During one week, the salespeople in the audio-video department sold 17 of the Kraco deluxe 8-track stereo tape players shown in the advertisement. If they receive commissions at the rate of 7%, calculate the total amount of the commissions paid to them.

SAVE 79.99

KRACO

**DELUXE 8 TRACK or CASSETTE
IN DASH with AM/FM and
40 WATT POWER BOOSTER**

99.99 OUR REGULAR 179.98

Acknowledgments: (appliances): reprinted by courtesy of TViews, a Division of CIS, New Milford, CT; *(tape players):* reprinted by courtesy of AMES Department Stores, Inc.

5. Ms. Cordero selected the wall-to-wall carpet advertised below to cover a floor whose dimensions are 9 feet by 12 feet. Mr. Willis, a salesperson in this department, receives commission at the rate of 12% of his sales.

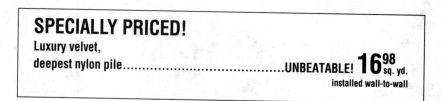

SPECIALLY PRICED!
Luxury velvet,
deepest nylon pile...UNBEATABLE! **16**⁹⁸ sq. yd.
installed wall-to-wall

a. Change each dimension from feet to yards.
b. Calculate the area of the room in square yards.
c. Calculate the price of the carpet needed to cover the floor of this room wall to wall.
d. Calculate the amount of Mr. Willis's commission on this sale.

INSTALLMENT BUYING

Many stores, including the More-for-Less Department Store, offer customers the option of paying for their purchases in *installments* instead of with immediate cash payments. This is a great convenience in the case of many or expensive purchases. However, a charge is made for installment buying, in consideration of the fact that the store is actually making a *loan* to the customer until he or she completes the payments for the purchase. This charge, sometimes called a "service charge" or a "carrying charge," is actually an *interest* charge for the loan. Interest charges are commonly calculated at the rate of 1% or $1\frac{1}{2}$% per month (12% or 18% per year) on the unpaid balance of the loan.

Customers may ask salespersons about the advantage or disadvantages of installment purchasing over direct cash purchasing. Each salesperson must be able to calculate the additional money the customer must pay if the installment plan is chosen.

EXAMPLE: Mr. Carson wishes to purchase the camera shown in the advertisement, but he is unable to buy it for cash. He learns from Mr. Anthony, his salesperson, that he can pay for it over a period of 1 year by making 12 equal monthly installments of $25 each.

SUPERAUTOMATIC 35mm CAMERA

• Automatic metering system
• f 1.7 lens $259

How much more will Mr. Carson pay the store if he uses the installment plan rather than paying cash for the camera?

SOLUTION:

$$
\begin{array}{r}
\$\ 25.00 \\
\times\quad 12 \\
\hline
50\ 00 \\
250\ 0\quad \\
\hline
\$300.00 \\
\end{array}
$$
Total installment price

Since the cash price is $259, the difference between these two prices is the actual cost of the credit extended to Mr. Carson:

$$
\begin{array}{r}
\$300 \\
-259 \\
\hline
\$\ 41 \\
\end{array}
$$
Cost of credit

Thus:

Cost of credit = installment price – cash price

EXAMPLE: Ms. Kent wishes to purchase the 50-inch big-screen TV shown in the advertisement. Instead of paying cash, she is given the option, or choice, of paying $109 per month for 24 months after making a *down payment* of $100. How much is she being charged for credit?

The One and Only Now at Ampitron

NOW ON DISPLAY
A Gift for the Entire Family

50"
Color
TV

Now
$2250

Giant 50" screen...
Like having a movie screen at home!
Stop in and see it.

SOLUTION: Making a *partial payment*, called a "down payment," reduces the cost of Ms. Kent's credit because she will owe less money on the purchase. Instead of owing $2250, she owes only $2250 – $100 = $2150, since she paid $100 "down." She is charged interest only for the loan of $2150.

$$
\begin{array}{r}
\$\ 109 \\
24 \\
\hline
436 \\
218 \\
\hline
\$2616 \\
+\ 100 \\
\hline
\$2716
\end{array}
$$

Total installment payments
Down payment
Total cost of purchase

Thus:

Cost of credit = installment price – cash price
$$C = \quad \$2716 \quad - \quad \$2250$$
$$C = \$466$$

PRACTICE EXERCISE 9

1. Paula Parson purchased the grandfather clock shown in the advertisement. Instead of paying cash, she agreed to make 12 equal monthly payments of $50 each. Calculate the amount she would have saved if she had paid cash for the clock.

> **GRANDFATHER CLOCKS**
> **559** were 750. Walnut finish, moon dial, chimes every half-hour.

2. Mr. Nicholls decided to buy a new stereo system, priced at $227. He agreed to pay $22 per month for 1 year for this purchase instead of paying cash. Calculate the cost of his credit.

3. Ms. Lane bought the wall-to-wall carpet advertised below for her entire apartment. She needed 78 sq. yd. to carpet her bedrooms, living room, and dining area.

> **$4 OFF EVERY SQ. YD.!**
> Saxony plush
> polyester pile. Reg. 20.49 **16⁴⁹** sq. yd.
> installed wall-to-wall

    a. Calculate the cash price of this carpet.

    b. Calculate the cost of credit if Ms. Lane decides to pay for her purchase by making a down payment of $125 and 12 equal monthly payments of $75 each.

4. Ms. Cartagena decided to buy a word processor to replace her typewriter. The one she chose sold for $549.95, and she was told that she could pay cash or pay $49.50 per month for 1 year to make this purchase. What is the cost of credit if she decides to pay in installments?

5. Some customers need more than 1 year to pay for very expensive purchases. Naturally, the cost of credit is higher. Calculate the cost of credit for Mr. and Mrs. Ryan, who bought a set of living room furniture priced at $1099.95, if they made a $100 down payment and paid the balance in equal monthly installments of $50 for 2 years (24 months).

COMPARISON SHOPPING—UNIT PRICING

Price of One Item

The "unit price" of any merchandise is the price of *one item* of that merchandise or the price of *one unit* of that merchandise. Since many items are packed in containers of varying sizes, customers are often confused when they try to determine which package gives them the best buy for their money. If you are the customer or the salesclerk, you can make decisions on best buys if you know how to calculate unit prices. Let's start with the unit price of *one item* of merchandise.

These calculations are easily made by dividing the total price of the merchandise by the number of units being sold at that price, or

$$\text{Unit price} = \frac{\text{total price}}{\text{number of units}}$$

EXAMPLE: In the men's clothing department of the More-for-Less Department Stores, shirts are on sale at a price of two for $31, and socks are selling in packages of three pairs for $5.49. What is the unit price of the shirts and socks, that is, the price of one shirt and one pair of socks?

SOLUTION: Since the

$$\text{Unit price} = \frac{\text{total price}}{\text{number of units}}$$

the

$$\text{Unit price} = \frac{\$31.00}{2} = \$15.50 \text{ per shirt}$$

and the

$$\text{Unit price} = \frac{\$5.49}{3} = \$1.83 \text{ per pair of soc}$$

PRACTICE EXERCISE 10

1. Mr. George purchases one can of tennis balls for $2.07 and one box of golf balls at $12.72. Tennis balls are usually sold in cans of three balls, and golf balls are usually sold in boxes of one dozen each. Calculate the price of one tennis ball and one golf ball.

2. Ms. Chase wishes to purchase one dozen ladies' handkerchiefs for use as gifts. The handkerchiefs come in packages of three for $4.50.

    a. What is the price of one handkerchief?
    b. How much will Ms. Chase pay for one dozen handkerchiefs?
    c. How many packages must she buy to have one dozen handkerchiefs?

3. **a.** Calculate the price of one pair of the socks advertised as shown.

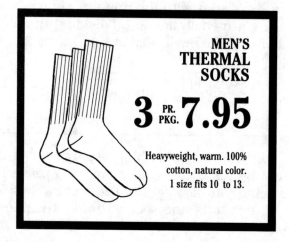

**MEN'S
THERMAL
SOCKS**

3 PR. PKG. **7.95**

Heavyweight, warm. 100%
cotton, natural color.
1 size fits 10 to 13.

b. How many packages would a customer need to buy to have one **dozen** pairs of socks?

c. How much would these packages cost?

4. Mr. Green purchased six neckties, which are on sale at a price of four for $30.

a. What is the cost of one necktie?

b. How much will Mr. Green pay for his six neckties?

c. What part of a dozen is six neckties?

Price of One Unit of That Item

The method of finding the unit price or the price of *one unit* of the item is the same as shown before. The difference is the size of the unit. For instance, if a quart of milk sells for 64¢, the unit price of 1 ounce of milk is 2¢ since there are 32 ounces in a quart and 64¢ ÷ 32 = 2¢. Likewise, the unit price of 1 pint of milk is 64¢ ÷ 2 = 32¢ since there are 2 pints in a quart.

What is the unit price per gallon of milk? Since there are 4 quarts in a gallon, the unit price per gallon of milk is $64¢ ÷ \dfrac{1}{4} = 64 × \dfrac{4}{1} = 256¢$ or $2.56.

Common "length" units are inches, feet, and yards; "weight" units are ounces and pounds; "liquid volume" units are fluid ounces, pints, quarts, and gallons; and so on.

Acknowledgment: Reprinted by courtesy of AMES Department Stores, Inc.

EXAMPLE: Ms. Murphy saw these advertisements for bulk shopping and decided to try it as a money-saving idea:

Gleamy Shampoo	**Honey Herbal Shampoo**	**Brown Balsam**
Concentrate	Concentrate	Concentrate Coconut
$24.00 a gallon	**$14.00 a gallon**	**$12.50 a gallon**
(1 gallon makes 16 gallons)	(1 gallon makes 16 gallons)	(1 gallon makes 12 gallons)

SOLUTION: She had to figure the unit price per gallon for each of the three shampoos so that she could decide which, if any, was the best buy.

$$\text{Unit price} = \frac{\text{total price}}{\text{number of units}}$$

Gleamy Shampoo

$$\text{U.P.} = \frac{\$24}{16 \text{ gallons}}$$

U.P. = $1.50 per gallon

Honey Herbal Shampoo

$$\text{U.P.} = \frac{\$14}{16 \text{ gallons}}$$

U.P. = $0.875 or $0.88 rounded off to the nearest cent

Brown Balsam

$$\text{U.P.} = \frac{\$12.50}{12 \text{ gallons}}$$

U.P. = $1.04\frac{2}{12}$ = $1.04 rounded off to the nearest cent

Clearly, the most expensive shampoo is Gleamy, the middle-priced shampoo is Brown Balsam, and the least expensive is Honey Herbal. Ms. Murphy knows that price is not the only measure of a good shampoo, so she must decide which she should purchase, taking all factors into consideration, and whether it is wise to buy such a large quantity of shampoo.

EXAMPLE: Mr. Daniels is invited to dinner and decides to bring a gift with him. In the More-for-Less Department Store he sees boxes of chocolates in three different sizes: a 1-pound box selling for $5.00, a $1\frac{1}{2}$-pound box for $7.00, and a 2-pound box for $9.00. Which box of chocolates is the best buy?

SOLUTION: The unit price per pound for the 1-pound box of chocolates is clearly $5.00 since the U.P. $= \dfrac{\$5.00}{1} = \$5.00$. The 2-pound box of chocolates costs $\dfrac{\$9.00}{2} = \$4.50$ per pound. The unit price of the $1\frac{1}{2}$-pound box of chocolates is $\dfrac{\$7.00}{1\frac{1}{2}}$. This involves more difficult computation, so let's look at it a little more closely.

We can calculate the answer in two different ways, with fractions or with decimals, as shown below.

Calculating with Fractions

$$\frac{\$7.00}{1\frac{1}{2}} = 7.00 \div 1\frac{1}{2}$$

$$= 7.00 \div \frac{3}{2}$$

$$\left(Remember: 1\frac{1}{2} = \frac{3}{2}. \right)$$

$$= \frac{7.00}{1} \times \frac{2}{3}$$

$$= \frac{14.00}{3}$$

$$= 4.66\frac{2}{3}$$

$$= \$4.67 \, per \, pound$$

Calculating with Decimals

$$1\frac{1}{2} = 1\frac{5}{10} \text{ since } \frac{1}{2} \times \frac{5}{5} = \frac{5}{10}$$

and $\dfrac{5}{10}$ written as a decimal is 0.5.

Therefore $1\frac{1}{2} = 1.5$.

$$\frac{\$7.00}{1\frac{1}{2}} = \frac{7.00}{1.5}$$

$$
\begin{array}{r}
4.66 \\
1.5\overline{)7.000} \\
6\,0xx \\
\hline
1\,00 \\
90 \\
\hline
100 \\
90 \\
\hline
10
\end{array}
$$

$$= \$4.66\frac{10}{15}$$

$$= \$4.66\frac{2}{3}$$

$$= \$4.67 \text{ per pound}$$

Clearly, the 2-pound box of chocolates is the best buy.

In the procedure headed "Calculating with Decimals," it was necessary to divide by the decimal 1.5. It is easier to divide by a *whole number*, so we multiplied 1.5 by 10, resulting in the answer 15. The same result could be obtained by moving the decimal point *one* place to the right. If we multiplied by 100, the decimal point would move two places to the right. Since the divisor, 1.5, was multiplied by 10, the dividend, 7.00, must also be multiplied by 10, resulting in the answer 70.0. This is shown in the division on page 200 by using arcs and arrows to indicate the movement of the decimal point. The decimal point is then placed directly above itself in the quotient, resulting in 4.66\frac{2}{3}$.

An additional 0 was placed after 70.0 in the dividend so that the division could be carried out to two decimal places. Zeros can always be placed to the right of the decimal point without changing the value of a number. That's what was done in this example.

PRACTICE EXERCISE 11

1. Sara purchased $2\frac{1}{2}$ yards of fabric to make a shirt and vest. The purchase price of this material is $7.25. Calculate the unit price per yard.

2. A customer in the carpet department wishes to purchase a long, narrow carpet. She finds two carpets, each 1 yard wide. One is 5 yards long and sells for $23; the other, 6 yards long, sells for $27. If each carpet is of equal quality, then:
    a. find the unit price for 1 yard of each carpet.
    b. which is the better buy?

3. Toothpaste is sold in tubes of three sizes, according to weight. During a special sale, the price of a 5-ounce tube was $1.50, the price of a 7-ounce tube was $1.96, and the price of an 8-ounce tube was $2.40. Which was the best buy?

4. In the drug and cosmetic department of the More-for-Less Department Store, rubbing alcohol is sold in bottles containing 1 pint for $0.65 and in bottles containing 1 quart for $1.25. Which bottle is the better buy?

COMPARING UNITS OF DIFFERENT SIZES

Salespersons and other employees of department stores like the More-for-Less need to know the relationships existing among units of length, weight, liquid measure, and dry measure to help customers determine unit prices, best buys, relationships between containers of different sizes, and so on. Stockclerks must be familiar with this information to determine shelf capacity when filling shelves with merchandise.

The table below is provided for your reference. It shows equivalent measures of length, weight, and liquid and dry measure.

Length
1 foot = 12 inches
1 yard = 3 feet = 36 inches

Weight
1 pound = 16 ounces

Liquid Measure
1 pint = 16 fluid ounces
1 quart = 2 pints = 32 fluid ounces
1 gallon = 4 quarts = 128 fluid ounces

Dry Measure
1 dozen = 12 units
1 gross = 12 dozen = 144 units

Note: Do not confuse fluid ounces with ordinary ounces. *Fluid ounces* are used to measure *liquid capacity*. *Ounces* are used to measure *weight*.

EXAMPLE: Ms. Kent, a stockclerk in the garden department of the More-for-Less, is asked to fill one of the shelves with boxes of grass seed. The shelf will hold 3 layers of boxes. Each layer can hold 4 rows of boxes, and each row has room for 6 boxes of seeds. How many boxes of seeds will she need to fill the shelf?

SOLUTION: Since there are 6 boxes in each row and 4 rows, she needs $6 \times 4 = 24$ boxes in each layer. There are 3 layers, so $24 \times 3 = 72$ boxes of seed.

EXAMPLE: The grass seed comes in larger cartons, which contain 1 dozen boxes. How many cartons will Ms. Kent have to bring up from the supply room to fill the shelf?

SOLUTION: To change from smaller units (single boxes) to larger units (dozens), you *divide*. Since 1 dozen = 12 units, you divide by 12: $\dfrac{72}{12} = 6$ cartons.

EXAMPLE: In the automobile supply department of the store, motor oil is on sale when purchased by the case. A case contains 24 quart cans of oil and sells for $28.80. How many fluid ounces are in 24 quarts?

SOLUTION: To change from larger units to smaller units, quarts to fluid ounces, you *multiply*. Since there are 32 fluid ounces in 1 quart, you multiply by 32: $24 \times 32 = 768$ fluid ounces.

How many pints are in 24 quarts?

Once again, you are changing from larger units (quarts) to smaller units (pints), and you multiply by the number of pints in a quart: $24 \times 2 = 48$ pints.

How many gallons are in 24 quarts?

This time you are changing from a smaller unit (quarts) to a larger one (gallons), so you divide. There are 4 quarts in a gallon, so you divide by 4: $24 \div 4 = 6$ gallons.

What is the unit price per quart for this motor oil?

$$U.P. = \frac{\$28.80}{24}$$

$$U.P. = \$1.20 \text{ per quart}$$

If this oil was sold in gallon cans at the same price of $1.20 per quart, what would the unit price per gallon be?

$$U.P. = \frac{\$28.80}{24 \text{ quarts}} \quad \text{and} \quad 24 \text{ quarts} = ? \text{ gallons}$$

$$\frac{24}{4} = 6 \text{ gallons}$$

$$U.P. = \frac{\$28.80}{6 \text{ gallons}}$$

$$U.P. = \$4.80 \text{ per gallon}$$

EXAMPLE: The candy and baked goods department is selling layer cakes in two sizes, 1 pound 4 ounces and 26 ounces. Which is heavier?

SOLUTION: One pound contains 16 ounces, so $16 + 4 = 20$ ounces. The other cake weighs 26 ounces so it is the heavier one.

EXAMPLE: If the smaller cake costs $3.98 and the larger one costs $5.10, which is the better buy?

SOLUTION: Calculate the unit price per ounce for each cake:

$$\text{U.P.} = \frac{\$3.98}{20} \qquad\qquad \text{U.P.} = \frac{\$5.10}{26}$$

$$\text{U.P.} = \$0.19\frac{18}{20} \qquad\qquad \text{U.P.} = \$0.19\frac{16}{26}$$

$$\frac{18 \div 2}{20 \div 2} = \frac{9}{10} \qquad\qquad \frac{16 \div 2}{26 \div 2} = \frac{8}{13}$$

$$\text{U.P.} = \$0.19\frac{9}{10} \qquad\qquad \text{U.P.} = \$0.19\frac{8}{13}$$

Which fraction, $\frac{8}{13}$ or $\frac{9}{10}$, is the larger one? Fractions like these often appear in problems in which unit prices are being calculated, and it is necessary to determine which is the larger in order to decide on lowest unit prices and best buys. The decision can sometimes be made just by inspection, as in the case of fractions like $\frac{1}{3}$ and $\frac{2}{3}$. Since these fractions have the same denominators, it is easy to see that $\frac{1}{3}$ is smaller than $\frac{2}{3}$.

When fractions do not have the same denominator, as in the case of $\frac{8}{13}$ and $\frac{9}{10}$, it is best to change each fraction to its decimal equivalent to help with the decision.

To change a fraction to its decimal equivalent, you divide the denominator into the numerator. Thus you divide 13 into 8 and divide 10 into 9:

$$\frac{8}{13} = 13\overline{)8.000} \qquad \text{and} \qquad \frac{9}{10} = 10\overline{)9.0}$$

with the long division showing $\frac{8}{13} = 0.615$ (steps: 78, 20, 13, 70, 65) and $\frac{9}{10} = 0.9$.

This shows that 0.615 is less than 0.9, and clearly $\frac{8}{13}$ is the smaller fraction and $\frac{9}{10}$ is the larger. This may be written as follows:

$$\frac{8}{13} < \frac{9}{10} \text{ or } \frac{9}{10} > \frac{8}{13}$$

The symbol "<" means "is less than," and the symbol ">" means "is greater than."

The larger cake has a slightly smaller unit price and is therefore the better buy.

PRACTICE EXERCISE 12

1. In the yard goods department, a bolt (roll) of cloth has 20 yards of material remaining on it. A customer requests 63 inches to make a dress. The material sells for $4.85 per yard.

   a. Find the number of yards in 63 inches.
   b. How many yards of material will be left on the bolt after the customer makes her purchase?
   c. How much did the customer pay for her purchase?

2. Change each of the following to the units indicated. Use the table on p. 202 for reference.

   a. 60 units = ___ dozen
   b. $3\frac{1}{2}$ dozen = ___ units
   c. 2 gross = ___ dozen
   d. 64 ounces = ___ pounds
   e. $2\frac{1}{4}$ pounds = ___ ounces
   f. 72 inches = ___ yards

   g. 5 feet = ___ inches
   h. $3\frac{1}{3}$ yards = ___ feet
   i. 96 fluid ounces = ___ pints
   j. 5 quarts = ___ fluid ounces
   k. 8 quarts = ___ gallons
   l. 3 gallons = ___ pints

3. A clerk in the hardware department was asked to fill a shelf with boxes of nails. He arranged the nails in 5 layers, with each layer containing 6 rows of boxes and each row containing 8 boxes of nails.

   a. How many boxes of nails were needed to fill the shelf?
   b. How many dozen boxes is this amount?
   c. Would a carton containing 2 gross boxes of nails contain enough to fill this shelf? If not, how many extra or how many more would be needed?

4. In the drug and cosmetics department, one brand of spray cologne is sold in a 2-fluid-ounce bottle for $15.00.

   a. What is the unit price per fluid ounce of this cologne?
   b. How many bottles, each containing 2 fluid ounces, will it take to make 1 pint?
   c. What is the unit price per pint of this cologne?

5. If a customer in the carpet department buys a runner 1 yard wide by 4 yards long and pays $53.75 for it, how much is he paying for each square foot of carpeting?

6. The baked goods department sells marble loaf cakes, weighing 14 ounces and selling for $1.26, and carrot loaf cakes, weighing 1 pound 2 ounces and selling for $1.44.

    a. Which cake is heavier?
    b. Calculate the unit price per ounce for each cake.
    c. Calculate the unit price per pound for each cake.

7. Change each fraction below to a decimal to compare the fractions in each pair. Then insert the symbol ">" ("is greater than") or the symbol "<" ("is less than") between the two fractions in each pair.

    a. $\dfrac{3}{5}$ $\dfrac{2}{3}$ f. $\dfrac{7}{8}$ $\dfrac{9}{10}$

    b. $\dfrac{1}{2}$ $\dfrac{3}{8}$ g. $\dfrac{4}{5}$ $\dfrac{3}{4}$

    c. $\dfrac{1}{2}$ $\dfrac{7}{16}$ h. $\dfrac{3}{8}$ $\dfrac{13}{32}$

    d. $\dfrac{2}{7}$ $\dfrac{1}{3}$ i. $\dfrac{3}{4}$ $\dfrac{5}{7}$

    e. $\dfrac{3}{5}$ $\dfrac{5}{8}$ j. $\dfrac{1}{4}$ $\dfrac{13}{64}$

USING GRAPHS TO SHOW INFORMATION

    Graphs are pictures that can be used to show such information as sales in different departments of a store, profits, or comparative sales in a particular department on different days or weeks. The More-for-Less Department Store uses a variety of graphs to check on the amount of business it does and on its earnings (profits). It uses bar graphs, line graphs, and pictographs. Each type of graph is used to represent different data.

Bar Graph

    This *bar graph* shows the total sales (in dollars) made by several departments of the More-for-Less Department Store during the week of July 23:

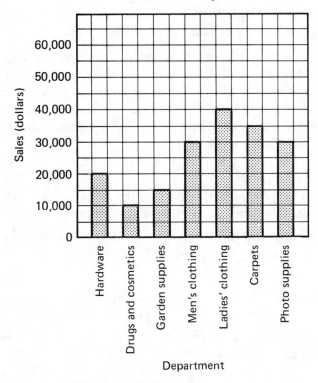

The names of the departments are shown at the bottom of the graph, along a horizontal line called the "horizontal axis." The amount of sales is shown at the left side of the graph, along the "vertical axis." Note that every two squares on the graph in the vertical direction represent $10,000 in sales. How much does one square represent?

From the bar graph it is easy to see that the hardware department made sales totaling $20,000 during this week. Which department had sales totaling $35,000? Of course, you see that it was the carpet department.

Bar graphs are often drawn with horizontal bars instead of the vertical bars shown in the graph here. This can easily be done by listing the departments along the vertical axis on the left and the amount of sales (dollars) along the horizontal axis at the bottom of the graph, the reverse of what was done here.

PRACTICE EXERCISE 13

1. Using the graph showing the weekly sales in several major departments for the week of July 23, answer the following questions:

   a. What was the amount of sales made by the garden supplies department?
   b. What was the amount of sales made by the photo supplies department?
   c. Which department made the least amount of sales during the week? How much did it sell, in dollars?
   d. Which department made the greatest amount of sales during the week? How much did it sell, in dollars?
   e. Which two departments had the same amount of sales during the week? What was the amount?
   f. How much more did the photo supplies department make in sales during the week than the garden supplies department?

2. This *horizontal bar graph* shows the weekly sales in several major departments during the week of November 4:

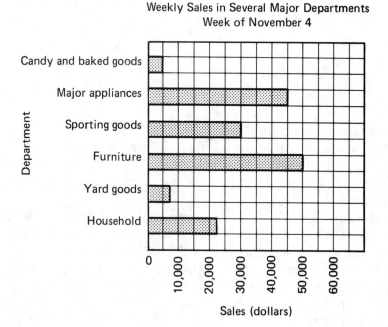

   a. What was the amount of sales made by the major appliances department?
   b. What was the amount of sales made by the yard goods department?
   c. Which department made the greatest amount of sales during the week? How much did it sell, in dollars?
   d. What department made the least amount of sales during the week? How much did it sell, in dollars?
   e. Which department had sales totaling $22,500?
   f. How much more did the sporting goods department make in sales during the week than the yard goods department?

Line Graph

A *line graph* is used to present data of a variable nature that change with some other variable, usually time. If you wanted to see how sales change from week to week in a single department, you would use a line graph.

The graph below shows the weekly sales during February in the furniture department.

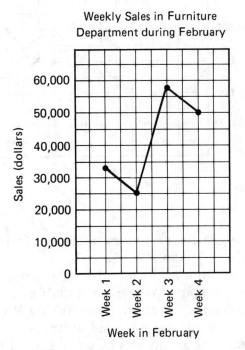

Weekly Sales in Furniture Department during February

Which week showed the greatest amount of sales? The third week showed the greatest amount. The amount was $57,500, since each square represents $5000 and the point for the third week is midway between $55,000 and $60,000.

Which week showed the least amount of sales? The second week did: only $25,000.

PRACTICE EXERCISE 14

1. Use the graph above to answer the following questions:

   a. What was the total amount of sales in the furniture department during week 1?

   b. What was the total amount of sales in the furniture department during week 4?

   c. How many more dollars' worth of furniture was sold during week 4 than during week 2?

   d. What was the total value of all furniture sold during the month of February?

   e. How many more dollars' worth of furniture was sold during the best week of February than during the poorest week?

2. This line graph depicts the sales in the hardware department during a 7-month period of time. Use it to answer the following questions.

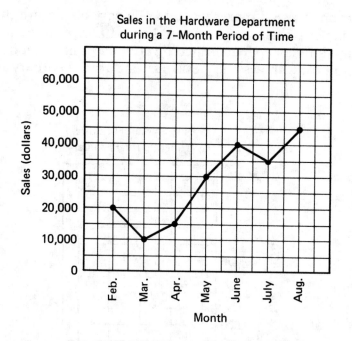

a. Which month was the best for sales in the hardware department?
b. Which month was the poorest for sales?
c. What was the total amount of sales during February?
d. What was the total amount of sales during June and July?
e. How many more dollars' worth was sold during May than during March?
f. The total of the three largest months, June, July, and August, surpassed the total of the other four months by what sum of money?
g. How many more dollars' worth of hardware was sold during the best month than during the poorest month?
h. What was the total amount of sales in the hardware department during this 7-month period of time?

Pictograph

The *pictograph* is a graph that uses pictures to represent items. Each picture may represent one or many items, and parts of a picture may be depicted as well. For instance, if a picture represents 10 items, then one-half of a picture represents 5 items. Two pictures represent 20 items, $3\frac{1}{2}$ pictures represent 35 items, and so on.

The pictograph below shows the soda sales in the foods department of the More-for-Less Department Store for one particular weekday period. (On another day, the number of cans of each kind of soda sold would have been different.)

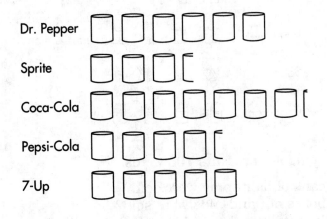

Note: Each can represents 40 cans.

Which soda had the best sales on this particular day? The largest number of cans shown is $7\frac{1}{4}$ for Coca-Cola.

How many Coca-Colas were sold?

$$7\frac{1}{4} \text{ cans} \times 40 \quad \text{or} \quad \frac{29}{4} \times 40 = \frac{29}{4} \times \frac{\overset{10}{\cancel{40}}}{1} = 290 \text{ cans}$$

PRACTICE EXERCISE 15

1. Use the pictograph above to answer the following questions:
   a. How many cans of Pepsi-Cola were sold?
   b. How many more cans of 7-Up were sold than cans of Sprite?
   c. What was the total number of sodas sold during this weekday period?
   d. How many cans would it be necessary to draw to show this total?
   e. What was the total number of cans of Coca-Cola and 7-Up sold?

2. This pictograph shows the sales for certain items in the drug department:

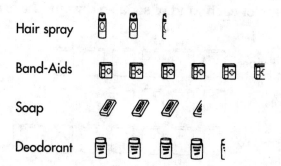

Note: Each symbol represents 100 items.

a. How many cans of hair spray were sold?
b. How many boxes of Band-Aids were sold?
c. What was the total number of bars of soap and containers of deodorant sold?
d. What was the total number of all four items sold?
e. How many more boxes of Band-Aids were sold than containers of deodorant?

FINDING THE AVERAGE

Mr. Thomas, the assistant buyer in the store, was asked to conduct a survey among the female employees to determine their *average* height. He discovered that there are *three* methods for determining the average. The most common of all of the three is the "mean." The other two are called the "median" and the "mode." Averages are referred to as "measures of central tendency." This indicates that each of them tends toward the center of all the numbers when they are arranged in order of size.

You find these three types of averages as follows:

Mean: Add all the values, and divide by the number of values.
Median: Arrange the values in ascending or descending order, and find the value that is in the middle of all the numbers.
Mode: Find the value that occurs most frequently among all the numbers. There may or may not be a mode.

The following example will help you to see how these measures of central tendency operate.

Mr. Thomas discovered in his survey that the heights of the 9 female employees in the drug and cosmetics department are 57 in., 66 in., 59 in., 61 in., 71 in., 58 in., 60 in., 65 in., and 61 in. He found the mean, median, and mode in the following way.

The *data* are the listed heights. He arranged the data in ascending size before he began.

Mean: 57 in. + 58 in. + 59 in. + 60 in. + 61 in. + 61 in. + 65 in. + 66 in.

+ 71 in. = 558 in. There are 9 values, so $9)\overline{558}$ = 62 in. (quotient 62)

Median: 57 in., 58 in., 59 in., 60 in., (61 in.) 61 in., 65 in., 66 in., 71 in. The middle value is the fifth one, 61 in.

Mode: The value 61 in. occurs twice while all the others occur only once, so 61 in. is the mode.

Thus Mr. Thomas discovers that the   *mean* height is 62 in.
median height is 61 in.
mode height is 61 in.

The three values may be the same or they may be different, as just shown. Each of them tends toward the center or middle value of all the data.

PRACTICE EXERCISE 16

1. The sales in the auto department of the More-for-Less Department Store for a particular Monday amounted to $4158.92 for 384 sales. What was the mean sale for the day?

2. Sales for the hardware department during one particular hour were $4.19, $6.83, $12.19, $22.98, $7.01, $1.19, and $1.19.

    a. Find the mean sale during this hour.
    b. What was the median sale during the hour?
    c. What was the mode sale for the hour?

3. Mr. Stern, an office worker in the store, timed himself at various times of the day to determine the average number of bills typed in 1 hour. He found that as the day progressed he typed fewer bills. These were his findings: 73, 73, 71, 70, 67, 63, 59 bills per hour.

    a. What was the average (mean) number of bills typed per hour?
    b. What was the median number of bills typed per hour?
    c. Was there a mode? What was it?

INTERPRETING DATA IN TABLES

Certain employees in the More-for-Less Department Store must be capable of reading tables and transmitting this information to the customer. This is a little difficult since the employee usually has to work down rows and across to columns to arrive at the correct information. At times he or she may have to consult many tables before finding the correct data.

Mr. Ross works in the mail-order department and receives phone calls for information and orders. Usually the orders come from advertisements that the customer sees in the daily newspaper.

BLACK and DECKER
VARIABLE SPEED DRILL
SAVE 5.00   #7127
34.99 OUR REGULAR 39.99
• ³⁄₈ speed drill with reverse action
• Double insulated and locking button

BLACK and DECKER
7¹⁄₄" CIRCULAR SAW
SAVE 10.00   #7383
59.99 OUR REGULAR 69.99
• Depth, bevel adjustment and double insulated; wrap around metal shoe

BLACK and DECKER
5" BENCH GRINDER
SAVE 10.00   #7900
39.99 OUR REGULAR 49.99
• Polishing and grinding wheels
• Includes tool rest, eye shield

BLACK and DECKER
WORKMATE
SAVE 10.00   #79-020
49.99 OUR REGULAR 59.99
• 16" tilt top, mounting clamps
• A must for do it yourselfers

Mr. Ross received an order from Mr. Tobin, who ordered one each of the variable speed drill, the circular saw, the bench grinder, and the Workmate, and asked that the order be mailed to him. Mr. Ross looked up the information in the catalog, as shown below, and made up his bill. He had to include a mailing charge as well as the cost of the merchandise.

What was the total bill that Mr. Ross completed?

Catalog No.	Product	Regular Price	Sales Price	Weight
10280G	Variable Speed Drill	$ 39.99	$ 34.99	$6\frac{1}{2}$ lb.
75020G	Drill Set	14.95	12.95	$\frac{1}{4}$ lb.
85020G	Bench Grinder	49.99	39.99	18 lb.
25090H	15" Chain Saw	186.99	146.99	21 lb.
44020H	Workmate	59.99	49.99	13 lb.
16180H	Circular Saw	69.99	59.99	15 lb.
94600J	Tool Chest	24.99	18.99	4 lb.
58320J	Soldering Gun	22.99	14.99	$2\frac{1}{4}$ lb.
44960J	Socket Wrench Set	12.99	10.99	$7\frac{1}{2}$ lb.

Acknowledgment: Reprinted by courtesy of AMES Department Stores, Inc.

The merchandise cost Mr. Tobin:

$$\begin{array}{r} \$34.99 \\ 59.99 \\ 39.99 \\ \underline{49.99} \\ \$184.96 \end{array}$$

Mr. Ross used the fourth-class (parcel post) rates shown below to compute the mailing charge that was added to the bill. First he added the weights of the four items:

$$\begin{array}{r} 6\frac{1}{2} \text{ lb.} \\ 15 \\ 18 \\ \underline{+\ 13} \\ 52\frac{1}{2} \text{ lb.} \end{array}$$

FOURTH-CLASS MAIL

Inter-BMC / ASF Parcel Post Rates
Nonmachinable Parcels
Surcharge Included

Weight Not Over (lbs.)	Zone 1 & 2	Zone 3	Zone 4	Zone 5	Zone 6	Zone 7	Zone 8
2	$3.69	$3.82	$3.96	$4.24	$4.35	$4.35	$4.35
3	3.79	3.99	4.20	4.62	5.04	5.50	5.55
4	3.89	4.15	4.44	5.00	5.56	5.85	6.10
5	3.99	4.31	4.67	5.38	6.08	6.70	6.90
6	4.09	4.48	4.91	5.76	6.60	7.83	10.05
7	4.18	4.64	5.15	6.14	7.12	8.56	11.10
8	4.28	4.81	5.39	6.52	7.64	9.28	12.15
9	4.38	4.97	5.62	6.90	8.17	10.01	13.20
10	4.48	5.13	5.86	7.28	8.69	10.74	14.25
11	4.58	5.30	6.10	7.66	9.21	11.47	15.25
12	4.68	5.46	6.33	8.04	9.73	12.19	16.30
13	4.75	5.58	6.49	8.29	10.07	12.67	17.35
14	4.82	5.69	6.66	8.54	10.42	13.15	18.40
15	4.88	5.78	6.77	8.73	10.67	13.49	19.45
16	4.93	5.86	6.89	8.90	10.90	13.81	20.50
17	4.98	5.94	6.99	9.06	11.12	14.11	21.41
18	5.03	6.01	7.10	9.22	11.33	14.40	21.88
19	5.08	6.09	7.19	9.37	11.53	14.67	22.33
20	5.13	6.15	7.29	9.51	11.72	14.93	22.76
21	5.18	6.22	7.38	9.65	11.90	15.18	23.16
22	5.22	6.29	7.47	9.78	12.07	15.41	23.55
23	5.27	6.35	7.55	9.90	12.24	15.64	23.93
24	5.31	6.41	7.63	10.02	12.40	15.86	24.28
25	5.35	6.47	7.71	10.14	12.55	16.07	24.63
26	5.39	6.53	7.79	10.26	12.70	16.27	24.96
27	5.43	6.59	7.86	10.37	12.85	16.47	25.28
28	5.47	6.64	7.94	10.47	12.99	16.66	25.59
29	5.51	6.70	8.01	10.58	13.13	16.84	25.89
30	5.55	6.75	8.08	10.68	13.26	17.02	26.18
31	5.59	6.80	8.15	10.78	13.39	17.19	26.46
32	5.63	6.86	8.21	10.87	13.51	17.36	26.73
33	5.67	6.91	8.28	10.97	13.64	17.52	27.00
34	5.70	6.96	8.34	11.06	13.76	17.68	27.25
35	5.74	7.01	8.41	11.15	13.87	17.84	27.51
36	5.78	7.05	8.47	11.24	13.99	17.99	27.75

Inter-BMC / ASF Parcel Post Rates
Nonmachinable Parcels
Surcharge Included

Weight Not Over (lbs.)	Zone 1 & 2	Zone 3	Zone 4	Zone 5	Zone 6	Zone 7	Zone 8
37	5.81	7.10	8.53	11.32	14.10	18.14	27.99
38	5.85	7.15	8.59	11.41	14.21	18.29	28.22
39	5.88	7.19	8.65	11.49	14.31	18.43	28.45
40	5.92	7.24	8.70	11.57	14.42	18.57	28.67
41	5.95	7.28	8.76	11.65	14.52	18.70	28.89
42	5.98	7.33	8.82	11.73	14.62	18.83	29.10
43	6.02	7.37	8.87	11.81	14.72	18.97	29.31
44	6.05	7.42	8.93	11.88	14.82	19.09	29.51
45	6.08	7.46	8.98	11.96	14.91	19.22	29.71
46	6.12	7.50	9.03	12.03	15.01	19.34	29.91
47	6.15	7.54	9.08	12.10	15.10	19.46	30.10
48	6.18	7.59	9.14	12.17	15.19	19.58	30.29
49	6.22	7.63	9.19	12.24	15.28	19.70	30.47
50	6.25	7.67	9.24	12.31	15.37	19.82	30.65
51	6.28	7.71	9.29	12.38	15.45	19.93	30.83
52	6.31	7.75	9.34	12.45	15.54	20.04	31.01
53	6.34	7.79	9.39	12.51	15.62	20.15	31.18
54	6.37	7.83	9.43	12.58	15.71	20.26	31.35
55	6.41	7.87	9.48	12.65	15.79	20.36	31.52
56	6.44	7.91	9.53	12.71	15.87	20.47	31.68
57	6.47	7.94	9.58	12.77	15.95	20.57	31.84
58	6.50	7.98	9.62	12.84	16.03	20.68	32.00
59	6.53	8.02	9.67	12.90	16.11	20.78	32.16
60	6.56	8.06	9.71	12.96	16.18	20.88	32.31
61	6.59	8.09	9.76	13.02	16.26	20.97	32.47
62	6.62	8.13	9.80	13.08	16.33	21.07	32.62
63	6.65	8.17	9.85	13.14	16.41	21.17	32.77
64	6.68	8.21	9.89	13.20	16.48	21.26	32.91
65	6.71	8.24	9.94	13.26	16.55	21.36	33.06
66	6.74	8.28	9.98	13.31	16.63	21.45	33.20
67	6.77	8.31	10.02	13.37	16.70	21.54	33.34
68	6.80	8.35	10.07	13.43	16.77	21.63	33.48
69	6.83	8.39	10.11	13.48	16.84	21.72	33.62
70	6.86	8.42	10.15	13.54	16.91	21.81	33.75

U.S. POSTAL SERVICE

OFFICIAL ZONE CHART
For Determining Zones From All Postal Units Having
ZIP Codes 09001—10499

This zone chart lists the first three digits (prefix) of the ZIP Codes of the sectional center facility of address.

To determine the zone distance to a particular post office, ascertain the ZIP Code of the post office to which the parcel is addressed. The first three digits of that ZIP Code are included in this chart, and to the right thereof the zone.

Zip Code Prefixes	Zone	Zip Code Prefixes	Zone	Zip Code Prefixes	Zone	Zip Code Prefixes	Zone	Zip Code Prefixes	Zone	Zip Code Prefixes	Zone
006-009	7	100-118	*1	255-266	4	394-396	6	506-507	5	723-724	5
010-013	2	119	*2	267	3	397	5	508-516	6	725-738	6
014	3	120-123	2	268-288	4			520-539	5	739	7
015-018	2	124-127	*2	289	5	400-402	5	540	6	740-762	6
019	3	128-136	3	290-293	4	403-406	4	541-549	5	763-770	7
020-024	2	137-139	2	294	5	407-409	5	550-554	6	773	6
025-026	3	140-149	3	295-297	4	410-418	4	556-559	5	774-775	7
027-029	2	150-154	4	298-299	5	420-427	5	560-576	6	776-777	6
030-033	3	155	3			430-458	4	577	7	778-797	7
034	2	156	4	300-324	5	460-466	5	580-585	6	798-799	8
035-043	3	157-159	3	325	6	467-468	4	586-593	7		
044	4	160-162	4	326-329	5	469	5	594-599	8	800-812	7
045	3	163	3	330-334	6	470	4			813	8
046-049	4	164-165	4	335-338	5	471-472	5	600-639	5	814	7
050-051	3	166-169	3	339-340	6	473	4	640-648	6	815	8
052-053	2	170-171	2	350-364	5	474-479	5	650-652	5	816-820	7
054-059	3	172-174	3	365-366	6	480-489	4	653	6	821	8
060-067	2	175-176	2	367-374	5	490-491	5	654-655	5	822-828	7
068-069	1	177	3	376	4	492	4	656-676	6	829-874	7
070-077	*1	178-199	2	377-386	5	493-499	5	677-679	7	875-877	7
078	*2	200-218	3	387	6			680-692	6	878-880	8
079	*1	219	2	388-389	5	500-503	6	693	7	881-884	7
080-086	2	220-238	3	390-392	6	504	5			890-898	8
087	1	239-253	4	393	5	505	6	700-722	6	900-999	8
088-098	*1	254	3								

The local zone rate applies to all parcels mailed at a post office or on its rural routes for delivery at that office or on its rural routes.

The following are wholly within the indicated zone:

Alaska	8	District of Columbia..	3	Mariana Islands	8	Rhode Island	2
Arizona	8	Georgia	5	Marshall Islands	8	Samoa (American)	8
California	8	Guam	8	Nevada	8	Utah	8
		Hawaii	8	Ohio	4	Virgin Islands	7
Canton Island	8	Idaho	8	Oregon	8	Wake Island	8
Caroline Islands	8	Illinois	5	Puerto Rico	7	Washington	8
Delaware	2	Louisiana	6				

*DENOTES THE INTRA BMC RATE THAT APPLIES TO PARCEL POST MAILED AND DELIVERED WITHIN THE SAME BMC.

Then he consulted the Official Zone Chart on p. 216. Mr. Tobin's zip code prefix was 492, which is in zone 4. Since his purchases weighed $52\frac{1}{2}$ lb., he read down the rate chart to 53 because the weight was at least 52 but did not exceed 53. Reading across to zone 4, Mr. Ross found that the parcel post rate was $9.39.
Adding this to the bill, he figured the total cost to be:

$$\begin{array}{r} \$184.96 \\ +\quad 9.39 \\ \hline \$194.35 \end{array}$$

Mr. Tobin did not live in the same state as the More-for-Less Department Store and therefore did not have to pay a sales tax on his purchases.

PRACTICE EXERCISE 17

1. Using the same information as Mr. Ross used, calculate the total price of the purchases in each of the following.

   a. Three circular saws being mailed to zone 2.
   b. Two soldering guns and a 15″ chain saw purchased at the store, plus an 8% sales tax.
   c. Two drill sets, a variable speed drill, and three socket wrench sets plus mailing to zip code 387.
   d. Four tool chests and a Workmate being mailed to the District of Columbia.
   e. Ten drill sets and three variable speed drills mailed locally.

2. a. What is the parcel post rate for mailing a package weighing $23\frac{3}{4}$ lb. to a zone with zip code prefix 289?
   b. If the postage for a package to be mailed to zone 3 costs $6.70, what is the greatest possible weight of the package?
   c. What is the cost of mailing a package weighing 7 lb. 4 oz. to Puerto Rico?
   d. What is the greatest possible weight of a package mailed to zip code 879 costing $14.25?

3. What is the total saving if you purchase a circular saw and a tool chest at the sale price?

4. a. How much would all the items in the advertisement cost if purchased at the regular price?
   b. What would the same items cost if purchased at the sale price?
   c. What is the difference between the two totals?

5. Mr. Samson operates the insurance department for the store. He, along with other salespersons, sell insurance to customers. This year several companies were granted rate increases, and these increases were listed in a table sent to all insurance salespersons. The listed increases are for the territory covered by the store.

Company	Required Insurance			Optional Insurance	
	Bodily Injury	Basic No Fault	Property Damage	Compre-hensive	Collision
Allcity Ins. Co.	19.6%	19.6%	19.6%	60.8%	59.4%
Allstate Ins. Co.	55.0	40.0	12.5	30.0	30.0
Amica Mutual Ins. Co.	10.0	—	15.0	—	—
Empire Mutual Ins. Co.	21.2	15.5	21.2	45.3	54.9
Firemen's Ins. Co. of Newark	—	—	—	22.2	34.7
General Accident F & L	—	139.7	10.7	24.0	33.3
Government Employees Ins. Co.	56.5	41.5	17.8	100.0	42.0
Hartford Accident & Indemnity Co.	21.4	156.9	12.2	8.6	10.3
Ins. Co. of North America	16.0	16.0	16.0	0.8	10.0

a. Allstate Ins. Co. was permitted a rate increase for property damage. What was the percentage increase?
b. Which company in this partial list was not granted an increase for comprehensive insurance.
c. Which company was granted a 10% increase in collision insurance?
d. Which company received the highest rate increase for basic no-fault insurance? How much was this?
e. Bodily injury was increased 21.4% for which company? If that type of insurance cost $67.32 previously, what is the new rate?
f. Which company had its comprehensive insurance rate doubled after the increase was granted?
g. General Accident F & L was granted a rate increase for basic no-fault insurance.

   (1) What was the increased percentage?
   (2) If the cost of insurance was $128.56 previously, what is the new rate?

EXAM TIME

1. Mr. Michaels purchased a 35-mm camera for $89.95, a case for $9.95, and 5 rolls of 35-mm film at $3.79 per roll in the photo supply department. Calculate the total price of this merchandise.

2. Mr. and Mrs. Preston bought a 20″ girl's hi-rise bicycle in the sporting goods department for their daughter. The sale price of this bike was $88.88. Mrs. Preston gave the salesclerk two $50 bills to pay for the purchase. How much change did she receive?

3. A customer in the major appliances department purchased a heavy-duty clothes washing machine for $439 and a matching deluxe electric dryer for $349. The sales tax on these purchases was charged at the rate of 8%. Calculate the total amount the customer was charged, including the tax.

4. Change each of the following tax rate percentages to a decimal:

    a. 5%   **b.** 6.5%   **c.** 3%   **d.** $7\frac{1}{2}$%   **e.** 36.5%

5. Rose Brown bought 3 CD's @ $13.99, 2 CD's @ $12.95, and 1 tape selling for $8.24.

    a. Calculate her bill.
    b. Add 6% sales tax.
    c. Determine the amount of change she would receive if she gave the salesclerk four $20 bills to pay for her purchases.

6. An AM/FM portable radio is on sale for $24.95. If its regular price is $29.87, how much is saved by buying the radio during the sale?

7. During a sale in the gift department, many items were marked "1/3 off."

    a. If a flower vase originally sold for $7.95, how much discount would a customer receive?
    b. How much would she actually pay for this merchandise?

8. The jewelry department advertised a sale on 14-karat gold jewelry. The advertisement read "Save 25%." How much would a customer now pay for a gold bracelet originally selling for $59.50?

9. Senior citizens receive a discount of 15% on all merchandise purchased on Monday and Wednesday. How much would a senior citizen shopping on Monday pay for a 5-piece copper-clad stainless steel cookware set that normally sells for $59.60?

10. Earrings are on sale in the jewelry department at two pairs for $26 or $13.95 per pair if only one pair is purchased.

    a. How much does a customer save on each pair of earrings by buying two pairs?
    b. What is her total savings on both pairs?

11. Salesclerks in the luggage department receive a commission of 18% on all sales they make. Mr. Johns sold a shoulder strap tote bag for $24.50, a 24" weekend suitcase for $37.44, and a suit carrier for $46.27.

   a. Calculate the total amount of these sales.
   b. Calculate his commission.

12. A customer in the carpet department purchased a piece of carpet 9' × 12' (9 feet wide by 12 feet long) for a small room in his apartment. This carpet was selling for $12.98 per square yard. Salespersons in this department receive commissions on all sales at the rate of 12%.

   a. Change the dimensions of the carpet to yards.
   b. Calculate the area of the carpet in square yards.
   c. Calculate the price the customer paid for the carpet.
   d. Calculate the commission earned by the salesperson on this sale.

13. Ms. Donohoe purchased a portable dishwasher in the major appliances department. It had a selling price of $480. She had the choice of paying cash for the dishwasher or making a down payment of $60 and paying the balance in equal monthly installments of $40 each for 1 year. If she chose to buy on the installment plan, how much was she being charged for credit?

14. The Ruiz family purchased a new bedroom set in the furniture department. The cash price for this set was $1395.95. Instead of paying cash, they chose to pay in installments of $65 each month for 2 years. How much were they being charged for credit?

15. Film is on sale in the photographic supplies department, and three rolls of color film are selling for $9.87.

   a. What is the cost of one roll of film?
   b. How much would a customer pay if he wished to purchase five rolls of film for a vacation trip?

16. Sport socks are sold in the men's clothing department in packages of three pairs for $7.00.

   a. How many packages would a customer purchase if he wished to buy a dozen pair of socks?
   b. What would be the total cost of all these packages?
   c. Calculate the unit price of these sport socks.

17. In the candy and baked goods department, a 2-pound box of cookies sells for $7.00 and a 3-pound box of the same cookies sell for $10.

   a. Calculate the unit price per pound of each box.
   b. Which box is the better buy?

18. A customer in the yard goods department buys $1\frac{1}{2}$ yards of material for $5.79. Calculate the unit price of this material per yard.

19. The foods department sells bottles of soda in three sizes: 1 pint for $0.35, 1 quart for $0.63, and $\frac{1}{2}$ gallon for $1.19.

    a. Calculate the unit price per fluid ounce for each of these three sizes.

    b. Which is the best buy?

    c. How many pints are there in 1 quart?

    d. How many pints are there in $\frac{1}{2}$ gallon?

    e. Calculate the unit price per pint for each of the three sizes.

    f. Is the same size bottle as found in (b) above still the best buy?

20. In the drug and cosmetics department, toothpaste tubes are placed on shelves in 6 rows of 8 tubes each. They are stacked in 3 layers.

    a. How many tubes are placed on each shelf?
    b. How many dozen is this?
    c. Is this number more than, less than, or equal to 1 gross?

21. Toothpaste is sold in tubes of many sizes, according to weight. One of the common sizes is 4 ounces.

    a. What fraction of a pound is 4 ounces? Express the answer in lowest terms.
    b. How many 4-ounce tubes would be needed to have 1 pound?

22. Insert the symbol "<" ("is less than") or the symbol ">" ("is greater than") between the fractions shown in each of the following pairs to make the comparison correct:

    a. $\frac{2}{5}$ $\quad$ $\frac{3}{8}$ $\qquad$ c. $\frac{5}{8}$ $\quad$ $\frac{2}{3}$

    b. $\frac{1}{4}$ $\quad$ $\frac{3}{10}$ $\qquad$ d. $\frac{1}{2}$ $\quad$ $\frac{4}{5}$

23. Insert the symbol "<" or the symbol ">" between the measures in each of the following pairs to make the comparison correct.

    a. 46 units $\qquad$ 4 dozen

    b. 60 ounces $\qquad$ $3\frac{1}{2}$ pounds

    c. 40 inches $\qquad$ 1 yard

    d. $5\frac{1}{2}$ feet $\qquad$ 70 inches

    e. 3 pints $\qquad$ 50 fluid ounces

    f. 2 gallons $\qquad$ 6 quarts

24. The bar graph below shows the sales in the major appliances department. Using the graph, answer the following questions:

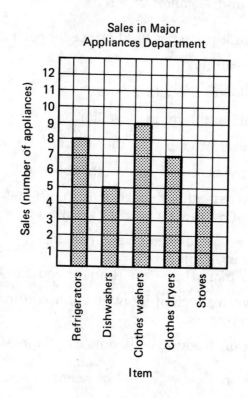

Sales in Major
Appliances Department

a. How many clothes dryers were sold?

b. How many more refrigerators were sold than dishwashers?

c. What was the total number of appliances sold during this period of time?

d. The number of refrigerators and clothes washers exceeds the number of dishwashers, clothes dryers, and stoves by what number?

e. Which item sold less than any of the others?

25. Study the line graph below to answer the questions about AM/FM radio sales during one week in the audio-video department.

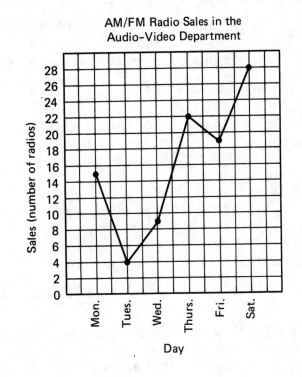

a. How many sales were made during the week?
b. How many radios were sold on Friday?
c. Which day had the greatest number of sales?
d. Which day had the least number of sales?
e. Between which two days did the greatest increase in sales occur?
f. How many more AM/FM radios were sold on Thursday than on Wednesday?
g. How many fewer radios were sold on Tuesday than on Saturday?

26. Use the table of rate increases below to answer the following questions:

| | Type of Insurance | | | | |
Company	Bodily Injury	Basic No Fault	Property Damage	Compre-hensive	Collision
Liberty Mut. Fire Ins. Co.	19.0%	100.0%	19.0%	—	—
National Grange Mut. Ins.	—	—	—	5.3%	5.1%
Nationwide Mut. Ins. Co.	—	—	22.8	42.3	38.9
Royal Globe Ins. Co.	17.2	17.2	7.2	31.9	27.7
State Farm Mut. Auto Ins. Co.	7.9	24.8	7.9	40.5	30.8
Travelers Indemnity Co.	33.0	33.3	25.0	34.2	30.0
Utica Mut. Ins. Co.	—	139.7	10.7	24.0	33.3

    a. Which company was granted a rate of increase double its previous rate for basic no-fault insurance?

    b. Which company wasn't granted any increase for comprehensive and collision insurance?

    c. If bodily injury insurance cost $32.91 with the Royal Globe Ins. Co. previously, what will it cost at the new rate?

    d. Which company was granted the least rate of increase for comprehensive insurance?

    e. Which company was granted the greatest rate of increase for basic no-fault insurance? If the previous cost was $56.78, what is the new cost?

27. In the ladies' shoes department sizes range from $4\frac{1}{2}$ to 9 for a certain type of shoe. On Monday, the following sizes in that style were sold: 5, 6, $7\frac{1}{2}$, $5\frac{1}{2}$, $6\frac{1}{2}$, $6\frac{1}{2}$, 7, $5\frac{1}{2}$, $6\frac{1}{2}$, 8, and 9.

    a. What was the mode?

    b. What was the median?

28. Find the mean sale for sales totaling $20,567.26 if there was 4739 sales on that day.

Now Check Your Answers

Now that you have finished the exam, check your answers against the correct ones beginning on page 228.

ANSWERS TO PRACTICE EXERCISES AND EXAM

PRACTICE EXERCISE 1

1. a. $233.32   b. $16.68   2. $8.82   3. $120.20
4. b. $25.96   c. Yes, $4.04   5. b. $24.95

PRACTICE EXERCISE 2

1. a. 0.06   b. 0.03   c. $0.05\frac{1}{2}$ d. 0.075   e. 0.15   f. 0.04

   g. $0.33\frac{1}{3}$ h. 0.60   i. 0.13   j. 0.048   k. 0.0625   l. $0.00\frac{1}{2}$

   m. 1.25   n. 0.01   o. 0.47   p. 1.10

2. a. $34.27   f. $432.75
    b. $18.87   g. $96.06
    c. $436.75   h. $80.29
    d. $29.90   i. $12.86
    e. $23.75   j. $65.90

3. a. $2.74   b. $1.51   c. $34.94   d. $2.39   e. $1.90

4. f. $19.47   g. $96.06   h. $80.29   i. $12.86   j. $65.90

5. $3.56

6. a. $129.97   b. $10.40   c. $140.37   d. Yes; $9.63

7. a. $89.98   b. $6.75   c. $96.73   d. $3.27

8. a. $32.87   b. $1.64   c. $34.51   d. No; she needs an additional
                                            $4.51.

PRACTICE EXERCISE 3

1. a. Adhesive bandages $0.80   b. $2.20   c. $5.16
       Satin Skin Cream   0.80
       Dental floss   0.60

2. a. $1498.00   b. $475.52   c. $6419.52

3. a. Small $5.00   b. $21.00   c. $61.94   d. $44.00
       Medium   7.00
       Large   9.00

PRACTICE EXERCISE 4

1. $80   2.   Both are honest advertisements.

PRACTICE EXERCISE 5

1. a. $8.74   b. $26.21   c. $27.52
2. a. $13.17   b. $30.73   c. $2.30   d. $33.03   3. $141.40

PRACTICE EXERCISE 6

1. a. $9.78   b. $0.98   c. $8.80   2. $32.87
3. a. $80.39   b. $589.51   c. $26.53   d. $616.04

PRACTICE EXERCISE 7

1. a. $13.44   b. $14.52   c. $0.75   d. $5.48
2. a. He saved $2.99 on the dickey shirt, $4.00 on the flannel shirts, and $1.00
   on the crew socks.
   b. $7.99
   c. $36.97

3. a. $90   b. $46.66
   c. (1) $ 44.95 $ 14.98
                19.95   6.65
                89.95   29.98
      (2) $44.95 $ 29.97
                19.95   13.30
                89.95   59.97
      (3) $103.24

4. a. $9.98, $15.98, and $19.00, respectively
   b. $14.97, $23.97, $28.50, respectively
   c. $67.44

5. a. Medium size $2.25   b. $4.80   c. $27.18
      Large   2.55

PRACTICE EXERCISE 8

1. $55.35

2. a. $9.00, $2.50, and $10.00, respectively
   b. $21.00

3. a. $92.89   b. $129.05

4. $118.99

5. **a.** 3 yd. by 4 yd.   **b.** 12 sq. yd.   **c.** $203.76   **d.** $24.45

PRACTICE EXERCISE 9

1. $41   2. $37.00   3. **a.** $896.22   **b.** $128.78

4. $44.05   5. $218.78

PRACTICE EXERCISE 10

1. Tennis ball $0.69
   Golf ball $1.06

2. **a.** $1.50   **b.** $18.00   **c.** 4 packages

3. **a.** $2.65   **b.** 4 packages   **c.** $31.80

4. **a.** $7.50   **b.** $45.00   **c.** one half-dozen

PRACTICE EXERCISE 11

1. $2.90   2. **a.** $4.60 per yard; $4.50 per yard, respectively

3. The 7-ounce tube   4. The quart bottle

PRACTICE EXERCISE 12

1. **a.** $1\frac{3}{4}$ yd.   **b.** $18\frac{1}{4}$ yd.   **c.** $8.49

2. **a.** 5   **b.** 42   **c.** 24   **d.** 4   **e.** 40   **f.** 2   **g.** 60
   h. 10   **i.** 6   **j.** 160   **k.** 2   **l.** 24

3. **a.** 240   **b.** 20   **c.** Yes 4 dozen extra

4. **a.** $7.50   **b.** 8 bottles   **c.** $120.00 per pint

5. $1.49

6. **a.** The carrot loaf cake
   b. marble cake, 9¢ per ounce; carrot cake, 8¢ per ounce
   c. marble cake, $1.44 per pound; carrot cake, $1.28 per pound

7. **a.** $0.6 < 0.66\frac{2}{3}$ **b.** $0.5 > 0.375$ **c.** $0.5 > 0.4375$ **d.** $0.29 < 0.33\frac{1}{3}$
   e. $0.6 < 0.625$ **f.** $0.875 < 0.9$ **g.** $0.8 > 0.75$ **h.** $0.375 < 0.41$
   i. $0.75 > 0.71$ **j.** $0.25 > 0.20$

PRACTICE EXERCISE 13

1. **a.** $15,000   **b.** $30,000   **c.** Drugs and cosmetics $10,000
   d. Ladies' clothing $40,000
   e. Men's clothing and photo supplies $30,000   **f.** $15,000

2. **a.** $45,000   **b.** $7,500   **c.** Furniture, $50,000
   d. Candy and baked goods, $5,000   **e.** Household   **f.** $22,500

PRACTICE EXERCISE 14

1. **a.** $32,500   **b.** $50,000   **c.** $25,000   **d.** $165,000   **e.** $32,500

2. **a.** August   **b.** March   **c.** $20,000   **d.** $75,000   **e.** $20,000
   f. $45,000   **g.** $35,000   **h.** $195,000

PRACTICE EXERCISE 15

1. **a.** 180   **b.** 50   **c.** 1060   **d.** $26\frac{1}{2}$ cans   **e.** 490

2. **a.** 250   **b.** 575   **c.** 775   **d.** 1600   **e.** 150

PRACTICE EXERCISE 16

1. **a.** $10.83   2. **a.** $7.94   **b.** $6.83   **c.** $1.19
3. **a.** 68   **b.** 70   **c.** Yes 73

PRACTICE EXERCISE 17

1. **a.** $186.05   **b.** $191.13   **c.** $107.12   **d.** $132.65   **e.** $239.69
2. **a.** $10.02   **b.** 29 lb.   **c.** $9.28   **d.** 10 lb.
3. $16   4. **a.** $219.92   **b.** $184.96   **c.** $34.96
5. **a.** 12.5%   **b.** Amica Mutual Ins. Co.   **c.** Ins. Co. of North America
   d. Hartford Accident & Indemnity   **e.** Hartford Accident & Indemnity
      Co. 156.9%   Co. $81.73
   f. Government Employees Ins. Co.   (1) 139.7%   (2) $308.16

EXAM TIME

1. $118.85   2. $11.12   3. $851.04

4. **a.** 0.05   **b.** 0.065   **c.** 0.03   **d.** $0.07\frac{1}{2}$ **e.** 0.365

5. **a.** $72.74   **b.** $4.36   **c.** $2.90

6. $4.92   **7. a.** $2.65   **b.** $5.30

8. $44.63   **9.** $50.66

10. **a.** $0.95   **b.** $1.90   **11. a.** $108.21   **b.** $19.48

12. **a.** 3 yd. by 4 yd.   **b.** 12 sq. yd.   **c.** $155.76   **d.** $18.69

13. $60   **14.** $164.05   **15. a.** $3.29   **b.** $16.45

16. **a.** 4 packages   **b.** $28   **c.** $2.33

17. **a.** 2 lb. = $3.50 per pound   **b.** 3 pounds
    3 lb. = $3.33 per pound

18. $3.86

19. **a.** 1 pint   = $0.021875
    1 quart   = $0.0196875

    $\frac{1}{2}$ gallon = $0.0185937

    b. $\frac{1}{2}$ gallon   **c.** 2 pints   **d.** 4 pints

    e. 1 pint   = $0.35   **f.** Yes
    1 quart   = $0.315

    $\frac{1}{2}$ gallon = $0.2975

20. **a.** 144   **b.** 12   **c.** Equal to

21. **a.** $\frac{1}{4}$ pound   **b.** 4

22. **a.** >   **b.** <   **c.** <   **d.** <

23. **a.** <   **b.** >   **c.** >
    d. <   **e.** <   **f.** >

24. **a.** 7   **b.** 3   **c.** 33   **d.** 1   **e.** Stoves

25. **a.** 97   **b.** 19   **c.** Saturday   **d.** Tuesday
    e. Wednesday and Thursday   **f.** 13   **g.** 24

26. **a.** Liberty Mut. Fire Ins. Co.   **b.** Liberty Mut. Fire Ins. Co.   **c.** $38.57
    d. National Grange Mut. Ins. Co.   **e.** Utica Mut. Ins. Co. $136.10

27. **a.** $6\frac{1}{2}$ **b.** $6\frac{1}{2}$ **28.** $4.34

How Well Did You Do?

0–51 **Poor.** Reread the unit, redo all the practice exercises, and retake the exam.

52–60 **Fair.** Reread the sections dealing with the problems you got wrong. Redo those practice exercises, and retake the exam.

61–68 **Good.** Review the problems you got wrong. Redo them correctly.

69–80 **Very good!** Continue on to the next unit. You're on your way.

THE GAS, GREASE, AND GADGETS SERVICE STATION AND AUTO SUPPLY SHOP

The Gas, Grease, and Gadgets Service Station is a busy garage and auto supply shop. It employs gasoline pump attendants, mechanics, and salespeople to serve its customers, who come in regularly to purchase gasoline, to have their cars serviced or repaired, and to buy parts and accessories for their cars. The station's employees must have many mathematical and mechanical skills to serve their customers well.

SELLING GASOLINE

The station sells gasoline in three grades: regular, extra, and premium. All cars manufactured since 1974 use only gasoline without lead. The price of each type of gasoline is different, for example:

> Regular gasoline costs 119.9¢ per gallon.
> Extra gasoline costs 129.9¢ per gallon.
> Premium gasoline costs 137.9¢ per gallon.

These prices include all taxes: the federal tax and state and city sales taxes. These taxes pay the cost of building and maintaining highways and roads, fighting air pollution, and serving the motorist in many other ways.

The gasoline pump automatically registers the price of each purchase, including all taxes, as gasoline is pumped into the tank of the car.

Selling Gas by the Gallon

Mr. Sanders is a regular customer who owns a 1992 Cadillac that uses premium gasoline (sometimes called "hi-test," since it has a higher octane rating). He drives in to buy 15 gallons of premium gas at 137.9¢ per gallon. What amount will the pump register as his bill?

First we change 137.9¢ to dollars by dividing by 100. Since we are changing from smaller (cents) units to larger (dollars) units, we divide by the number of cents in a dollar. Therefore:

$$
\begin{array}{r}
1.379 \\
100\overline{)137.900} \\
\underline{100\ \mathrm{xxx}} \\
37\ 9 \\
\underline{30\ 0} \\
7\ 90 \\
\underline{7\ 00} \\
900 \\
\underline{900}
\end{array}
$$

or $1.379 per gallon

Dividing by 100 resulted in moving the decimal point two places to the left.

Then we multiply the price per gallon by the number of gallons:

$1.379	3 decimal places
× 15	0 decimal places
6 895	
13 79	
$20.685	3 decimal places

Remember: When multiplying decimals, the number of decimal places in the product is the sum of the number of decimals places in the numbers being multiplied (3 + 0 = 3).

The answer $20.68(5) must be rounded off to the nearest cent and becomes $20.69 since the digit in the thousandths place (5) is 5 or more.

PRACTICE EXERCISE 1

1. Ms. Robinson drives her car to the regular gasoline pump (119.9¢ per gallon) and asks the attendant for 10 gallons of gas. How much change will she receive if she offers a $20 bill to pay for her purchase?

2. Mr. O'Brien drives into the Gas, Grease, and Gadgets Service Station in his Pontiac and asks for 13 gallons of premium gasoline at 137.9¢ per gallon.

   a. Will a $20 bill be enough to pay for this purchase?
   b. How much (if any) change will he receive?

3. What is the cost of 12.4 gallons of premium gasoline at 137.9¢ per gallon?

4. Ms. Smith needs only 7.9 gallons of regular gasoline to fill her tank. If regular costs 119.9¢ per gallon, what is her total bill?

Selling Gas by the Dollar

Ms. Rich owns a Toyota, which has a small gasoline tank and uses regular gasoline. She drives up to the pump and asks for $10 worth of gasoline. How many gallons will she get for this amount of money?

Since regular gas sells for 119.9¢ per gallon, we first change this price to dollars: $1.199. If the number of gallons × the price per gallon = the total cost, we can show this relationship as follows:

Number of gallons × price per
                 gallon = total cost

$$N \times 1.199 = \$10$$

or

$$1.199N = 10$$

Divide both sides of the equation by 1.199:

$$\frac{1.199N}{1.199} = \frac{10}{1.199}$$

$$N = 1.199\overline{)10.000\,000}\ \ \ 8.340$$

8.340	Multiply 1.199 by 1000
1.199)10.000 000	to convert it to a whole
9 592 xxx	number. Multiply 10.
4080	by 1000 also. This
3597	moves the decimal
4830	point three places to
4796	the right. Thus: 10 ×
340	1000 = 10000.
000	
340	

Ms. Rich will receive 8.340 . . . gallons of gasoline for $10. The dots indicate that the division may by continued further. The answer may also be rounded off to two decimal places, 8.34 gallons, or to one decimal place, 8.3 gallons.

This equation or formula may be written in the following way:

Number of gallons × price per gallon = total cost
      N × p = c

or

$$Np = c$$

because no symbol between N and p indicates multiplication.

Since $Np = c$, we can divide both sides of the equation by p:

$$\frac{N\overset{1}{p}}{\underset{1}{p}} = \frac{c}{p} \text{ or } N = \frac{c}{p}$$

As before,

$$\text{Number of gallons} = \frac{\text{total cost}}{\text{price per gallon}}$$

and it is necessary to divide to obtain the answer.

PRACTICE EXERCISE 2

1. Ms. Green has just purchased a new station wagon that uses only premium gasoline, which sells for 137.9¢ per gallon. She asks the attendant at the gasoline pump for $15 worth of premium gas. How many gallons of gas will be pumped into the tank of her car?

2. Mr. Chan has an Oldsmobile with a high-performance engine, which uses premium gasoline. The car has a 25-gallon gasoline tank that is almost empty. Mr. Chan asks for $25 worth of "hi-test" at 129.9¢ per gallon. Will his tank be able to hold all of this gasoline? How many gallons of gasoline will $25 buy?

3. Divide each of the following to two decimal places:

   a. $1.379\overline{)10}$ b. $1.489\overline{)20}$ c. $1.329\overline{)15}$

4. How many gallons of gasoline that costs 132.9¢ per gallon will you receive for $15.42?

5. Regular gasoline at a rival station costs 115.9¢ per gallon if you serve yourself. How many gallons of gasoline will you receive for $15?

6. Do the examples in Problem 3 on your calculator.

Selling Gas by the Fill-Up

Ms. Sanchez drives into the station in her new car, which uses only premium gasoline, selling for 137.9¢ per gallon. She asks the attendant to "fill it up." This means that she wants to buy as much gasoline as she needs to fill up the gasoline tank completely. The tank has a capacity of (will hold) 24 gallons of gasoline, and the gasoline gauge on the dashboard of the car reads as shown:

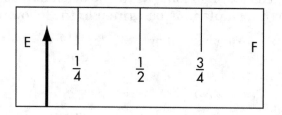

The "E" means "empty," and the "F" means "full."

Since the gasoline gauge reads half-way between "E" and "$\frac{1}{4}$," Ms. Sanchez has $\frac{1}{2}$ of $\frac{1}{4}$ of a tankful of gasoline, and

$$\frac{1}{2} \times \frac{1}{4} = \frac{1}{8}.$$

If the full tank contains 24 gallons, now the tank contains only

$$\frac{1}{\cancel{8}} \times \cancel{24}^{\,3} \quad \text{or} \quad 3 \text{ gallons.}$$

The car will have to receive 24 – 3 = 21 gallons of gas to "fill it up." The cost will be:

$$137.9¢ = \begin{array}{r} \$ 1.379 \\ \times \quad 21 \\ \hline 1\ 379 \\ 27\ 58 \\ \hline \$28.95(9) = \$28.96 \end{array}$$

If Ms. Sanchez pays for her purchase with a $20 and a $10 bill, how much change will she receive?

$$\begin{array}{r} \$30.00 \\ -28.96 \\ \hline \$\ 1.04 \end{array}$$

(*Remember:* Change = amount of money offered – amount of sale.)

PRACTICE EXERCISE 3

1. Ms. Brown has a car that uses regular gasoline, selling at 119.9¢ per gallon, and asks the attendant to "fill it up." Her gasoline gauge reads "$\frac{1}{2}$ full," and her gasoline tank has a capacity of 16 gallons.

   a. How many gallons of gas does she need to "fill it up"?
   b. What will this gasoline cost her?

2. Ms. Moss has a car with a gasoline tank whose capacity is 20 gallons. She uses extra gasoline, which sells for 129.9¢ per gallon. The pointer on her gasoline gauge shows that her tank is $\frac{1}{4}$ full.

   a. How many gallons of gasoline are left in the tank?
   b. How many gallons must she buy to "fill it up"?
   c. Will $20 be enough to pay for this purchase?
   d. How much (if any) change will Ms. Moss receive?

3. A car has a gasoline tank with a capacity of 24 gallons. Calculate the amount of gasoline remaining in the tank and how much must be bought for a "fill-up" if the gasoline gauge reads:

   a. $\frac{3}{4}$ full

   b. half-way between $\frac{1}{4}$ and $\frac{1}{2}$ full (*Hint:* Change each fraction to eighths and find the middle point between them.)

   c. half-way between $\frac{1}{2}$ and $\frac{3}{4}$ full

   d. half-way between $\frac{3}{4}$ and full

4. Calculate the total cost of regular gasoline needed for a "fill-up" in Problem 3.a–d if regular gasoline costs 119.9¢ per gallon.

SELLING MOTOR OIL, ANTIFREEZE, AND OTHER ENGINE FLUIDS

When motorists drive into the Gas, Grease, and Gadgets Service Station to purchase gasoline, it is the responsibility of the attendant at the gasoline pumps to wash their windshields (at no charge) and to ask them whether they would like a check made "under the hood." This means checking for the proper levels of motor oil, radiator fluid (water or antifreeze), and sometimes other fluids, such as windshield washer fluid, brake fluid, and power steering fluid. If any of these fluids is low, it is important for the attendant to advise the driver of the condition that needs attention. The driver then must decide whether he or she wishes to purchase the fluid that is needed. Water is free (for the windshield washer container, in warm weather, and for the radiators of some cars), but all other fluids must be paid for. The charges for these are added to the charge for gasoline, and the total is the driver's bill.

We Now Carry
AUTOMOBILE SUPPLIES

EXAMPLE: Mr. Benson drives into the station in his new Buick and asks for a "fill-up" with premium gas, which is selling for 137.9¢ per gallon. His gasoline tank holds 18 gallons, and the indicator on his gasoline gauge shows that the tank is $\frac{1}{4}$ full. Joe, the attendant, checks under the hood of the car and finds that the engine needs 2 quarts of oil. Mr. Benson asks for one of the best grades of oil, called 10W40, which sells for $1.75 per quart plus 8% sales tax. He gives Joe his credit card to pay for his purchases, and Joe must total the bill and enter the amount on Mr. Benson's sales slip. What was the total cost of both the gasoline and the oil?

SOLUTION: *The cost of the gasoline:* Since the gasoline tank is $\frac{1}{4}$ full:

$$\underset{2}{\frac{1}{\cancel{4}}} \times \frac{\overset{9}{\cancel{18}}}{1} = \frac{9}{2}$$

$$= 4\frac{1}{2} \text{ gallons are in the tank}$$

Therefore $18 - 4\frac{1}{2} = 13\frac{1}{2}$ gallons of gasoline are needed to fill the tank. At 137.9¢ per gallon, or $\$1.379 \times 13\frac{1}{2}$, we get

$$
\begin{array}{r}
\$ \ 1.379 \\
\times \ \ 13.5 \\
\hline
689 \ 5 \\
4 \ 137 \\
13 \ 79 \\
\hline
\end{array}
$$

$\$18.61(6)5 = \$18.62$ for gasoline

The cost of the oil: Oil sells for $1.75 per quart, and 2 quarts of oil are needed, so $\$1.75 \times 2 = \$3.50$. The sales tax rate is 8% or 0.08:

$$
\begin{array}{r}
\$3.50 \\
\times 0.08 \\
\hline
\end{array}
$$

$\$0.28(0)0 = \$0.28$ sales tax

Therefore the oil costs

$$
\begin{array}{r}
\$3.50 \\
+0.28 \\
\hline
\$3.78
\end{array}
$$

Mr. Benson's sales slip indicates that he received two items:

$13\frac{1}{2}$ gallons gasoline @ \$1.379 = \$18.62

$$
\begin{array}{lr}
2 \text{ quarts oil @ } \$1.75 = & 3.50 \\
\text{Sales tax (oil)} & +0.28 \\
\hline
& \$22.40
\end{array}
$$

EXAMPLE: Mr. Roy drives into the station and asks for 15 gallons of regular gasoline at 119.9¢ per gallon. A check under the hood of his car reveals that his engine needs 1 quart of oil and his radiator requires 1 gallon of antifreeze. Mr. Roy asks the attendant to add 1 quart of grade 30 oil, which costs $1.50 a quart, and 1 gallon of antifreeze, which costs $6.50 a gallon. He offers a $20 bill to pay for his purchases. Will this amount be sufficient to pay for all these items? If not, how much more will he be required to pay?

SOLUTION: *The cost of the gasoline:* 119.9¢ = $1.199 per gallon and

$$\begin{array}{r} \$\ 1.199 \\ \times\ 15 \\ \hline 5\ 995 \\ 11\ 99 \\ \hline \$17.985 \end{array} = \$17.99 \text{ for the gas}$$

The cost of the oil and the antifreeze:

1 quart of oil	=	$1.50
1 gallon of antifreeze	=	$6.50
		$8.00

and

$$\begin{array}{r} 8\% = 0.08 \text{ sales tax} \qquad \$8.00 \\ \times 0.08 \\ \hline \$0.64\cancel{0}\cancel{0} \end{array}$$

Then

$$\begin{array}{r} \$8.00 \\ +0.64 \\ \hline \$8.64 \end{array}$$

Thus the total bill for all three purchases is

$$\begin{array}{r} \$17.99 \\ 8.00 \\ 0.64 \\ \hline \$26.63 \end{array}$$

Surely, $20 will not cover the cost so Mr. Roy must add $6.63 to pay for his purchases.

PRACTICE EXERCISE 4

1. Mr. Ryan drives up to the gasoline pump and asks for a "fill-up" with premium gasoline, selling for 137.9¢ per gallon. His gasoline tank holds 22 gallons, and the indicator on the gas gauge shows that the tank is $\frac{1}{2}$ full. A check of the engine shows that he needs a quart of oil and some brake fluid. He asks for the most expensive oil, premium grade 10W20W50, selling for $2.25 per quart. The charge for the brake fluid is $2.25. Sales tax is charged only on the oil and brake fluid and is at the rate of 8%.

   a. Calculate the number of gallons of gasoline Mr. Ryan needs to fill his tank.
   b. Calculate the cost of this gasoline.
   c. Find the total cost of the oil and the brake fluid.
   d. Find an 8% sales tax on the oil and the brake fluid.
   e. Find Mr. Ryan's total bill.

2. Ms. Green drives up to the extra gasoline pump in her new Ford and asks for 8 gallons of gasoline, which sells at 129.9¢ per gallon. She also needs 2 quarts of oil at $1.85 a quart and some windshield washer fluid at $1.25.

   a. Calculate the cost of the gasoline.
   b. Calculate the total cost of the oil and the windshield washer fluid.
   c. Find the sales tax on the oil and the windshield washer fluid at the rate of $7\frac{1}{2}\%$.
   d. Calculate Ms. Green's bill.

3. Ms. Smith drives into the station to purchase $10 worth of gasoline. An engine and radiator check shows that she needs 2 quarts of oil and 1 gallon of antifreeze. Oil costs $1.75 per quart, antifreeze sells for $6.50 per gallon, and the tax rate on these items is 8%.

   a. Find the cost of the oil and the antifreeze.
   b. Calculate the sales tax on these items.
   c. Find the total bill for Ms. Smith, including the gas.

4. a. Find the total amount of this sale:
       10 gallons of regular gasoline @ 119.9¢ per gallon
       2 quarts of 10W-40 motor oil @ $1.75 per quart
       1 gallon of antifreeze @ $6.50 per gallon
       Fluid for the power-steering unit @ $1.95
       Add an 8% sales tax for the oil, antifreeze, and power steering fluid.
   b. Determine the amount of change to be returned to the customer, who offers $30 in payment.

5. a. A customer has only $10 with him to pay for 5 gallons of regular gasoline at 119.9¢ per gallon and 2 quarts of oil at $1.65 per quart. If the sales tax rate is 8%, will he have enough to pay for his purchases?
   b. Will there be any change? If so, how much?

SERVICING CARS IN THE SERVICE BAYS

Some employees of the Gas, Grease, and Gadgets Service Station are assigned by the manager to work in the service bays instead of at the gasoline pumps. These employees lift cars on the racks to lubricate (grease) them, change their engine oil, change or rotate tires, charge weak batteries, and make minor repairs whenever they can. The station's mechanics are responsible for making major repairs, tuning engines, and so on.

EXAMPLE: Mr. Davidson drives his car in for servicing. He asks for a lubrication, an oil change, and the installation of a new oil filter. The charge for lubrication (greasing points at which wear occurs because of friction) is $5.50; a new oil filter costs $7.50. Mr. Davidson's car requires 5 quarts of oil, and he asks for oil selling at $2.25 per quart. Sales tax at the rate of 8% is added to the total cost of materials and labor. What is Mr. Davidson's total bill for service?

SOLUTION:

5 quarts @ $2.25   =

$11.25   Total cost of oil
 5.50   Lubrication
 7.50   Oil filter
$24.25   Total cost of materials and labor

A sales tax of 8% amounts to

$ 24.25
× 0.08
$1.94ØØ = $1.94

Therefore:

$24.25   Total cost of materials and labor
+ 1.94   Sales tax
$26.19   Total bill

Different drivers ask for different grades of motor oil when they bring their cars in for oil changes or when oil must be added to engines that are low on oil. These different grades are of varying qualities and viscosities (the thickness or thinness of an oil is called its "viscosity"), and they sell at different prices. The Gas, Grease, and Gadgets Service Station sells oil as follows:

Single-grade oil, grade 30   @ $1.25 per quart
Grade 10W30   @   1.50 per quart
Grade 10W40   @   1.75 per quart
Grade 10W20W50   @   2.25 per quart

All but the first of these oils are multigrade (they are mixtures of thick and thin oils). Calculate the average price per quart of oil sold by this station.

The *average* price of 1 quart of oil is found by adding the prices of 1 quart of all grades of oil sold by the station and dividing by 4 since there are four different grades of oil. Therefore:

$1.25   per quart of grade 30
1.50   per quart of grade 10W30
1.75   per quart of grade 10W40
2.25   per quart of grade 10W20W50
$6.75   Total cost of 1 quart of all oils

$$\frac{\$6.75}{4} = \$1.68\frac{3}{4} = \$1.69 \text{ Average price per quart of oil}$$

The *average* price of any number of items is calculated by finding the total price of all the items and dividing by the number of items.

EXAMPLE: What is the average price per quart of:

5 quarts of oil selling at $1.25 per quart,
6 quarts of oil selling at   1.50 per quart,
4 quarts of oil selling at   1.75 per quart,
3 quarts of oil selling at   2.25 per quart?

SOLUTION:

$1.25	$1.50	$1.75	$2.25
× 5	× 6	× 4	× 3
$6.25	$9.00	$7.00	$6.75

Their sum is:

$ 6.25
9.00
7.00
6.75
$29.00

and $29 must be divided by 18 since there are 18 quarts of oil:

```
        1.611
18)29.000
   18 xx
   11 0
   10 8
      20
      18
      20
```

The average price per quart of oil is $1.61(1) . . ., which is rounded off to $1.61 per quart.

PRACTICE EXERCISE 5

1. Ms. Gonzalez drives a small car that she brings in for lubrication (a "grease job"), an oil change, and a new oil filter. Lubrication costs $5.50, and an oil filter for her small car sells for $5.25. Her car's engine requires 5 quarts of oil, and she asks for 10W40 oil, which sells for $1.75 per quart. On the day Ms. Gonzalez comes in for service, the Gas, Grease, and Gadgets Service Station is running a special offer:

> **OIL - LUBE & FILTER SPECIAL**
> Now includes most
> American and Foreign Cars
> **Featuring 10W40** **$17.95**
> **Quaker State Deluxe**

How much money will Ms. Gonzalez save, compared to the regular cost of the service, if she takes advantage of the price offered in the advertisement?

2. Mr. Moy brings his car into the station for a brake job. This service involves replacing the brake linings front and rear, refacing all drums and rotors, and repacking inner and outer wheel bearings. The station normally charges $135 for parts and $60 for labor for a brake job. How much would Mr. Moy save by taking advantage of the special price shown in the advertisement below?

> **FRONT WHEEL DRIVE & MAC PHERSON SUSPENSION ALIGNMENT**
> Special $28⁸⁸
>
> **TEXACO**
> **CIRCLE SERVICE**
> **BEAR.**
>
> **DRUM BRAKE OR DISC/DRUM COMBO**
> Backed By FREE brake adjustment for life OF LINING
> • REPLACE BRAKE LININGS FRONT & REAR
> • REFACE ALL DRUMS & ROTORS
> • REPACK INNER & OUTER WHEEL BEARINGS
> **$169⁹⁵**
> * most popular cars
> **10% discount on all exhaust system work**

Acknowledgment: Reprinted by courtesy of TViews, a Division of CIS, New Milford, CT.

3. Monday, Ms. Ramos drove her car into the station for a lubrication, which usually costs $5.50, and an oil change. She wanted 5 quarts of 10W30 oil for her engine, and this grade of oil usually sells for $1.50 per quart. How much did Ms. Ramos save by coming into the station on a day when the special shown in the advertisement below was in effect?

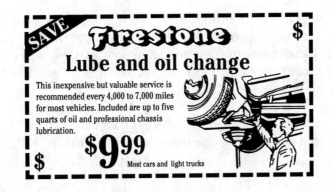

4. One morning many cars were brought into the Gas, Grease, and Gadgets Service Station for oil changes. During the day, 25 quarts of oil selling at $1.25 per quart, 40 quarts of oil at $1.50 per quart, 55 quarts of oil at $1.75, and 30 quarts of oil at $2.25 were sold. Calculate the average price per quart of the oil sold that day.

5. During a busy hour in the late afternoon, the Gas, Grease, and Gadgets Service Station sold 275 gallons of gasoline of all grades. Total receipts for all of these sales amounted to $362.52. Calculate the average price per gallon of the gasoline sold during that hour, and express the answer to the nearest tenth of a cent.

6. Gasoline is usually sold by the *gallon*, but oil is usually sold by the *quart*. Since there are 4 quarts in a gallon, to change gallons to quarts you multiply by 4 because you are changing from larger units (gallons) to smaller units (quarts). To change quarts to gallons you divide by 4 because you are changing smaller units (quarts) to larger units (gallons).

In problem 5, 275 gallons of gasoline were sold. What is this amount in quarts? $275 \times 4 = 1100$ quarts of gasoline.

In Problem 4, 25 quarts of oil selling at $1.25 per quart were sold. This amount can be changed to gallons as follows: $\dfrac{25}{4} = 6\dfrac{1}{4}$ gallons of oil. Using Problem 4, find:

    a. The number of gallons in 40 quarts.
    b. The number of gallons in 55 quarts.
    c. The number of gallons in 30 quarts.
    d. The total number of gallons of oil sold on that day.

Acknowledgment: Reprinted by courtesy of TViews, a Division of CIS, New Milford, CT.

7. a. Calculate the number of quarts of antifreeze in 13 containers of the fluid if each container holds 1 gallon.

  b. If a driver purchased 8.5 gallons of gasoline, how many quarts of gasoline did he receive?

  c. A driver bought 1 gallon 3 quarts of antifreeze for his car's radiator. How many quarts was this?

8. Changing from gallons to quarts and from quarts to gallons often involves working with decimals, proper fractions, improper fractions (the numerator is greater than the denominator), and mixed numbers. It is important to know how to change from any one of these forms of a given number to any other form. For example:

To change $3\frac{5}{8}$ to a *decimal:* $\frac{5}{8}$ means $8\overline{)5.000}$ and $3\frac{5}{8} = 3.625$

$$
\begin{array}{r}
0.625 \\
8\overline{)5.000} \\
\underline{4\,8} \\
20 \\
\underline{16} \\
40 \\
\underline{40}
\end{array}
$$

To change $3\frac{5}{8}$ to an *improper fraction:*

$$3\frac{5}{8} = 1 + 1 + 1 + \frac{5}{8}$$

$$= \frac{8}{8} + \frac{8}{8} + \frac{8}{8} + \frac{5}{8}$$

$$= \frac{29}{8}$$

Or, using a shortcut: $3\frac{5}{8} = 3\frac{5}{8} = \frac{8 \times 3 + 5}{8} = \frac{29}{8}$

To change $\frac{29}{8}$ to a *mixed number:* $\frac{29}{8}$ means $8\overline{)29}$

$$
\begin{array}{r}
3\frac{5}{8} \\
8\overline{)29} \\
\underline{24} \\
5
\end{array}
$$

a. Change each of the following mixed numbers to a two-place decimal:

(1) $5\frac{3}{4}$ (4) $6\frac{1}{4}$ (7) $10\frac{1}{8}$ (9) $4\frac{2}{5}$

(2) $7\frac{1}{2}$ (5) $5\frac{1}{5}$ (8) $1\frac{1}{7}$ (10) $3\frac{5}{8}$

(3) $2\frac{7}{8}$ (6) $4\frac{2}{3}$

b. Change each of the following mixed numbers to an improper fraction:

(1) $1\frac{3}{8}$ (4) $8\frac{3}{4}$ (7) $5\frac{5}{8}$ (9) $3\frac{1}{7}$

(2) $3\frac{1}{2}$ (5) $3\frac{1}{4}$ (8) $2\frac{2}{5}$ (10) $8\frac{3}{8}$

(3) $4\frac{1}{8}$ (6) $7\frac{1}{16}$

c. Change each of the following improper fractions to a mixed number:

(1) $\frac{23}{4}$ (4) $\frac{19}{8}$ (7) $\frac{23}{5}$ (9) $\frac{22}{3}$

(2) $\frac{13}{8}$ (5) $\frac{43}{8}$ (8) $\frac{15}{7}$ (10) $\frac{17}{6}$

(3) $\frac{9}{2}$ (6) $\frac{55}{16}$

9. Mr. Johnson drove in one day for the following services:

Charging a weak battery...	$7.50
State inspection...	17.00
Repairing a flat tire...	5.00
Lubrication..	5.50
Rotation of tires (for longer tire life)	12.00
Oil filter	5.75
Five quarts of 10W40 oil at $1.75 per quart	
Gasoline filter...	3.50
Air filter...	7.50
Subtotal $	_____
8% Sales tax	_____
Total bill	_____

a. Fill in the missing amount in this bill.

b. Determine the change Mr. Johnson would receive if he offered four $20 bills as payment.

WORKING ON MAJOR REPAIRS AND SERVICE

At the Gas, Grease, and Gadgets Service Station major repairs and service are done only by trained mechanics. These mechanics are able to tune engines, install new parts (such as fuel pumps, water pumps, generators, alternators, voltage regulators, and fan belts), align front ends of automobiles, repair brakes, repair steering, and make other major repairs. They must have many automotive skills to do this work properly and efficiently.

EXAMPLE: Mr. Seabrook owns an 8-cylinder automobile and drives into the Gas, Grease, and Gadgets Service Station for major service. He wishes to take advantage of the special prices shown below:

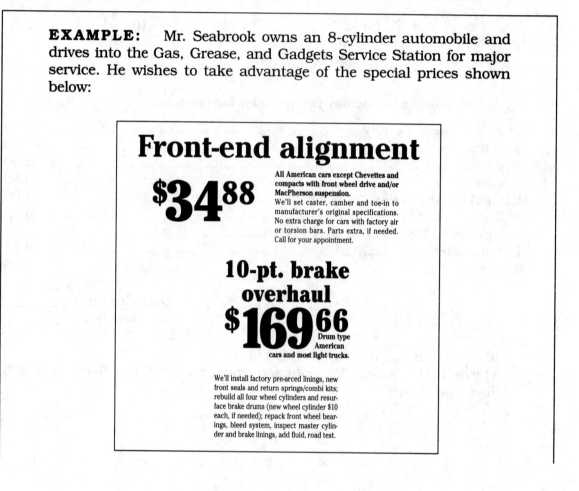

Front-end alignment

$34⁸⁸

All American cars except Chevettes and compacts with front wheel drive and/or MacPherson suspension.
We'll set caster, camber and toe-in to manufacturer's original specifications. No extra charge for cars with factory air or torsion bars. Parts extra, if needed. Call for your appointment.

10-pt. brake overhaul

$169⁶⁶
Drum type American cars and most light trucks.

We'll install factory pre-arced linings, new front seals and return springs/combi kits; rebuild all four wheel cylinders and resurface brake drums (new wheel cylinder $10 each, if needed); repack front wheel bearings, bleed system, inspect master cylinder and brake linings, add fluid, road test.

Acknowledgment: Reprinted by courtesy of TViews, a Division of CIS, New Milford, CT.

Mr. Seabrook wants to have a brake overhaul and the front end of his car aligned. The mechanic who inspects the car advises him that he needs a new muffler and tailpipe. The muffler will cost $54.95 and the tailpipe $39.95. There will be an additional charge of $6 for clamps and hangers needed to install these parts. Sales tax at the rate of $7\frac{1}{2}\%$ is charged on the entire bill. Calculate the cost to Mr. Seabrook for this major service.

SOLUTION: The advertisement says that the complete charge for a brake overhaul, including all parts and labor, is $169.66. The charge for the labor of a front-end alignment is $34.88. These charges must be added to the charges for the muffler, tailpipe, and clamps and hangers. Therefore:

$$
\begin{array}{r}
\$169.66 \\
34.88 \\
54.95 \\
39.95 \\
\underline{6.00} \\
\$305.44
\end{array}
$$

Sales tax of $7\frac{1}{2}\% = 0.07\frac{1}{2} = 0.075$

$$
\begin{array}{r}
\$305.44 \\
\times\ 0.075 \\
\hline
1\ 52720 \\
21\ 3808 \\
\hline
\$22.90(8)00 = \$22.91
\end{array}
$$

Adding in the sales tax gives:

$$
\begin{array}{r}
\$305.44 \\
\underline{+\ 22.91} \\
\$328.35
\end{array}
$$

EXAMPLE: Ms. Martin's car breaks down and is towed into the station for repairs. It is found that the car has a dead battery, caused by a bad alternator and voltage regulator. The battery will cost $65.95, the alternator $89.95, and the voltage regulator $24.95. There is an additional charge for labor to install these parts. The station charges $28 per hour for labor, and the work will require $1\frac{1}{2}$ hours to complete. Sales tax at the rate of 8% is charged on the entire bill. Calculate the total cost to Ms. Martin for this service.

SOLUTION:

Cost of parts:
$65.95
89.95
24.95
$180.85

Cost of labor: $1\frac{1}{2}$ hours @ $28 per hour.

$$1\frac{1}{2} \times 28 = \frac{3}{\cancel{2}} \times \frac{\cancel{28}^{14}}{1} = \frac{42}{1} = \$42$$

Total cost of parts and labor:

$180.85
42.00
$222.85

Adding in the sales tax of 8% gives:

$222.85
× 0.08
$17.82(8)0 = $17.83

$222.85
+ 17.83
$240.68   Total

PRACTICE EXERCISE 6

1. Mr. Salvado's car needs a brake job at the cost shown in the advertisement on page 242. He also needs to have a transmission tune-up, including changing the filter and transmission oil, checking the linkage, and adjusting the bands, for which a charge of $39.95, including labor, is made. Add a sales tax charge of 6% to Mr. Salvado's bill, and determine the amount of change, if any, he would receive from three $100 bills.

2. Ms. Park drives in to have a new fuel pump installed in her car at a cost of $62.95. The mechanic who checks her car sees that she has a worn fan belt which needs replacement, at a cost of $8.50, and two hoses which also need replacement, at a cost of $6.50 each. Total labor for the entire job will be $2\frac{1}{4}$ hours, at $28 per hour. Sales tax at the rate of 5% is charged on the entire bill. Calculate the total amount of Ms. Park's bill.

3. An advertisement states that the charge for tuning a car that has only 4 cylinders is $30.88, the charge for a 6-cylinder car is $36.88, and the charge for an 8-cylinder car is $40.88. Calculate the average charge for a tune-up.

4. Calculate the cost for labor on major repairs, at $28 per hour, for a job requiring each of the following amounts of time:

   a. $1\frac{3}{4}$ hours   c. 3 hours

   b. $\frac{1}{2}$ hour   d. 2 hours 15 minutes

5. If the total charge for labor is equal to the number of hours worked times the cost of 1 hour of labor ($28), this can be written as an equation:

$$\text{Charge} = \text{number of hours} \times \text{price per hour}$$
$$C \quad = \qquad N \qquad \times \qquad 28$$
$$C \quad = \quad 18N$$

Solving for N, you divide both sides of the equation by 18 to undo the multiplication of 18 and N:

$$\frac{C}{28} = \frac{28N}{28}$$
$$\frac{C}{28} = N$$

   a. The total amount of money collected by mechanics at the Gas, Grease, and Gadgets Service Station for labor was $602. How many hours of labor does this represent?

   b. A customer was charged $21.00 for labor costs on a small job. What fractional part of an hour does this amount pay for? How many minutes does this represent? (*Remember:* When changing from larger to smaller units, you multiply by the number of units in the relationship. In this case, 60 minutes = 1 hour.)

6. Ms. Finch needed to purchase a set of four new tires, size P225/75R15, for her station wagon. When purchasing new tires, Ms. Finch must pay the price of the tires, plus a charge of $1.50 for a new valve for each tire, plus a charge of $3.75 for high-speed balancing of each tire, and a $1.00 disposal fee for each old tire. In addition, there is a sales tax of 8% of the entire bill.

a. Read the advertisement shown below to find the cost of one P225/75R15 size tire.
b. Find the cost of four tires.
c. Find the amount of Ms. Finch's bill, including the valves, balancing, disposal fee, and 8% sales tax.

TIRE SALE

A QUALITY TIRE AT A GOOD PRICE

40,000-mile wearout warranty	
SIZE P155/80R13	**24.99**
P165/80R13	$32.99
P175/80R13	35.99
P185/80R13	37.99
P185/75R14	39.99
P195/75R14	40.99
P205/75R14	42.99
P205/75R15	43.99
P215/75R15	45.99
P225/75R15	46.99
P235/75R15	49.99

OUR BEST STEEL-BELTED RADIAL

60,000-mile wearout warranty	
SIZE 155/TR12	**34.99**
155/TR13	$45.99
165/TR13	51.99
175/TR14	61.99
175/70TR13	64.99
185/70TR13	65.99
175/70TR14	66.99
185/70TR14	69.99
195/70TR14	71.99
205/70TR14	75.99
185/65TR15	79.99
195/60TR14	78.99
195/60TR15	79.99

7. Mr. Newton drove into the Gas, Grease, and Gadgets Service Station to have new shock absorbers installed on his car. He needed four shock absorbers, two regular-strength ones for the front wheels at $21.99 each, and two heavy-duty ones for the rear wheels at $26.99 each. These prices did not include the cost of labor, nor did they include the sale tax at the rate of 8% on the entire bill. Mr. Newton received a bill of $120.92, which included $8.96 for sales tax. (The charge for labor is $28 per hour.)

   a. What was the total cost of all four shock absorbers?
   b. What was the total charge for parts and labor (before the sales tax was added to this amount)?
   c. How much of the bill was the charge for labor?
   d. How much working time does the charge represent?
      (1) Express the answer in terms of hours.
      (2) Express the answer in minutes.
   e. If Mr. Newton had chosen "Shock Special No. 2," as advertised below (the price includes labor), how much money would he have saved? (Include sales tax at the rate of 8%.)

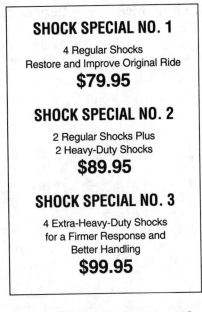

8. Some customer, unlike Ms. Finch in Problem 6, prefer to purchase lower-priced tires instead of the more expensive steel-belted radial tires, in order to save money.

a. Using the advertisement on page 250, determine the price of a 40,000-mile wearout-warranty tire, size P155/80R13. Add $1.00 disposal fee, $1.50 for a new valve, and a charge of $3.75 for high-speed balancing of each tire.

b. Calculate the total price of four tires.

c. Add sales tax at the rate of 5%.

d. Determine the amount of change the customer would receive from $150.

9. In one week, the Gas, Grease, and Gadgets Service Station sold 36 new tires of different sizes and qualities. If the total cost of these tires was $1665, calculate the average cost per tire.

10. Seven batteries were sold by the station in 1 week, at the following prices:

1 @ $39.95
2 @ $49.95
2 @ $59.95
1 @ $64.95
1 @ $79.50

Calculate the average selling price of a battery during that week.

WORKING WITH THE METRIC SYSTEM

The Gas, Grease, and Gadgets Service Station displays the following sign over the entrance to the service area:

AUTO REPAIRS:

- FRONT-END ALIGNMENT
- AUTOMATIC TRANSMISSION REPAIR
- TUNE-UPS
- BRAKE REPAIRS
- MAJOR ENGINE WORK

WE REPAIR *ALL* FOREIGN CARS AND TRUCKS

Since the station does a great deal of work on foreign cars and trucks, its employees must be familiar with the *metric* system. Metric measurements are used in the construction of all vehicles built in foreign countries, such as Japan, Germany, Italy, and Sweden, as well as France, where the metric system originated in the late eighteenth century.

There are four areas of metric measurement with which automobile mechanics must be familiar: temperature, length or distance, volume or capacity, and weight.

Metric Measurements of Temperature (Part I)

Temperature measurements in the English system are made in degrees *Fahrenheit* (F), and temperature measurements in the metric system are made in degrees *Celsius* (C). The English system is in common everyday use in the United States, but the metric system is used in all scientific measurements and in daily life in most other countries of the world. The United States has made some effort to introduce the metric system into daily use in this country.

An understanding of the Fahrenheit and Celsius systems of temperature measurement is important in order for the auto mechanic and service station attendant to provide proper guidance to the motorist. When the weather gets very cold and temperatures fall below freezing, antifreeze must be installed in automobile radiators instead of plain water to prevent freezing and cracking of the radiator. Antifreeze protects against freezing because it does not freeze until the temperature drops far below the freezing point of water.

Similarly, in very hot weather, automobile radiators containing plain water will overheat and engine damage may result. Radiators containing antifreeze do not overheat until the temperature is somewhat higher, since the boiling point of antifreeze is higher than that of water.

Temperature

	Fahrenheit	Celsius
Boiling point	212°F	100°C
Freezing point	32°F	0°C

As the table shows, there is a spread of 180° between the freezing and boiling points on the Fahrenheit scale: 212°F – 32°F = 180°F, and a spread of 100° on the Celsius scale: 100°C – 0°C = 100°C. Since 180°F = 100°C, we can divide both sides of the equation by 180, and arrive at:

$$\frac{180F}{180} = \frac{100C}{180} \quad \text{or} \quad 1F = \frac{5}{9}C$$

or 1° Fahrenheit is equal to 5°/9 Celsius. A Fahrenheit degree is smaller than a Celsius degree.

Similarly, dividing both sides of the equation by 100, we obtain the value of 1 Celsius degree:

$$\frac{180F}{100} = \frac{100C}{100} \quad \text{or} \quad \frac{9}{5}F = 1C$$

These results may be used to convert easily from one system to the other.

EXAMPLE: Many automobiles contain thermostats that are set at 180°F to maintain a proper operating temperature of the fluid in the radiator. Express this temperature in Celsius degrees.

SOLUTION: Two steps are needed to find the solution:

a. Subtract 32°: 180°F – 32°F = 148°F.

b. Multiply by $\frac{5}{9}$: $\frac{5}{9} \times 148 = \frac{740}{9} = 82\frac{2}{9}°$ C.

There is also an *algebraic formula* that can be used to convert from a Fahrenheit to a Celsius temperature. It is written as

$$C = \frac{5}{9}(F - 32)$$

In the example above, F was 180°, so:

$$C = \frac{5}{9}(180 - 32)$$

$$C = \frac{5}{9}(148)$$

$$C = \frac{740}{9}$$

$$C = 82\frac{2}{9}°$$

EXAMPLE: A flashing sign on a bank near the service station displays alternately the time of the day and the temperature in Fahrenheit degrees and in Celsius degrees. An attendant looks up while he is pumping gasoline and notices that the temperature reading is 30°C. Is the season winter or summer?

SOLUTION: Two steps are required to find the answer. They are the reverse of the steps shown in the other example.

a. Multiply by $\frac{9}{5}$: $\quad \frac{9}{5} \times 30 = \frac{9}{\overset{}{\underset{1}{5}}} \times \frac{\overset{6}{\cancel{30}}}{1} = \frac{54}{1} = 54°.$

b. Add 32°: $\quad 54 + 32 = 86°F.$

It appears that the season is summer.

The *algebraic formula* that can be used to convert temperature written as Celsius to Fahrenheit is

$$F = \frac{9}{5}C + 32$$

If C = 30°, then

$$F = \frac{9}{5}(30) + 32$$

$$F = \frac{9}{\underset{1}{5}} \cdot \frac{\overset{6}{\cancel{30}}}{1} + 32$$

$$F = 54 + 32$$

$$F = 86°$$

Therefore 30°C = 86°F.

PRACTICE EXERCISE 7

1. **a.** 100°C = _____°F   **b.** 32°F = _____°C

2. The average temperatures at the Gas, Grease, and Gadgets Service Station during one week in the middle of the winter were as follows: Monday, 37°F; Tuesday 25°F; Wednesday, 19°F; Thursday, 23°F; Friday, 8°F; Saturday, 29°F; Sunday, 34°F.

    a. If an automobile radiator contained only plain water, would freezing occur on Monday?

    b. If not, how many degrees Fahrenheit would the temperature have had to drop on Monday before the water in the radiator would have frozen?

    c. Express Monday's average temperature in degrees Celsius.

    d. Which day of the week was the warmest?

    e. Which day of the week was the coldest?

    f. How many degrees Fahrenheit difference was there between the average temperature on the warmest day of the week and the average temperature on the coldest day?

    g. On which days of the week would the fluid in a car radiator have frozen if it had been just plain water instead of an antifreeze mixture?

    h. How many days of the week was the average temperature *above* freezing?

    i. Calculate the average temperature for the entire week in degrees Fahrenheit.

    j. In degrees Fahrenheit, how much did the average temperature *fall* between Monday and Wednesday?

    k. In degrees Fahrenheit, how much did the average temperature *rise* between Friday and Saturday?

DIRECTED NUMBERS

Sam, one of the garage attendants, attempted to express Thursday's temperature of 23°F in degrees Celsius and ran into difficulty. Why? Try it.

When he attempted to follow the instructions: "Subtract 32 and then multiply by $\frac{5}{9}$," he found that he was unable to subtract 32 from the smaller number 23. He discussed the problem with his friend Hazel, who told him that you can subtract a number having a larger value from one which has a smaller value. She explained that you get an answer that is *negative*. She said that subtracting 32°from 23°, or 23°C – 32°C, = –9°C. The minus sign preceding the 9 indicates that it is a negative 9°C as opposed to a +9°C, which is a positive 9°. The negative 9°C means that it is a temperature of 9°below freezing or below 0°C.

Sam became interested in these new types of numbers and decided to read more about them. He found a math book that told him all about negative and positive numbers, which are called "directed" or "signed" numbers.

He found that these numbers were probably discovered or invented for the same reason he became interested in them. Numbers can be placed on a *number line* so that each number can be associated with a point on the line. For example:

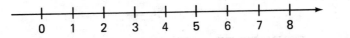

Sam found that addition can be performed on this number line. For instance, suppose that he wanted the sum of 3 and 2; 3 + 2 = ? Starting at 0, he moved 3 spaces to the right and then moved 2 more spaces to the right. He landed on 5, which he knew was the correct answer.

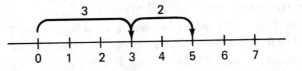

You can subtract on the number line too. You move to the left instead of to the right as in addition. For example, 4 − 3 = ?

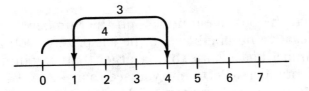

Sam then attempted to subtract a number with a larger value from one of a smaller value: 3 − 5 = ? He placed this on the number line as follows:

What was his answer? He did not land anywhere on the number line. He should land somewhere to the left of zero. but there are no numbers there.

This is where the book helped Sam discover the answer. He found that there are other numbers which he did not know about, negative numbers. These numbers could be placed on the number line to the *left* of zero, like this:

The problem that puzzled Sam, $3 - 5 = ?$, can now be done:

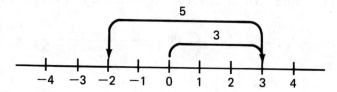

Moving 3 spaces to the right and then 5 spaces to the left, you land on -2 or negative 2.

The numbers to the *right* of zero are called positive numbers. When you moved 3 spaces to the right, you landed on $+3$ or positive 3.

Instead of right and left, you may go up and down—deposit and withdrawal, above sea level and below sea level—to signify directed numbers. "Above" and "up" signify positive numbers, while "down" and "below" are represented by negative numbers. Values above $0°C$ are represented by positive numbers; values below $0°C$ are written as negative numbers.

ORDERING

You see that the positive numbers on the number line are to the right of zero, while the negative numbers are to the left of zero. You also notice, as Sam did, that each number has a value that is larger than any number to its left. This is called *ordering* numbers. Certainly, you see that $+2 > +1$, since $+2$ is to the right of $+1$. Is -3 greater or less than -5? Looking at the number line, Sam saw that -3 is to the right of -5, so $-3 > -5$:

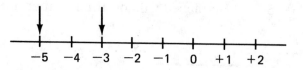

PRACTICE EXERCISE 8

Use either $>$ or $<$ to order the following numbers:

a.	$+6$	$+8$	f.	0	-3
b.	-3	-4	g.	-4	-2
c.	$+6$	-2	h.	-5	$+1$
d.	-3	0	i.	$+4$	-2
e.	$+4$	$+2$	j.	-1	$+1$

ADDING

Sam also found that these new numbers, like the old ones, can be added, subtracted, multiplied, and divided. The answers were more difficult to find, however, since the sign of the number played a big role in the computation.

He found that adding became easier when he used the number line. Suppose that he wanted to add: $(+3) + (+2) = ?$

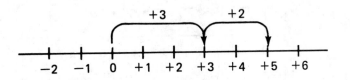

He moved 3 spaces to the right and then moved 2 more spaces to the right, landing on $+5$. Therefore: $(+3) + (+2) = +5$.

Add: $(+5) + (-7) = ?$

Move 5 spaces to the right for $+5$, then move 7 spaces (in what direction?) to the left, corresponding to -7. Where do you land? Therefore: $(+5) + (-7) = -2$.

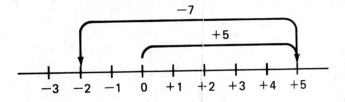

Add: $(-3) + (-4) = ?$

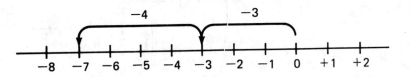

Therefore: $(-3) + (-4) = -7$.

PRACTICE EXERCISE 9

Use the number line to add each of the following:

a. $(+4) + (-3)$ f. $(+3) + (+1)$
b. $(+5) + (+3)$ g. $(+6) + (-1)$
c. $(-2) + (-2)$ h. $(-5) + (-1)$
d. $(+3) + (-4)$ i. $(-3) + (+5)$
e. $(-1) + (+5)$ j. $(0) + (-2)$

MULTIPLYING

Multiplication is repeated addition, and Sam investigated, not the product of the numbers, but the sign that must be placed in the answer. He started with the product of two positive numbers: $(+3)(+2) = ?$ (*Remember:* No sign between two numbers indicates multiplication.) He used the number line for help:

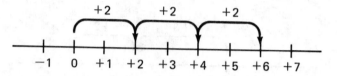

The product $(+3)(+2)$ is the same as adding $+2$ three times, so he showed that on the number line above, or

$$\begin{array}{r} +2 \\ +2 \\ +2 \\ \hline +6 \end{array}$$

Thus: $(+3)(+2) = +6$.

And $(+3)(-2) = ?$ This indicates adding three -2's, or

$$\begin{array}{r} -2 \\ -2 \\ -2 \\ \hline -6 \end{array}$$

Using the number line, you get:

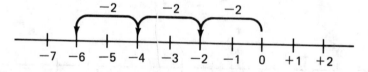

Thus: $(+3)(-2) = -6$.

The product of the same two numbers is the same regardless of how they are multiplied together; for example, $7 \times 5 = 5 \times 7$, so $(+3)(-2) = (-2)(+3)$. Multiplication of two numbers with unlike signs will result in the same signed answer, *negative.*

We saw that the result was positive when multiplying two positive numbers together, but what occurs when we multiply two negative numbers?

Let's try: $(-3)(-2) = ?$

Sam knows that $(-3)(+2) = -6$, so he continued from this point in the following manner. He worked out this pattern of values to arrive at the solution:

$$(-3)(+2) = -6$$
$$(-3)(+1) = -3$$
$$(-3)\ (0) = \ \ \ 0$$
$$(-3)(-1) = \ \ \ ?$$

He knew that a positive number multiplied by a negative number results in a negative signed answer, and he knew that any number times 0 is always 0. Therefore he followed the pattern of answers and saw that -6, -3, and 0 were increasing by 3 and that the next answer in this pattern must be $+3$. In this way he discovered that $(-3)(-1) = +3$ and that $(-3)(-2) = +6$.

Multiplying two numbers with the same sign, positive or negative, results in a signed answer that is always positive.

PRACTICE EXERCISE 10

Multiply each of the following pairs of numbers. Be sure to include the sign in each answer.

a. $(+4)(+2)$
b. $(-3)(+6)$
c. $(-3)(-5)$
d. $(-4)(-4) = (-4)^2 = ?$
e. $(-2)(+2)$

f. $(-5)(-2)$
g. $(+2)(+8)$
h. $(-2)(+6)$
i. $(+1)(+4)$
j. $(+5)(-3)$

DIVIDING

Division is the inverse (direct opposite) of multiplication, so Sam used the same results he found for multiplication. Since $\frac{6}{2} = 3$ and $2 \times 3 = 6$, he used this idea to find the four signed answers for the four possible combinations of division examples.

By the Laws of Multiplication

$\frac{+6}{+2} = ?$ and $(+2)(?) = +6$		$? = +3$	$\frac{+6}{+2} = +3$
$\frac{-6}{-2} = ?$ and $(-2)(?) = -6$		$? = +3$	$\frac{-6}{-2} = +3$
$\frac{+6}{-2} = ?$ and $(-2)(?) = +6$		$? = -3$	$\frac{+6}{-2} = -3$
$\frac{-6}{+2} = ?$ and $(+2)(?) = -6$		$? = -3$	$\frac{-6}{+2} = -3$

Therefore, to divide *like* signed numbers, you:

1. Divide the numbers.
2. Place a positive sign in the answer.

To divide *unlike* signed numbers, you:

1. Divide the numbers.
2. Place a negative sign in the answer.

PRACTICE EXERCISE 11

Divide:

a. $(+8) \div (-2)$

b. $(-6) \div (-2)$

c. $(+10) \div (+2)$

d. $\dfrac{-9}{+3}$

e. $(-9) \div (-3)$

f. $(10) \div (-2)$

g. $\dfrac{-15}{+3}$

h. $\dfrac{+9}{-3}$

i. $\dfrac{-8}{+2}$

j. $(-2) \div (-1)$

SUBTRACTING

Subtraction is the inverse operation of addition. Webster defines *inverse* as "the direct opposite," and subtraction is the opposite operation of addition. Instead of adding on, you are taking away.

If you want to subtract 5 from 8, you write the problem like this:

$$\begin{array}{r} 8 \\ -5 \\ \hline ? \end{array} \quad \text{Minuend} \quad \text{or, horizontally, } 8 - 5 = ?$$

Minuend   or, horizontally, 8 − 5 = ?
Subtrahend
Remainder

Since subtraction is the opposite of addition, you mean

$$\begin{array}{r} 5 \\ +\,? \\ \hline 8 \end{array}, \text{ which is the same as } \begin{array}{r} 8 \\ -5 \\ \hline ? \end{array}$$

Therefore you subtract signed numbers by finding a number that, when added to the subtrahend, gives you the minuend. For example:

$$\begin{array}{r} -3 \\ -\,-2 \\ \hline \end{array} \text{ becomes } \begin{array}{r} -2 \\ +\ ? \\ \hline -3 \end{array} \text{ and } \quad ? = -1 \text{ since } \begin{array}{r} -2 \\ +\ -1 \\ \hline -3 \end{array}$$

Therefore:

$$\begin{array}{r} -3 \\ -\ -2 \\ \hline -1 \end{array}$$

Or

$$\begin{array}{r} +5 \\ -\ -3 \\ \hline ? \end{array} \quad \text{becomes} \quad \begin{array}{r} -3 \\ +\ \ ? \\ \hline +5 \end{array} \quad \text{and} \quad ? = +8 \text{ since } \begin{array}{r} -3 \\ +\ +8 \\ \hline +5 \end{array}$$

Therefore:

$$\begin{array}{r} +5 \\ -\ -3 \\ \hline +8 \end{array}$$

Or

$$\begin{array}{r} -3 \\ -\ +2 \\ \hline ? \end{array} \quad \text{becomes} \quad \begin{array}{r} +2 \\ +\ \ ? \\ \hline -3 \end{array} \quad \text{and} \quad ? = -5 \text{ since } \begin{array}{r} +2 \\ +\ -5 \\ \hline -3 \end{array}$$

Therefore:

$$\begin{array}{r} -3 \\ -\ +2 \\ \hline -5 \end{array}$$

PRACTICE EXERCISE 12

Subtract:

a. $(+4) - (-3)$
b. $(+5) - (+3)$
c. $(-2) - (-2)$
d. $(+3) - (-4)$
e. $(-1) - (+5)$

f. $(+3) - (+1)$
g. $(+6) - (-1)$
h. $(-5) - (-1)$
i. $(-3) - (+5)$
j. $(-2) - (+1)$

Metric Measurements of Temperature (Part II)

Sam was now able to tackle the problem of changing Thursday's temperature of 23°F to degrees Celsius. He followed the instructions given previously:

1. Subtract 32 from the Fahrenheit temperature.
   23°F −32° becomes either $(23) + (-32) = -9°$
   or $(23) - (32) = -9°$.

Whichever way it is done, the answer is still −9°.

2. Multiply by $\frac{5}{9}$:

$$\frac{5}{9} \times -9° = \frac{5}{\overset{1}{\cancel{9}}} \cdot \frac{(-\overset{1}{\cancel{9}})}{1} = \frac{5(-1)}{1} = -5°C$$

The Celsius temperature is negative (below zero) because it is below freezing, and the freezing point of water on the Celsius scale is 0°C. Any temperature below freezing on the Celsius scale is a negative number.

PRACTICE EXERCISE 13

1. a. Express Friday's temperature (see p. 256) in degrees Celsius.
    b. Is the result a negative temperature? Why or why not?

2. During the winter months, a 0°F temperature reading is not unusual. Convert that temperature to degrees Celsius.

3. a. Is a temperature of −8°F colder than a temperature of −5°F?
    b. How much lower is −8°F than −5°F?
    c. What is the *difference* between +13°F and −5°F? (*Hint:* (+13) − (−5) = ?)

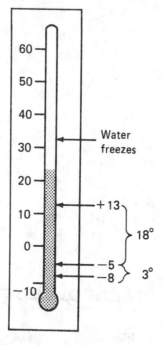

4. If the temperature went from −8°F to 18°F, how many degrees Fahrenheit did it rise?

5. If the temperature went from 23°F to −3°F, how many degrees Fahrenheit did it fall?

6. How many degrees did the temperature rise if it went from 9° below zero F to 4° below zero F?

7. How many degrees did the temperature fall if it went from −1°C to −6°C?

8. **a.** If a thermometer read 0°F at 2 P.M. and 3°F below zero at 4 P.M., did the temperature rise or fall between 2 P.M. and 4 P.M.?

   b. How many degrees did it rise or fall?

9. During a warm week in the middle of the summer, the average temperatures at the Gas, Grease, and Gadgets Service Station were as follows: Monday, 28°C; Tuesday, 25°C; Wednesday, 32°C; Thursday, 30°C; Friday, 33°C; Saturday, 35°C; Sunday, 27°C.

   a. Which day of the week was the warmest?

   b. Which day of the week was the coolest?

   c. How many degrees Celsius difference was there between the average temperature on the warmest day of the week and the average temperature on the coolest day of the week?

   d. Express Tuesday's temperature in degrees Fahrenheit.

   e. Calculate the average temperature for the entire week in degrees Celsius.

   f. In degrees Celsius, how much did the average temperature rise between Monday and Friday?

   g. In degrees Celsius, how much did the average temperature fall between Friday and Sunday?

   h. Express Friday's average temperature in degrees Fahrenheit.

10. The normal body temperature of a person is 98.6°F. Express this temperature in degrees Celsius.

11. If a person's body temperature was 40°C, would he have a fever? Express this temperature in degrees Fahrenheit.

12. The thermostat on a foreign car indicates that it has been set at 90°C, in order to maintain the proper operating temperature for the engine. Express this temperature in degrees Fahrenheit.

13. Would a Celsius thermometer be more likely to read 32° in the summer or in the winter? Why? (*Hint:* Express 32°C in degrees Fahrenheit.)

14. In the pictures of the thermometers shown, replace the question marks with the proper values of the Fahrenheit or Celsius temperature.

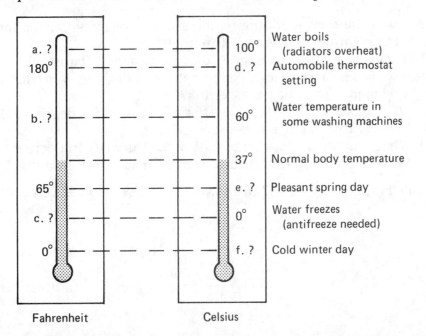

15. The graphs drawn below show the same data in degrees Fahrenheit and in degrees Celsius.

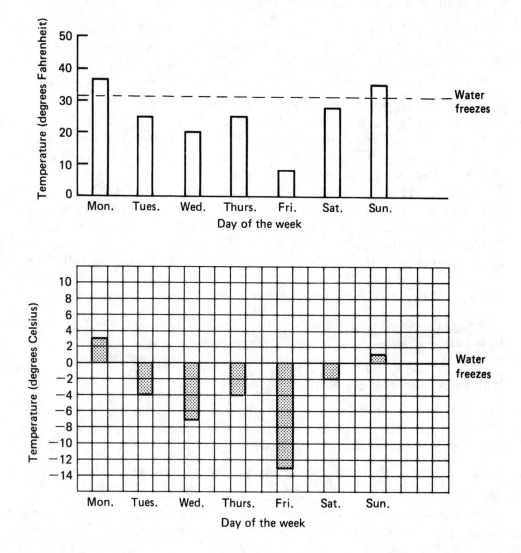

a. What was the corresponding temperature in degrees Celsius for Wednesday's temperature in degrees Fahrenheit?
b. What was the temperature for Friday, in degrees Celsius?
c. What was the corresponding temperature in degrees Fahrenheit for Monday's temperature in degrees Celsius?
d. (1) What was the average temperature for the week in degrees Fahrenheit?
   (2) Convert that average to degrees Celsius.
e. (1) What was the average temperature for the week in degrees Celsius?
   (2) Does this value check with your answer to question d.(2) above?

16. Use the line graph on p. 267 to answer the following questions:

a. Between which two days did the greatest drop in temperature occur?

b. Which was the coldest day of the week?

c. Which day of the week was warmer, Monday or Sunday?

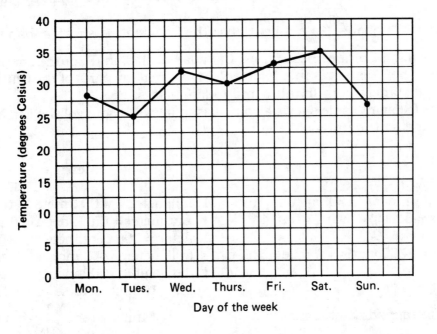

Metric Measurements of Length or Distance

Automobile mechanics who make repairs to foreign cars must use a *metric set of wrenches* and other *metric tools*. These tools are graduated in *centimeters* and *millimeters*, rather than in fourths, eighths, sixteenths, and thirty-seconds of an inch, as tools made for American cars are machined. Therefore the mechanic must be familiar with units of length in the metric system.

The basic unit of length, called the "meter," is a little longer than a yard. One yard is *exactly* 36 inches long, and one meter is *approximately* 39.37 inches long. Smaller lengths are called "centimeter" and "millimeter." The prefix "centi" means "one hundredth" or $\frac{1}{100}$. Since a centimeter is $\frac{1}{100}$ or 0.01 of a meter and a meter is approximately 39.37 inches,

$$1 \text{ centimeter} = \frac{1}{100} \times 39.37 = \frac{39.37}{100} = 0.3937 \text{ or } 0.4 \text{ inch approximately.}$$

The prefix "milli" means "one thousandth" or $\frac{1}{1000}$. Since a millimeter is $\frac{1}{1000}$ or 0.001 of a meter and a meter is approximately 39.37 inches,

$$1 \text{ millimeter} = \frac{1}{1000} \times 39.37 = \frac{39.37}{1000} = 0.03937 = 0.04 \text{ inch}$$
approximately (a very small length).

Larger lengths or distances are called "kilometers." Since a kilometer is 1000 meters,

1 kilometer = 1000 × 39.37 inches = 40,000 inches approximately.

This distance is about $\frac{5}{8}$ of a mile. Hence the kilometer is shorter than our mile.

Road signs in all European countries, Canada, Mexico, and South America are all designed to show distances in kilometers. Signs in many parts of the United States now show distances both in miles and in kilometers.

The following table shows a comparison between some of the English units of length and metric units of length:

English	Metric
12 inches = 1 foot ≈ 0.305 meter	1 millimeter = 0.001 meter ≈ 0.04 inch
36 inches = 3 feet = 1 yard ≈ 0.914 meter	1 centimeter = 0.01 meter ≈ 0.4 inch
5280 feet = 1760 yards = 1 mile ≈ 1.609 kilometers	1 meter = 100 centimeters = 1000 millimeters ≈ 40 inches
	1 kilometer = 1000 meters
	Note: 10 millimeters = 1 centimeter
≈ means is approximately equal to	1 millimeter = $\frac{1}{10}$ centimeter

The units shown above are commonly abbreviated as follows:

inch = in.	millimeter = mm
foot = ft.	centimeter = cm
yard = yd.	meter = m
mile = mi.	kilometer = km

Except in scientific writing a period is usually placed after each English abbreviation, but a period is never placed after a metric abbreviation.

The table above was provided to help you to change from one metric unit of length to another and to change from English units of length to metric units of length and vice versa.

REMINDER! *When changing from units of larger length to those of smaller length, MULTIPLY by the number of the smaller unit of length that is contained in the larger unit.*

For example, you know that a kilometer is a larger unit than a meter, so to change from kilometers to meters you multiply by the number of meters in a kilometer. Since 1000 m = 1 km, you multiply by 1000.

To change from meters to centimeters, you multiply by 100 since a meter is larger than a centimeter and 100 cm = 1 m.

To change from centimeters to millimeters, you multiply by 10 since the centimeter is the larger unit and 10 mm = 1 cm.

REMINDER! *When changing from units of smaller length to those of larger length, DIVIDE by the number of the smaller unit of length that is contained in the larger unit.*

For example, you know that a meter is a smaller unit than a kilometer and that 1000 m = 1 km, so to change from meters to kilometers, you divide by 1000.

To change from centimeters to meters, you divide by 100 since there are 100 cm = 1 m and the centimeter is a smaller unit than the meter.

To change from millimeters to centimeters, you divide by 10, since the millimeter is a smaller unit than the centimeter and 10 mm = 1 cm.

Changing from the metric system to the English system or vice versa requires the use of the table provided above for your reference. For example to change from kilometers to miles, you multiply by $\frac{5}{8}$; to change from miles to kilometers, you divide by $\frac{5}{8}$. To change from centimeters to inches, you multiply by 0.4; to change from inches to centimeters, you divide by the same quantity, 0.4.

To change from millimeters to inches, you multiply by 0.04; to change from inches to millimeters, you divide by 0.04.

EXAMPLE: A mechanic at the Gas, Grease, and Gadgets Service Station is repairing a Volkswagen, which is manufactured in Germany and whose measurements are completely metric. To loosen a nut, he needs a 13-mm wrench.

a. What is the measurement of this wrench in centimeters?

SOLUTION: To change from millimeters to centimeters is to change from a smaller unit to a larger one, so you divide by the number of millimeters contained in a centimeter: 10 mm = 1 cm, that is, you divide by 10. Therefore:

$$\frac{13}{10} = 1\frac{3}{10} = 1.3 \text{ cm}$$

b. What size English wrench would most closely correspond to this metric wrench?

SOLUTION: To change from centimeters to inches, you multiply by 0.4: 1.3 cm × 0.4 in. = 0.52 in. Since 0.52 is just a little more than $0.50 = \frac{50}{100} = \frac{1}{2}$, a 13-mm metric wrench corresponds closely to an English wrench of $\frac{1}{2}$-inch size. Metric wrenches commonly range from 4.5 mm to 17 mm, including sizes of 4.5, 5, 6, 7, 8, 9, 10, 11, 12, 13, 15, 17 mm in some standard sets.

EXAMPLE: A driver of a car with Canadian license plates stopped at the Gas, Grease, and Gadgets Service Station for gasoline. He remarked that he had just driven 560 km from the province of Quebec. How great a distance is this when expressed in miles?

SOLUTION: To change from kilometers to miles, you multiply by $\frac{5}{8}$:

$$560 \times \frac{5}{8} = \frac{\overset{70}{\cancel{560}}}{1} \times \frac{5}{\underset{1}{\cancel{8}}} = 350 \text{ miles}$$

PRACTICE EXERCISE 14

1. A mechanic working on a Fiat automobile, manufactured in Italy, needed a 6-mm socket wrench to loosen a bolt.

   a. Express this measurement in centimeters.
   b. Determine the approximate size of an English wrench, in inches, that would most closely correspond to this metric wrench.

2. A replacement part on a Datsun sedan, manufactured in Japan, measured 15 cm in length.

   a. Express this measurement in millimeters.
   b. Express this measurement approximately in inches.

3. A common wrench size in a standard English set of tools is $\frac{3}{4}$ inch.

   a. Express this measurement approximately in centimeters.
   b. Express this measurement approximately in millimeters.

4. A rod connecting two parts of an automobile steering system measures 1 foot 6 inches in length. Express this measurement in:

   a. inches   b. centimeters   c. millimeters

5. Change each measurement *from millimeters to centimeters*:

   a. 100 mm   b. 25 mm   c. 7.5 mm

6. Change each measurement *from centimeters to millimeters*:

   a. 100 cm   b. 50 cm   c. 2.5 cm   d. .1 cm

7. Change each measurement *to inches*, approximately:

   a. 100 cm   b. 50 mm   c. 150 cm   d. 500 mm   e. 25 cm

8. Change each measurement to its approximate equivalent *in centimeters:*

   a. 8 in.   **b.** 2 ft.   **c.** 1 yd.   **d.** 20 in.   **e.** $\frac{1}{4}$ in. (0.25 in.)

9. Change each measurement in Problem 8 to its approximate equivalent *in millimeters.*

10. A road sign in Italy indicates that the distance to Rome is 640 km. Approximately how many miles is this?

11. A foreign-car driver asks the attendant at the Gas, Grease, and Gadgets Service Station how far a certain city is from the station. The attendant responds that it is 25 miles away. Approximately how many kilometers is this?

12. Change each distance *from miles to kilometers:*

    a. 75 mi.   **b.** 125 mi.

13. Change each distance *from kilometers to miles:*

    a. 240 km   **b.** 96 km

14. A Peugeot automobile, manufactured in France, has a speedometer that registers speed in kilometers per hour (km/hr.) instead of miles per hour (mi./hr.). What is the speed of a car in miles per hour if its speedometer registers 80 km/hr.?

15. If the speed limit on a road in France is shown as 100 km/hr. and an American car is being driven down the road at a speed of 60 mi./hr., is the driver exceeding the speed limit?

16. In the Olympic Games, there are races over many distances, both long and short.

    a. How long in kilometers is a 10,000-meter race?
    b. What part of a kilometer is a 100-meter dash?

17. Many newly built cars have speedometers that register top speeds of only 80 miles per hour (mph). Express this speed in kilometers per hour.

18. Change each distance *from kilometers to meters:*

    a. 12 km   **b.** 0.8 km

19. Change each distance *from meters to kilometers:*

    a. 3000 m   **b.** 500 m

Metric Measurements of Volume or Capacity

The basic unit of volume in the metric system is the *cubic centimeter,* abbreviated as cc or cm^3. The 3 in cm^3 is an "exponent" and is read as "cubed." Therefore cm^3 is read as "centimeters cubed," and of course you remember that it means that cm is a factor three times, or cm · cm · cm.

One cubic centimeter is the volume of a cube whose length, width, and height each measure 1 cm. (See the accompanying diagram.)

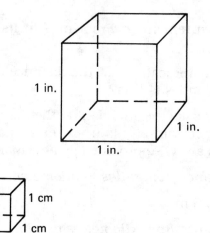

In automobile engines the displacement (volume) of the cylinders may be given in cubic inches (in.^{3}) in an American-made car or in cubic centimeters (cm^3) in a foreign car. A cubic centimeter is very small in comparison to a cubic inch. Since 1 cm = 0.4 in. (approx.) and 1 cm · 1 cm · 1 cm = 0.4 in. · 0.4 in. · 0.4 in.,

$$1 \text{ cm}^3 = (0.4 \text{ in.})^3$$
$$1 \text{ cm}^3 = 0.064 \text{ cubic inch}$$

Dividing both sides of this equation by 0.064, you get

$$\frac{1 \text{ cm}^3}{0.064} = \frac{\overset{1}{\cancel{0.064} \text{ cubic inch}}}{\underset{1}{\cancel{0.064}}}$$

or

$$
\begin{array}{r}
15.625 \\
0.064{\overline{)1.000\ 000}} \\
\underline{64\text{x xxx}} \\
360 \\
\underline{320} \\
400 \\
\underline{384} \\
160 \\
\underline{128} \\
320 \\
\underline{320}
\end{array}
$$

1 cubic inch = 15.625 or 16 cm^3 approximately

Thus it takes almost 16 cubic centimeters to equal 1 cubic inch, which is the volume of a cube whose length, width, and height each measure 1 inch.

Laboratory equipment in all scientific work is almost always marked in cubic centimeters, and this unit is used also in the automotive industry. Since 1 cm^3 is so small, a larger unit of volume is also needed, especially for everyday measurement of liquids. This larger unit is called a *liter*, and 1 liter is equal to 1000 cm^3. This is equal to the volume of a cube whose length, width, and height each measure 10 cm, since

$$10 \text{ cm} \cdot 10 \text{ cm} \cdot 10 \text{ cm} = 1000 \text{ cm}^3$$

One liter (abbreviated as L) is just a little larger than 1 quart; 1 L = 1.06 qt. (approx.). Since 1 quart = 32 fluid ounces, 1 L actually contains a little more than that, about 33.5 fluid ounces. Gasoline is sold in the United States by the gallon, which is equal to 4 quarts or 128 fluid ounces, and oil is sold by the quart ($\frac{1}{4}$ gallon or 32 fluid ounces). These items are all sold in European countries by the liter.

As you can see, the automobile mechanic should be familiar with the relationship between English units of volume and capacity and the corresponding metric units so that he can properly service foreign cars brought into the station for repairs and service.

The following table is provided for reference:

English	**Metric**
16 fluid ounces = 1 pint	1000 cubic centimeters = 1 liter (L) =
32 fluid ounces = 2 pints = 1 quart	1.06 quarts (approx.)
8 pints = 4 quarts = 1 gallon	

To change from liters to cubic centimeters, you multiply by 1000.
To change from cubic centimeters to liters, you divide by 1000.
Liters may be interchanged with quarts for fair approximations.

To change from cubic inches to cubic centimeters, you multiply by 16 (approx.).

To change from cubic centimeters to cubic inches, you divide by 16 (approx.).

EXAMPLE: Large American cars built in the 1960s were available with powerful 400-cubic-inch engines.

a. Express this volume in cubic centimeters.
b. Express this volume in liters.

SOLUTION:

a. To change from cubic inches to cubic centimeters, you multiply by 16, so 400 × 16 = 6400 cm^3.
b. To change cubic centimeters to liters, you divide by 1000, so $\frac{6400}{1000} = 6.4$ liters.

EXAMPLE: Gasoline is sold in a certain European country for the equivalent of 83¢ per liter.

a. What would be its approximate price per gallon?
b. How much would it cost a motorist to fill a gasoline tank whose capacity was 20 gallons?

SOLUTION:

a. Since 1 liter is approximately equal to 1 quart and 1 gallon = 4 quarts, 1 gallon is approximately equal to 4 liters. Therefore $0.83 × 4 = $3.32 per gallon, approximately.
b. The approximate cost of 20 gallons is $3.32 × 20 = $66.40.

PRACTICE EXERCISE 15

1. A driver brings his automobile into the Gas, Grease, and Gadgets Service Station for service, and Johnny, the mechanic, notices that it has a 304-cubic-inch engine of the V-8 type.

   a. Express this volume in cubic centimeters.
   b. Express this volume in liters.

2. A small foreign car driven into the station displays the following in chrome on its front fender: 3.2 liters. This is a description of the total capacity of the cylinders of its engine.

   a. Express this amount in cubic centimeters.
   b. Express this amount in cubic inches.

3. Regular gasoline sells at the Gas, Grease, and Gadgets Service Station for 119.9¢ per gallon. Calculate the approximate price of 1 liter of regular gasoline, and express the answer rounded off to the nearest tenth of a cent.

4. The gasoline tank of a medium-sized car has a capacity of 16 gallons.

   a. Determine the approximate capacity of this tank in liters.
   b. If gasoline sells for the equivalent of 73¢ per liter in a European country, calculate the cost of filling the tank.

5. While traveling in Europe, Ms. Kelly stopped at a service station and spent the equivalent of $20 for gasoline. If the selling price of the gasoline was the equivalent of 80¢ per liter,

   a. how many liters were purchased?
   b. how many gallons (approximately) was this?

Metric Measurements of Weight

The basic unit of weight (actually mass, but commonly called "weight") in the metric system is the *gram* (g). It is a very small unit, approximately the weight of one paper clip. It takes more than 28 grams to equal 1 ounce (actually 28.35 grams), and about 453.6 grams to equal 1 pound. Since the gram is so small, a larger unit is needed for everyday use for weighing goods. This larger unit, called the *kilogram* (kg), is equal to 1000 grams or approximately 2.2 pounds.

Very large weights, such as the weight of an automobile or a truck, are expressed in metric *tonnes*, which are equal to 1000 kilograms each. Since 1 kilogram = 2.2 pounds approx. and 1 tonne = 1000 kilograms, $1000 \times 2.2 = 2200$ pounds. You see that a metric tonne is 200 pounds larger than an English ton, since an English ton is equal to 2000 pounds. Note the difference in spelling between "tonne" and "ton." Very small weights, such as the weights of chemicals and drugs, are expressed in *milligrams* (mg); $1 \text{ mg} = \dfrac{1}{1000} \text{ g} = 0.001 \text{ g}$.

The following table is provided for reference.

English	Metric
16 ounces (oz.) = 1 pound (lb.) $\approx$ 0.45 kilograms	1 milligrams (mg) = 1 gram (g)
2000 lb. = 1 ton $\approx$ 909.1 kilograms	1000 g = 1 kilogram (kg) $\approx$ 2.2 lb.
	1000 kg = 1 tonne $\approx$ 2200 lb.
	28 g $\approx$ 1 oz.
	454 g $\approx$ 1 lb.

REMINDER! *When changing from larger units to smaller ones, you multiply, so:*
    To change from grams to milligrams, you multiply by 1000.
    To change from kilograms to grams, you multiply by 1000.
    To change from tonnes to kilograms, you multiply by 1000.

REMINDER! *When changing from smaller units to larger ones, you divide, so:*
    To change from milligrams to grams, you divide by 1000.
    To change from grams to kilograms, you divide by 1000.
    To change from kilograms to tonnes, you divide by 1000.

Also:
    To change from kilograms to pounds, you multiply by 2.2.
    To change from pounds to kilograms, you divide by 2.2.

EXAMPLE: A heavy American car drives into the Gas, Grease, and Gadgets for service and is raised on the lift in the service bay. The car weighs 4400 pounds. Express the weight of the car in:

a. English tons.
b. Metric weights: kilograms, tonnes, and grams.

SOLUTION:

a. Since there are 2000 pounds in a ton, you divide by 2000, so

$$\frac{4400}{2000} = 2000\overline{)4400.0}^{\;2.2} = 2.2 \text{ tons}$$
$$\underline{4000} \text{ x}$$
$$400\ 0$$
$$\underline{400\ 0}$$

b. To change from pounds to kilograms, you divide by 2.2, so

$$\frac{4400}{2.2} = 2.2\overline{)4400.0}^{\;200\ 0.} = 2000 \text{ kg}$$
$$\underline{44}$$

To change from kilograms to tonnes, you divide by 1000, so

$$\frac{2000}{1000} = 2 \text{ tonnes}$$

To change from kilograms to grams, you multiply by 1000, so

$$2000 \times 1000 = 2{,}000{,}000 \text{ g}$$

PRACTICE EXERCISE 16

1. A medium-size American car weighs 3300 pounds. Express the weight of this car in:

   a. English tons   c. tonnes
   b. kilograms   d. grams

2. One of the lifts in a service bay at the Gas, Grease, and Gadgets Service Station is marked "Capacity: 11,000 pounds." This is the maximum weight it can raise safely. Express this weight in:

   a. English tons   b. kilograms   c. metric tonnes

3. A small foreign car weighs 1000 kg. Express this weight in:

   a. metric tonnes   **b.** pounds   **c.** English tons

4. An attendant uses a jack to raise one corner of a car in order to change a flat tire. The capacity of this jack is 650 kg. Express this weight in:

   a. pounds   **b.** grams

5. Change each weight from grams to milligrams:

   a. 10 g   **b.** $\frac{1}{2}$ g

6. Change each weight from milligrams to grams:

   a. 5 mg   **b.** 150 mg

SELLING AUTOMOTIVE SUPPLIES

The Gas, Grease, and Gadgets Service Station includes an auto supply shop, in which customers may purchase a variety of supplies and accessories for their cars. The shop employs sales personnel, who must be able to give advice to customers on the proper materials to purchase, total their bills, calculate the sales tax, and make change properly.

> *WE OFFER*
> *COMPLETE*
> *LINES OF PARTS*
> *FOR FOREIGN AND*
> *DOMESTIC CARS,*
> *AS WELL AS*
> *AUTOMOBILE PAINTS*
> *AND SUPPLIES*

EXAMPLE: One week the shop ran a sale on battery booster cables, battery chargers, and motor oil, as shown in the advertisements below. A customer bought one set of cables, one 10 amp battery charger, and five quarts of STP motor oil. Note: The rebates mentioned in the ads are sent directly to the customer by mail and do not affect the price charged by the shop.

a. How much did the customer save on the cables and the battery charger, compared to their regular prices?
b. Calculate the total purchases.
c. Add sales tax at the rate of 8%.
d. Determine the amount of change the customer received from four $20 bills.

Acknowledgment: Reprinted by courtesy of AMES Department Stores, Inc.

SOLUTION:

a. List price of cables   = $23.49
   Sale price of cables   =   −19.49
            Savings   = $ 4.00

   List price of battery charger   = $ 54.99
   Sale price of battery charger   =   − 39.99
            Savings   = $ 15.00

   Total savings:

$15
+ 4
$19

b. Cost of five quarts of motor oil $6.45

   Total cost of merchandise:

$19.49
 39.99
  6.45
$65.93

c. Sales tax at the rate of 8%:

$ 65.93
×   0.08
$5.27(4)4

d. Total bill:

$65.93
+ 5.27
$71.20

   Change:

$80.00
−71.20
$ 8.80

PRACTICE EXERCISE 17

1. Dan comes into the Gas, Grease, and Gadgets Service Station and Auto Supply Shop to buy a lock de-icer, a snow brush/scraper, and *two* windshield wiper blades for his car. These items are on sale, as shown in the advertisement on p. 280.

a. Calculate the cost of the merchandise.
b. Add the sales tax at the rate of 8%.
c. Determine the total bill.

2. A customer purchases a set of front and rear all-vinyl car mats and a 2-ton hydraulic floor jack for her car at the prices shown below.

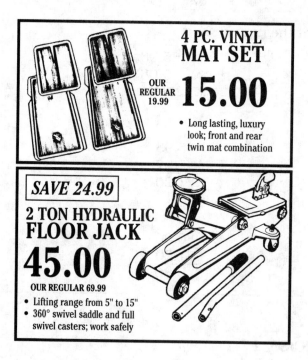

a. Calculate the total purchase.
b. Add sales tax at the rate of 8%.
c. Calculate the amount of change she would receive from a $100 bill.

Acknowledgment: Reprinted by courtesy of AMES Department Stores, Inc.

3. Some customers of the Gas, Grease, and Gadgets Service Station change the engine oil in their cars themselves, instead of having it done in the service bays, to save money. Oil sold by the case, containing 12 quarts of oil, is much less expensive than oil installed by an attendant. A customer bought a case of the 10W40 oil shown in the advertisement at $1.05 per quart. The station's regular price for 1 quart of 10W40 oil installed is $1.75.

CASTROL
MOTOR OIL

1.05/qt.

Case of 12 12.60

• 10W30, 5W30, 10W40 OR 20W50.

 a. How much did the customer save on each quart from the installed cost?
 b. How much did he save on 12 quarts?
 c. How many gallons of oil are there in 12 quarts?
 d. The customer's car has a large engine that requires 6 quarts of oil at every oil change. How many times will the customer be able to change his oil, using the 12 quarts in the case?
 e. What is the total cost of the case of oil, including sales tax at the rate of 8%?

4. Some automobile owners like to decorate their cars with special equipment, like the items advertised below.

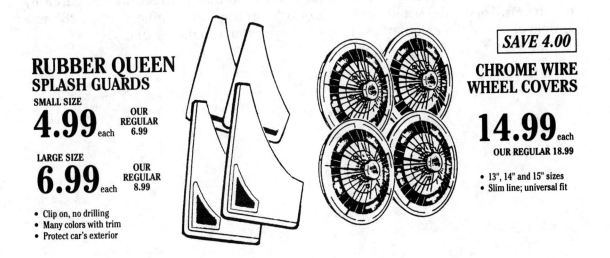

RUBBER QUEEN
SPLASH GUARDS

SMALL SIZE
4.99 each   OUR REGULAR 6.99

LARGE SIZE
6.99 each   OUR REGULAR 8.99

• Clip on, no drilling
• Many colors with trim
• Protect car's exterior

SAVE 4.00

CHROME WIRE
WHEEL COVERS

14.99 each

OUR REGULAR 18.99

• 13", 14" and 15" sizes
• Slim line; universal fit

Acknowledgment: Reprinted by courtesy of AMES Department Stores, Inc.

a. Calculate the total cost to Mr. Manuel, who purchased a pair of small-size splash guards and a set of four wheel covers.

b. Include 8% sales tax on all merchandise.

c. Determine the amount of change Mr. Manuel will receive from the $100 bill he offers as payment.

5. Using all the advertisements in the section headed "Selling Automotive Supplies," complete the following bill for Mr. Ching:

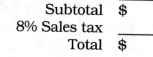

1	Set of booster cables ....$	
1	Pair of wiper blades	
1	4-Piece vinyl mat set..	
6	Quarts of 10W40 motor oil................................	
	Subtotal $	
	8% Sales tax	
	Total $	

RECORD KEEPING IN THE SERVICE STATION OFFICE

An important job in every service station is keeping records of the income and the expenses. The income is derived from various sources: sales of gasoline and oil, repairs, and sales in the auto supply shop. Expenses include the cost of supplies, gasoline, and oil; rent; fuel; utilities; taxes; and wages. Income and expenses must be carefully calculated. Money remaining after all expenses are paid becomes the "earnings" or "profit" of the station.

At the Gas, Grease, and Gadgets Service Station the records are carefully kept by Mr. Lopez, the bookkeeper, who works in the station office. On p. 283 is a sample of one of the calculations that Mr. Lopez must make regularly.

EXAMPLE: The station employs 7 people: 3 gasoline pump attendants, 2 mechanics, 1 salesclerk, and 1 bookkeeper. Each person works 40 hours of regular time per week (not counting overtime), and each is paid according to the amount of skill and responsibility required by the job performed and the length of time he or she has been employed at the station. The mechanics have the greatest degree of skill and make major repairs on cars; therefore, they are the highest paid employees of the station.

a. Calculate each person's earnings for a 40-hour week, using the hourly pay scale printed below.

b. Determine the total payroll for the station for the week.

Gasoline pump attendants:	L. Amoroso	$6.90
	J. Pierre	5.75
	H. Gross	5.65
Mechanics:	H. Brink	12.75
	H. Suarez	15.00 (manager)
Salesclerk:	C. Miller	5.50
Bookkeeper:	M. Lopez	7.50

SOLUTION:

a.
L. Amoroso	40 hours	@	$6.90	=	$ 276
J. Pierre	" "	@	5.75		230
H. Gross	" "	@	5.65		226
H. Brink	" "	@	12.75		510
H. Suarez	" "	@	15.00		600
C. Miller	" "	@	5.50		220
M. Lopez	" "	@	7.50		300

b. The total payroll is $ 2362

PRACTICE EXERCISE 18

1. The station was very busy one Saturday, and Mr. Suarez, the manager, asked Mr. Amoroso to come to work although Saturday was a regular day off for him. Mr. Suarez was required to pay Mr. Amoroso time-and-a-half for the extra 8 hours of work he performed on that day (*overtime* work). Mr. Amoroso's regular rate of pay is $6.90 per hour. Calculate:

   a. the number of overtime hours he earned at time-and-a-half.
   b. his total earnings for the 8 hours of overtime.
   c. his total wages for the week (5 regular days plus 1 day of overtime).

2. **a.** The Gas, Grease, and Gadgets Service Station pays an average of 80¢ per gallon for the gasoline it *buys*. Remember that a large sum of money is added to the price per gallon of gasoline for the federal tax, state tax, city sales tax, and tax on leaded gasoline. If the station buys 12,000 gallons of gasoline each week, how much does it pay for the gasoline it buys?

    b. If the station sells the 12,000 gallons of gasoline each week and makes a profit of 12¢ per gallon, calculate the profit.

3. During 1 week, Brink and Suarez worked on customers' cars for a total of 72 hours. If the charge is $38 per hour for labor, how much did the station earn from labor during that week?

4. During 1 week in July, the station sold 160 quarts of oil at an *average* price of $1.45 per quart, 25 gallons of antifreeze at $6.50 per gallon, 24 tires at an *average* price of $49.75 per tire, 4 batteries at a *total* price of $168.50, a number of oil, gasoline, and air filters at a total price of $140, and various accessories in the auto supply shop at a *total* price of $254.75. Determine the total income for the station from the sale of all these items.

5. The Gas, Grease, and Gadgets Service Station has an *average* gross income of $33,000 per week. *Gross* income is total money collected for sales and labor. After expenses are deducted, it is called *net* income. One expense is rent, and the station must set aside 9% of its gross income to pay for rent. What is the rent?

6. The records for the Gas, Grease, and Gadgets Service Station show that its *net* income each week is derived from the following sources:

Sale of gasoline	25%
Sale of oil, antifreeze, other fluids, batteries, tires, filters	20%
Major and minor repairs, including lubrication and oil change servicing, tune-ups, alignments, brake repairs, repairs to exhaust systems	40%
Miscellaneous sources: auto supply shop, towing, etc.	15%
Total	100%

The percentages in an *entire* operation must always total 100%. Information given in percentage form, such as this, can best be shown on a *circle graph* like the one on p. 285.

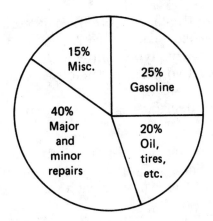

The entire circle represents 100%, so gasoline, which is 25% of the total net income, is

$$\frac{25}{100} \text{ or } \frac{\overset{1}{\cancel{25}}}{\underset{4}{\cancel{100}}} = \frac{1}{4} \text{ of the circle}$$

The 20% that represents sale of oil, tires, etc., is

$$\frac{20}{100} = \frac{1}{5} \text{ of the circle}$$

and so on.

a. What fraction of the circle is major and minor repairs?

b. What fraction of the circle is 15%?

c. What is the sum of the four fractions representing the percents in the circle graph?

d. Which three items together account for $\frac{3}{4}$ of the station's net income?

(*Hint:* Change $\frac{3}{4}$ to a decimal and then write the decimal as a percent.)

A circle contains 360°, and the portion represented by gasoline, 25%, contains a certain number of degrees in the circle above. How many degrees is 25% or $\frac{1}{4}$ of a circle? 25% = 0.25 and 25% of 360° = ?

$$\frac{1}{4} \text{ of } 360° = \qquad \text{or} \qquad 0.25 \times 360° =$$

$$\frac{1}{\cancel{4}} \times \frac{\overset{90}{\cancel{360}}}{1} = 90° \qquad\qquad \begin{array}{r} 360 \\ \times 0.25 \\ \hline 18\ 00 \\ 72\ 0 \\ \hline 90.(0)0 = 90° \end{array}$$

    e. What is the name of this or any other angle containing 90°?

    f. How many degrees is 20% of a circle?

    g. What is the name for this type of angle?

    h. Represent the other two parts of the circle graph, and name the type of angle.

    i. Add the number of degrees that represent the four parts of the graph. What is their sum?

7. The *difference* between the *gross* income and the *net* income of the Gas, Grease, and Gadgets Service Station is the cost of doing business, or the expenses, of the station. The expenses each week at the station are as follows:

    Cost of gasoline...35%
    Cost of oil, antifreeze, batteries, tires, parts,
        supplies, accessories..30%
    Cost of labor (payroll)...10%
    Cost of rent and electricity 15%
    Cost of taxes 10%

    a. Find the number of degrees each percent represents in a circle.
       35% = _____°   15% = _____°
       30% = _____°   10% = _____°
       10% = _____°

    b. Find the sum of all the degrees.

    c. Name the type of angle each of the percents will represent in a circle graph.

    d. Using a *protractor*, draw each of the percents as a part of the circle to construct the circle graph of the expenses of the station.

    e. Which item(s) was (were) *least* expensive?

    f. Which item was *most* expensive?

    g. Which *two* items were *equally* expensive?

    h. Which *two* items taken together totaled one-half the cost of doing business?

    i. Which *three* items taken together totaled one-half the cost of doing business?

8. In drawing a circle graph to show sources of gross income for the station, the bookkeeper filled in items representing 37%, 26%, and 22% of the total gross income. One item was omitted. What percent of the total gross income did this last item represent?

EXAM TIME

1. a. Calculate the price of 18 gallons of premium gasoline, at 137.9¢ per gallon.

    b. Determine the amount of change a customer would receive if he offered $30 in payment for this purchase.

2. If extra gasoline is selling for 129.9¢ per gallon, and a customer asks for $20 worth of this gasoline, determine the number of gallons she will receive. Round off the answer to two decimal places.

3. **a.** A car has a gasoline tank with a capacity of 16 gallons, and the fuel tank indicates that the tank is $\frac{3}{8}$ full. Calculate the number of gallons needed to fill this tank.

    b. How much will this cost if the customer asks for regular gasoline, selling at 119.9¢ per gallon?

4. A car has a gasoline tank with a capacity of 24 gallons. How many gallons of gasoline are needed to fill this tank if the fuel gauge indicates that the tank is:

    a. $\frac{1}{4}$ full? **b.** $\frac{1}{2}$ full? **c.** $\frac{5}{8}$ full?

5. If the price of 1 gallon of regular gasoline is 119.9¢, the price of extra gasoline is 129.9¢ per gallon, and premium gasoline costs 137.9¢ per gallon, calculate the average price of 1 gallon of gasoline at the station.

6. **a.** Every gallon of regular gasoline, selling at 119.9¢, includes in its price 9¢ federal tax, 8¢ state tax, and 5¢ city sales tax. Calculate the total tax included in the price of each gallon.

    b. If a customer buys 20 gallons of this gasoline, how much is he actually paying in taxes?

7. **a.** Henry buys 12 gallons of premium gasoline at 137.9¢ per gallon and 2 quarts of 10W30 grade engine oil at $1.50 per quart. Calculate his total bill.

    b. How much change will he receive if he offers a $20 bill as payment?

8. If sales tax is charged at the rate of 8%, determine the amount of tax that must be collected when a battery costing $35.95 is sold. Round off the answer to the nearest cent.

9. **a.** A customer purchases 2 gallons of antifreeze for her car radiator, at a price of $6.50 per gallon. Sales tax is added to the bill at the rate of 7%. Determine the total bill.

    b. Determine the amount of change she will receive if she offers a $20 bill in payment.

10. On a certain day, the station sold 12 quarts of 30 grade oil at $1.25 per quart, 20 quarts of 10W30 grade oil at $1.50 per quart, 24 quarts of 10W40 grade oil at $1.75 per quart, and 8 quarts of 10W20W50 grade oil at $2.25 per quart. Calculate, to the nearest cent, the average price of 1 quart of oil sold that day.

11. Calculate the total bill for Sam, who orders the following: lubrication, $5.50; oil filter, $5.50; 5 quarts of 10W40 oil at $1.75 per quart. Include a charge of $7\frac{1}{2}$% in the bill for sales tax.

12. Determine the bill for each of the following transactions:

    a. A new set of four tires, at $56.50 per tire, plus 8% sales tax.
    b. A tune-up of a 4-cylinder engine at a cost of $46.88, a new battery for $59.95, a front-end alignment for $34.88, and a new muffler and tailpipe installed at a cost of $84.95 for both, all plus 8% sales tax.
    c. Which of the bills was the higher?
    d. By how much was it higher than the other?

13. If the cost of labor at the station is $28 per hour, determine the amount a customer is charged for labor for a repair job that requires:

    a. 2 hours   **b.** $1\frac{1}{4}$ hours   **c.** $\frac{1}{2}$ hour

14. A customer purchased a set of battery booster cables for $16.99, a set of windshield wiper blade refills for $3.59, and a can of car wax for $3.29 in the auto supply shop. Calculate his total bill, including sales tax at the rate of 5%.

15. Oil is sold by the quart, but gasoline is sold by the gallon.

    a. Change each of the following from quarts to gallons:
       (1) 16 quarts   (2) 10 quarts
    b. Change each of the following from gallons to quarts:
       (1) 8 gallons   (2) $4\frac{1}{2}$ gallons

16. How much did the temperature rise on a winter afternoon if it was 3°F below zero at 1 P.M. and 8°F above zero at 4 P.M.?

17. Using the symbol > or <, order these pairs of numbers:

    a. −4   +2   **b.** +7   +3   **c.** −4   −3

18. Add:

    a. (−5) + (−4)   **b.** (−2) + (+6)   **c.** (+2) + (−8)

19. Subtract:

    a. (−5) − (−4)   **b.** (−2) − (+6)   **c.** (+2) − (−8)

20. Multiply:

    a. (−5)(−4)   **b.** (−2)(+6)   **c.** (+2)(−8)

21. Divide:

    a. (−8) ÷ (−4)   **b.** (+6) ÷ (−2)   **c.** (−8) ÷ (+2)

22. **a.** An automobile radiator containing a proper mixture of antifreeze and water will not overheat until its temperature reaches at least 230°F. Express this temperature in degrees Celsius.
    b. The same radiator will not freeze until its temperature drops below −30°C. Express this temperature in Fahrenheit degrees.

23. A standard wrench in a metric set measures 9 mm.

    a. Express this measurement in centimeters.

    b. Convert it to the nearest $\frac{1}{8}$ inch.

24. a. A road sign shows that a certain city is 480 km away. Express this distance in miles.

    b. Express a speed of 45 mph in kilometers per hour.

25. a. Change each distance from meters to kilometers:
    (1) 2500 meters   (2) 250 meters

    b. Change each distance from kilometers to meters:

    (1) 3.5 kilometers   (2) $\frac{3}{4}$ kilometer

26. a. One liter is approximately equal to 1 quart. How many liters of gasoline will be required to fill the gasoline tank of a car if it has a capacity of 24 gallons?

    b. If a liter of gasoline sells for 97¢ in a European country, how much will it cost for a fill-up of this tank?

27. The limit that a lift in the station can support is stamped on the controls: 6600 pounds. Express this limit in:

    a. tons   b. kilograms   c. tonnes

28. The records of Mr. Lopez, the Gas, Grease, and Gadgets Service Station's bookkeeper, indicate that Mr. Brink, who earns $12.75 per hour, worked 40 hours, 37 hours, 35 hours, and 32 hours, respectively, for the 4 weeks in a month. How much did he earn for the month of work?

29. The station buys 12,000 gallons of gasoline each week from the oil company that supplies it, including regular, extra, and premium grade gasoline, and it pays $9600 for this. Calculate the average price it pays for each gallon of gasoline.

30. If 55% of the sales of gasoline at the station are of regular gasoline and 30% are of extra gasoline, determine the percent of premium gasoline sold.

31. A major gasoline company spends its money in various ways. The money allotted for this purpose is shown in the circle graph.

Where Does The Money Go?

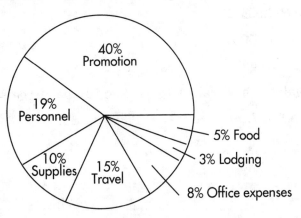

a. Find the number of degrees for each percent shown in the graph.
b. What is the name of the angle that represents the 40% sector?
c. Find the sum of the angles for all the percents listed in the graph.
d. Which item is the *least* expensive?
e. Which item is the *most* expensive?
f. What item, when added to personnel, office expenses, lodging, and food, accounts for half of the expenditures?

Now Check Your Answers

Now that you have completed the exam, check your answers against the correct ones, which follow the answers to the practice exercises below.

ANSWERS TO PRACTICE EXERCISES AND EXAM

PRACTICE EXERCISE 1

1. $8.01   2. a. Yes   b. $2.07   3. $17.10   4. $9.47

PRACTICE EXERCISE 2

1. 10.9 gal.   2. Yes. 19.25 gal.
3. a. 7.25   b. 13.43   c. 11.29
4. $11.60   5. 12.9 gal.   6. a. 7.25   b. 13.43   c. 11.29

PRACTICE EXERCISE 3

1. a. 8 gal.   b. $9.59
2. a. 5 gal.   b. 15 gal.   c. Yes   d. 51¢
3. a. 6 gal., 18 gal.   b. 9 gal., 15 gal.   c. 15 gal., 9 gal.   d. 21 gal., 3 gal.
4. a. $7.19   b. $17.99   c. $10.79   d. $3.60

PRACTICE EXERCISE 4

1. a. 11 gal.　b. $15.17　c. $4.50　d. $0.36　e. $20.03
2. a. $10.39　b. $4.95　c. $0.37　d. $15.71
3. a. $10.00　b. $0.80　c. $20.80
4. a. $24.90　b. $5.10
5. a. Yes　b. Yes, $0.65

PRACTICE EXERCISE 5

1. $1.55　2. $25.05　3. $3.01　4. $1.70　5. $1.318

6. a. 10 gal.　b. $13\frac{3}{4}$ gal.　c. $7\frac{1}{2}$ gal.　d. $37\frac{1}{2}$ gal.

7. a. 52 qt.　b. 34 qt.　c. 7 qt.

8. a. (1) 5.75　(2) 7.50　(3) 2.88　(4) 6.25　(5) 5.20
   (6) 4.67　(7) 10.13　(8) 1.14　(9) 4.40　(10) 3.63

   b. (1) $\frac{11}{8}$　(2) $\frac{7}{2}$　(3) $\frac{33}{8}$　(4) $\frac{35}{4}$　(5) $\frac{13}{4}$

   (6) $\frac{113}{16}$　(7) $\frac{45}{8}$　(8) $\frac{12}{5}$　(9) $\frac{22}{7}$　(10) $\frac{67}{8}$

   c. (1) $5\frac{3}{4}$　(2) $1\frac{5}{8}$　(3) $4\frac{1}{2}$　(4) $2\frac{3}{8}$　(5) $5\frac{3}{8}$

   (6) $3\frac{7}{16}$　(7) $4\frac{3}{5}$　(8) $2\frac{1}{7}$　(9) $7\frac{1}{3}$　(10) $2\frac{5}{6}$

9. a. Oil $8.75　　b. $1.70

   Subtotal　=　$72.50
   Sales tax　=　　5.80
   Total　=　$78.30

PRACTICE EXERCISE 6

1. $77.51　2. $154.82　3. $36.21

4. a. $49.00　b. $14.00　c. $84.00　d. $63.00

5. a. $21\frac{1}{2}$ hr.　b. $\frac{3}{4}$ hr. = 45 min.

6. a. $46.99　b. $187.96　c. $230.00

7. a. $97.96　b. $111.96　c. $14.00　d. (1) $\frac{1}{2}$ hr.　(2) 30 min.
   e. $34.57

8. a. $31.24　b. $124.96　c. $124.96 + 6.25 = $131.21　d. $18.79

9. $29.75　10. $57.74

PRACTICE EXERCISE 7

1. **a.** 212°F   **b.** 0°C

2. **a.** No   **b.** 5°F   **c.** 2.8°C   **d.** Sunday
   e. Friday   **f.** 26°F   **g.** Tuesday, Wednesday, Thursday, Friday, Saturday
   h. 2   **i.** 25°F   **j.** 18°F   **k.** 21°F

PRACTICE EXERCISE 8

a. <   **b.** >   **c.** >   **d.** <   **e.** >
f. >   **g.** <   **h.** <   **i.** >   **j.** <

PRACTICE EXERCISE 9

a. +1   **b.** +8   **c.** −4   **d.** −1   **e.** +4
f. +4   **g.** +5   **h.** −6   **i.** +2   **j.** −2

PRACTICE EXERCISE 10

a. +8   **b.** −18   **c.** +15   **d.** +16   **e.** −4
f. +10   **g.** +16   **h.** −12   **i.** +4   **j.** −15

PRACTICE EXERCISE 11

a. −4   **b.** +3   **c.** +5   **d.** −3   **e.** +3
f. −5   **g.** −5   **h.** −3   **i.** −4   **j.** +2

PRACTICE EXERCISE 12

a. +7   **b.** +2   **c.** 0   **d.** +7   **e.** −6
f. +2   **g.** +7   **h.** −4   **i.** −8   **j.** −3

PRACTICE EXERCISE 13

1. **a.** $-13\frac{1}{3}°$ C   **b.** Yes. It's below freezing.   2. −17.8°C

3. **a.** Yes   **b.** 3°F   **c.** 18°F   4. 26°F   5. 26°F   6. 5°F

7. 5°C   8. **a.** It fell   **b.** 3°F

9. **a.** Saturday   **b.** Tuesday   **c.** 10°C   **d.** 77°F
   e. 30°C   **f.** 5°C   **g.** 6°C   **h.** 91.4°F

10. 37°C   11. Yes 104°F   12. 194°F   13. 89.6°F Summer

14. **a.** 212°F   **b.** 140°F   **c.** 32°F
    d. 82.2°C   **e.** 18.3°C   **f.** −17.8°C

15. a. −7°C   b. −13°C   c. 37°F

d. (1) 25°F   (2) $-3\frac{8}{9}°$C   e. (1) $-3\frac{8}{9}°$C   (2) Yes

16. a. Saturday-Sunday   b. Tuesday   c. Monday

PRACTICE EXERCISE 14

1. a. 0.6 cm   b. $\frac{1}{4}$ in.   2. a. 150 mm   b. 6 in.

3. a. 1.875 cm   b. 18.75 mm

4. a. 18 in.   b. 45 cm   c. 450 mm

5. a. 10 cm   b. 2.5 cm   c. 0.75 cm

6. a. 1000 mm   b. 500 mm   c. 25 mm   d. 1 mm

7. a. 40 in.   b. 2 in.   c. 60 in.   d. 20 in.   e. 10 in.

8. a. 20 cm   b. 60 cm   c. 90 cm   d. 50 cm   e. 0.63 cm

9. a. 200 mm   b. 600 mm   c. 900 mm   d. 500 mm   e. 6.25 mm

10. 400 miles   11. 40 km   12. a. 120 km   b. 200 km

13. a. 150 mi.   b. 60 mi.

14. 50 mph   15. No 96 km/hr.

16. a. 10 km   b. $\frac{1}{10}$ km

17. 128 km/hr.   18. a. 12,000 m   b. 800 m

19. a. 3 km   b. 0.5 km

PRACTICE EXERCISE 15

1. a. 4864 cm³   b. 4.864 L
2. a. 3200 cm³   b. 200 in.³
3. $0.30 per liter   4. a. 64 L   b. $46.72
5. a. 25 L   b. 6 gal.

PRACTICE EXERCISE 16

1. a. 1.65 tons   b. 1500 km   c. 1.5 tonnes   d. 1,500,000 g
2. a. 5.5 tons   b. 5000 kg   c. 5 tonnes
3. a. 1 tonne   b. 2200 lb.   c. 1.1 tons
4. a. 1430 lb.   b. 650,000 g
5. a. 10,000 mg   b. 500 mg
6. a. 0.005 g   b. 0.15 g

PRACTICE EXERCISE 17

1. a. $9   b. $0.72   c. $9.72
2. a. $60.   b. $64.80   c. $35.20
3. a. $0.70   b. $8.40   c. 3 gal.   d. 2 times
   e. $12.60 + 1.01 = $13.61
4. a. $69.94   b. $26.91   c. $73.03
5. $47.79 + 3.82 = $51.61

PRACTICE EXERCISE 18

1. a. 12 hr.   b. $82.80   c. $358.80
2. a. $9600   b. $1440
3. $2736   4. $2151.75   5. $2970
6. a. $\frac{2}{5}$ b. $\frac{3}{20}$ c. 1   d. oil, tires, etc., major and minor repairs, and miscellaneous

   e. a right angle   f. 72°   g. acute angle

   h. misc. 54°
   (acute angle), major
   and minor repairs 144°
   (obtuse angle)

   i. 360°

7. a. 35% = 126°
      30% = 108°
      10% = 36°
      15% = 54°
      10% = 36°

   b. 360°

   c. 126°, obtuse
      108°, obtuse
      36°, acute
      54°, acute
      36°, acute

   d.

   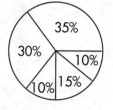

   e. Labor; taxes   f. Gasoline   g. Labor; taxes
   h. Gasoline + rent and electricity   i. Oil + labor + taxes

8. 15%

EXAM TIME

1. **a.** $24.82   **b.** $5.18   **2.** 15.40 gal.   **3. a.** 10 gal.   **b.** $11.99

4. **a.** 18 gal.   **b.** 12 gal.   **c.** 9 gal.   **5.** $1.292 = $1.29

6. **a.** 22¢   **b.** $4.40   **7. a.** $19.55   **b.** $0.45

8. $2.88   **9. a.** $13.91   **b.** $6.09   **10.** $1.64

11. $21.23

12. **a.** $226 + 18.08 = $244.08   **b.** $226.66 + 18.13 = $244.79

   c. (b)   **d.** $0.71

13. **a.** $56   **b.** $35   **c.** $14   **14.** $25.06

15. **a.** (1) 4 gal.   (2) $2\frac{1}{2}$ gal.   **b.** (1) 32 qt.   (2) 18 qt.

16. 11°F   **17. a.** <   **b.** >   **c.** <

18. **a.** –9   **b.** +4   **c.** –6

19. **a.** –1   **b.** –8   **c.** +10

20. **a.** +20   **b.** –12   **c.** –16

21. **a.** +2   **b.** –3   **c.** –4

22. **a.** 110°C   **b.** –22°F   **23. a.** 0.9 cm   **b.** $\frac{3}{8}$ in.

24. **a.** 300 mi.   **b.** 72 km/hr

25. **a.** (1) 2.5 km   (2) 0.25 km   **b.** (1) 3500 m   2) 750 m

26. **a.** 96 L   **b.** $93.12   **27. a.** 3.3 tons   **b.** 3000 kg   **c.** 3 tonnes

28. $1836   **29.** 80¢   **30.** 15%

31. **a.**   40% = 144°   8% = 28.8°
         19% = 68.4°   3% = 10.8°
         10% = 36°   5% = 18°
         15% = 54°
   b. Obtuse   **c.** 360°   **d.** Lodging
   e. Promotion   **f.** Travel

How Well Did You Do?

0–48   **Poor.** Reread the unit, redo all the practice exercises, and retake the exam.

49–57   **Fair.** Reread the sections dealing with the problems you got wrong. Redo those practice exercises, and retake the exam.

58–65   **Good.** Review the problems you got wrong. Redo them correctly.

66–76   **Very good!** Continue on to the next unit. You're on your way.

THE MONEY BAGS SAVINGS BANK

The Money Bags Savings Bank is a large bank with many employees. It offers a variety of services to its customers.

BANKING HOURS

The bank is open from 9 A.M. to 3 P.M. each day from Monday to Friday. For the convenience of its users, it is also open Friday from 5 P.M. to 8 P.M. and Saturday from 9 A.M. to 1 P.M. Since there are 3 hours from 9 A.M. to 12 noon and 3 more hours from 12 noon to 3 P.M., it is easy to see that the bank is normally open for business a total of 6 hours each day from Monday to Friday. This does not take into account the extra hours that the bank is open on Friday and Saturday.

PRACTICE EXERCISE 1

1. How many hours is the bank open on Friday evenings?

2. What is the total number of hours that the bank is open on Friday?

3. How many hours is the bank open on Saturday?

4. What is the total number of hours that the bank is open each week from Monday through Saturday?

5. Approximately how many hours is the bank open for business during an entire year? (The answer will not be exact because banks are closed on legal holidays.)

DEPOSITING MONEY

Customers often come into the bank to put money into their savings accounts. They are called "depositors" because they are "depositing" money. Their accounts are "credited" or "added to" with the amount of money they deposit. They may deposit cash or checks or both.

Customers who wish to make deposits must fill out "deposit slips." A sample slip is shown here:

	DOLLARS	CENTS	
NAME _Mabel Quinn_			DEPOSIT
ADDRESS _549 First Avenue_			ACC'T. NO. _2565_
Anytown, USA			DATE _July 1, 19--_
CASH ____			
CHECKS ____	29	35	
" LIST SEPARATELY	3	89	
"	138	50	MONEY BAGS
"	30	00	SAVINGS BANK
"			Anytown, U.S.A.
TOTAL	201	74	
CASH RETURNED	−75	00	
DEPOSIT	126	74	

137   NOTE: PLEASE FILL OUT CARD FOR CHANGE OF ADDRESS

Notice that each check to be deposited must be listed separately. The amounts are then *added* to arrive at the total. If any cash is being deposited, that amount is also *added* in arriving at the total. Notice that Ms. Quinn deposited only checks (the "cash" line is left blank on the deposit slip). Also, she wanted to deposit only part of the total value of the checks and to receive $75 in cash. She showed this on the line marked "cash returned," and $75 was subtracted from the total of $201.74, leaving $126.74 to be deposited in Ms. Quinn's account.

Joe Miller is a teller at the Money Bags Savings Bank. It is his responsibility, and that of the other bank tellers, to check additions and subtractions shown by customers on their deposit slips before entering deposits into the bank's computers. Such checks help to prevent errors.

PRACTICE EXERCISE 2

1. Ms. Quinn gives Joe Miller her bankbook to have her deposit recorded by the bank's computers. If her handbook shows she already has $279.35 in the bank, what balance will it show after her new deposit is entered?

2. **a.** Ms. Rivera comes into the bank to deposit $45 in cash and a check for $16.75 in her savings account. What is her total deposit?
   b. If Ms. Rivera's balance before her deposit was $595.25, find her new balance after her deposit was recorded.

3. Mr. Chan deposits $735 in cash in his savings account. His old balance was $2194.17. What will his balance be after his deposit is recorded?

4. Mr. Smith wishes to deposit checks for the following amounts: $34.68, $225, and $9.55, and he wants $63 returned to him in cash. His bankbook shows a balance of $873.12 before this transaction.

   a. Prepare a deposit slip for Mr. Smith.
   b. What will his actual deposit be?
   c. Calculate Mr. Smith's balance after this transaction is completed.

		DEPOSIT	
NAME _____			
ADDRESS_____		ACC'T. NO. _____	
		DATE _____	
	DOLLARS	CENTS	
CASH _____			
CHECKS _____			
" LIST SEPARATELY			MONEY BAGS
" _____			SAVINGS BANK
" _____			
" _____			Anytown, U.S.A.
TOTAL			
CASH RETURNED			
DEPOSIT			

137   NOTE: PLEASE FILL OUT CARD FOR CHANGE OF ADDRESS

5. After all deposits have been entered on the bank's records, calculate the total amount of money that Ms. Quinn, Ms. Rivera, Mr. Chan, and Mr. Smith have on deposit, all together, in the Money Bags Savings Bank.

6. **a.** Which of these four has the least amount of money on deposit at the bank?
   b. Which has the greatest amount of money on deposit?
   c. What is the difference between the least amount and the greatest amount of money that these four persons have on deposit at the bank?

WITHDRAWING MONEY

People must sometimes take money out of their savings account to meet personal expenses or to pay bills. This is called "withdrawing" money. Their accounts are "debited" or "subtracted from" by the amount of the withdrawal. Money taken from a savings account may be withdrawn as cash or in one or more checks, or partly in cash and partly in checks. The bank issues its own checks to customers who request them. There is usually no charge made for issuing official bank checks (sometimes called "teller's checks").

Customers who wish to make withdrawals must fill out and sign "withdrawal slips." A sample slip is shown here:

```
┌─────────────────────────────────────────────────────────────────────┐
│                                                    ┌──────────────┐   │
│   ACCT. NO. ___2565_____                 │   DOLLARS    │   │
│                                                    ├──────────┬───┤   │
│   DATE __May 12, 19-- _____                 │  $115    │37 │   │
│                                                    └──────────┴───┤   │
│            MONEY BAGS SAVINGS BANK                          CENTS │   │
│  W                                                                    │
│  I   Pay to __Bell Fuel Oil Company_____ or Bearer          │
│  T                                                                    │
│  H   _One Hundred Fifteen_____ Dollars _37_ Cents             │
│  D              (AMOUNT IN WORDS)                                     │
│  R   and charge to my account.                                       │
│  A          Signature__Mabel Quinn_____                    │
│  W            (In Ink)              (Name in Full)                    │
│  A          Present                                                   │
│  L          Address__549 First Avenue_____                    │
│                                                                       │
│                     __Anytown, USA_____                    │
│  ┌──┬──┬──┐              (CITY)     (STATE)   (ZIP CODE)             │
│  └──┴──┴──┘                                                           │
│   B  S  S                                                            │
└─────────────────────────────────────────────────────────────────────┘
```

The amount of money withdrawn is written in words as well as in numerals. This duplication is often required by banks to make certain that there is no error or that no change has been made in the numerals. Although the number of dollars (one hundred fifteen) is written in words, it is permissible to write the number of cents (37) in numerals (37/100 is also correct since cents are hundredths of a dollar).

It is important for Jane and the other tellers to check each withdrawal slip to be certain that the same amount is written in numerals and in words. The withdrawal slip must be *signed* by the person making the withdrawal, and the bank book must also be presented.

Ms. Quinn withdrew $115.37 from her savings account to pay for her fuel bill. She did not take this money in cash; instead, she asked the bank to issue an official check, made payable to the Bell Fuel Oil Company, for the amount withdrawn. Ms. Quinn had the choice of taking her money in cash or having a check made payable to anyone she wished, including herself. All that is needed is to write the name of the person or company to whom the check should be made payable on the line marked "Pay to."

Ms. Quinn's account is "debited" for $115.37 when she makes this withdrawal. This amount is *subtracted* from her old balance to leave her new balance. If her old balance was $406.09, her balance after her withdrawal is made is $406.09 – $115.37 = $290.72.

REMINDER! *When adding or subtracting decimals, line up the decimal points under each other. The decimal point in the answer must line up in the same column as the other decimal points.*

PRACTICE EXERCISE 3

1. Mr. Brown needed a check for $295.50, made payable to the Fancy Furniture Company to pay for two living room chairs he bought. His savings account had a balance of $1230.25 before he made this withdrawal. Calculate Mr. Brown's new balance after this transaction is completed.

2. Ms. Rosen withdrew $495 from her account to pay the monthly rent for her apartment, $23.65 to pay her gas and electric bill, and $75 for personal expenses. She had $945.30 on deposit in her savings account before she made these transactions.

   a. How much was the total withdrawal?
   b. What was her new balance?

3. Ms. Rodriquez had $357.85 on deposit in her savings account before she made a withdrawal of $125 in cash. How much was her balance after this withdrawal?

4. If Mr. Jones's bankbook showed a balance of $584.66 *after* he made a withdrawal of $200.97, how much was his balance *before* he made this withdrawal?

5. Mr. Kim had a balance of $3487.16 in his savings account at the beginning of a week. During that week he came into the bank often to make deposits and withdrawals. He deposited $150 in cash and $287.50 in checks, and withdrew $654.32 and $157.95 by writing two checks.

   a. How much were his total deposits?
   b. How much were his total withdrawals?
   c. What was his new balance at the end of the week?

SIMPLE INTEREST

The Money Bags Savings Bank, like all banks, pays its customers *interest* on their savings accounts. Interest is the money paid to depositors for leaving their money with the bank. To earn money to pay this interest and to meet its other expenses of operation, the bank invests some of the money deposited by customers in a variety of ways. It receives income on these investments and uses part of this income to pay interest to its customers. One of the ways in which the bank earns money is by making loans to customers who need money to buy or build a house (such a loan is called a "mortgage"), to make home repairs, to meet college costs, or to serve other useful purposes.

Customers pay interest to the bank when they take loans from the bank. The amount of interest paid depends on the amount of money borrowed, the rate of interest (which varies from bank to bank and also according to the purpose for which the money is borrowed), and the length of time for which the money is borrowed. A bank employee, usually an assistant to the vice-president, like Hector Ruiz, can help a depositor to estimate the actual cost of a loan (the total of the customer's interest payments) by using a formula or equation to calculate "simple interest." Simple interest is interest paid on the amount of money borrowed, called the "principal." The formula is:

$$\text{Interest} = \text{principal} \times \text{rate} \times \text{time}$$
$$I = P \times R \times T$$

or

$$I = PRT$$

The rate of interest is the rate per *year*, called the "annual rate," and it is expressed as a percentage. In making the calculation, it is necessary to change the percentage to a decimal. Time must always be measured in years or fractions of a year.

EXAMPLE: After talking with Mr. Ruiz, Mr. Ferguson borrowed $3000 for 1 year to help pay for his daughter's college tuition. If the interest rate was $8\frac{1}{2}\%$, calculate the amount of interest paid on this loan.

SOLUTION: Interest = principal × rate × time

or

$$I = PRT, \text{ where } P = \$3000$$
$$R = 8\frac{1}{2}\% = 8.5\% = 0.085$$
$$T = 1 \text{ year}$$
$$I = \$3000 \times 0.085 \times 1$$
$$I = \$255$$

In this calculation, $8\frac{1}{2}\%$ was changed to a decimal by first writing $8\frac{1}{2}\% =$ 8.5%, since $\frac{1}{2} = \frac{5}{10} = 0.5$ in decimal form. We could also have divided the denominator (2) into the numerator (1) to change $\frac{1}{2}$ into a decimal fraction. Then 8.5% was changed to a decimal by dividing by 100. This was done because percent means "hundredths," so

$$8.5\% = 8.5 \text{ hundredths} = \frac{8.5}{100} = 0.085$$

$$
\begin{array}{rl}
\$\quad 3000 & \text{0 decimal place} \\
\times 0.085 & \text{3 decimal places} \\
\hline
15\ 000 & \\
240\ 00 & \\
\hline
\$255.000 & \text{3 decimal places} \\
\times 1 & \\
\hline
\$255.000 & = \$255 \text{ interest paid}
\end{array}
$$

Borrowers usually repay borrowed money to the bank by making regular monthly payments until the loan is completely paid off. They are charged interest *only* on the amount of money they still owe each month. Over the time of a loan, this unpaid balance averages out to *approximately one-half of the original principal.* Therefore a good *estimate* of the interest cost of a loan can often be obtained by using one-half the original principal when making calculations using the interest formula. This amount is called the "average unpaid balance."

EXAMPLE: Mr. Orefanos figures that the average unpaid balance of his automobile loan is $3000. What is the amount of the original principal?

SOLUTION: If the average unpaid balance is approximately one half of the original principal, then the original principal is $6000.

EXAMPLE: Mr. Green borrowed $5000 for 5 years at 12% interest to make improvements in his home.

a. How much is his average unpaid balance?
b. Approximately how much interest will he be charged for this loan?
c. How much money will Mr. Green have to pay back to the bank?
d. How much is each payment if he repays this loan in monthly payments over a 3-year period?

SOLUTION:

a. Mr. Green's average unpaid balance for the 5-year term of the loan was approximately one-half of $5000 or $\frac{1}{2} \times 5000 = \$2500$.

b. We use the interest formula to calculate the total of his interest payments:

$$I = PRT, \text{ where } P = \$2500 \text{ (average unpaid balance)}$$
$$R = 12\% = 0.12$$
$$T = 5 \text{ years}$$
$$I = \$2500 \times 0.12 \times 5$$
$$I = \$1500$$

c. $5000 + $1500 = $6500

d.
```
         180.55(5)
   36)6500.000  = $180.56
      36xx xxx
      290
      288
       20 0
       18 0
        2 00
        1 80
          200
          180
           20
```

When using the interest formula, be sure that you:

1. Change the percent to a decimal by dividing by 100, for example, 12% = 0.12.

2. Multiply correctly, and place the decimal point in the product the number of places equal to the sum of the decimal places in the multiplier and the multiplicand.

3. Round off all answers involving money to two decimal places (for cents).

PRACTICE EXERCISE 4

1. Ms. Shea borrowed $700 for 2 years at 11% simple interest to buy some furniture. If she did not make any payments on this loan until the end of the 2-year period, what was the amount of interest charged for the loan?

2. Mr. Simpson borrowed $7000 at $9\frac{1}{2}$% for 3 years to remodel his house. He paid the loan back in equal installments each month for the entire 3 years.

   a. How much is the average unpaid balance?
   b. Use this amount in the interest formula to calculate the total of Mr. Simpson's interest payments.
   c. Calculate the total amount of money Mr. Simpson must pay back to the bank.
   d. How many monthly installments did he pay during the 3-year period?
   e. Calculate the amount of each monthly payment.

3. Mr. Gonzalez borrowed $8000 for 4 years at 8% interest to help pay for his daughter's college expenses. He repaid the loan in equal monthly payments for the entire 4 years.

   a. How much was his average unpaid balance?
   b. Calculate the total of his interest payments.
   c. Calculate the total amount of money he must pay back to the bank.
   d. Find the amount of each monthly payment over the 4-year period.

4. Ms. Jones borrowed $850 at 9% simple interest for 18 months to buy a used car. If no payments were made on this loan until the end of the 18-month period, how much interest was charged for this loan? (*Remember:* Time must be expressed in years, and 12 months = 1 year.)

COMPOUND INTEREST

Interest paid by banks on savings accounts is *not* simple interest. It is called "compound interest." When interest is "compounded," this means that each time interest is calculated it is calculated not only on the original money the customer deposited (the principal) but also on all interest earned earlier and credited to the account. In other words, compound interest is interest paid on interest as well as on principal. Compounding may be done annually (once a year), semiannually (twice a year), quarterly (every 3 months), monthly (every month), or even more often. Many banks today, including the Money Bags Savings Bank, compound continuously (every minute of every hour of every day) by using computers, and credit the interest to customer's savings accounts every 3 months (quarterly).

Compounding helps to pay the customer a little more interest than he or she would get if banks paid only simple interest. The more frequent the compounding, the greater is the amount of interest the customer actually receives for saving.

The official rate of interest paid on a savings account, perhaps 2% or $2\frac{1}{4}$% is called the "nominal" rate (meaning the rate in name only). The actual rate paid is higher because of compounding and is called the "effective" rate. For example, a 2% nominal annual rate is actually equal to a 2.02% effective annual rate, and a 2.35% nominal annual rate is equal to a 2.38% effective annual rate, when compounding is done continuously.

The following table shows the nominal rate and the effective rate paid by the Money Bags Savings Bank on regular savings accounts and also on special accounts, called "savings certificates." Effective rates are paid only when the depositor allows all the interest earned to remain in the account until the end of the year. If the interest is withdrawn, the "nominal" rate of interest is the percent of interest earned.

Daily Compounding for Every Style of Savings

5.47% effective annual yield* on **5.25%** a year
6-year savings certificates (min. deposit $1000)

4.60% effective annual yield* on **4.50%** a year
4-year savings certificates (min. deposit $1000)

4.34% effective annual yield* on **4.25%** a year
2 ½-year savings certificates (min. deposit $500)

3.92% effective annual yield* on **3.85%** a year
1-year savings certificates (min. deposit $500)

Highest Passbook Rates

2.38% effective annual yield* on **2.35%** a year
Day of deposit to day of withdrawal passbook savings account

The table shows that interest rates are lowest for regular savings accounts, 2.38% effective annual rate on 2.35% nominal annual rate. The rates paid on these accounts are lower than the other rates because customers may make deposits into and withdrawals from regular savings accounts at any time. Interest is paid on money in the account from the day the money is deposited until the day it is withdrawn.

Higher rates of interest are paid on special "savings certificates." The reason for these higher rates is that the depositor promises to leave his or her money on deposit in the account untouched for a definite period of time. The interest on savings certificates may be withdrawn as it is earned, but if any part of the principal (the money originally deposited) is taken out before the end of the time period agreed upon, there is a penalty of the loss of some interest.

These accounts also require a minimum deposit of principal when they are opened, usually either $500 or $1000. Specified time periods vary; they may be 1 year, $2\frac{1}{2}$ years, 4 years, 6 years, or even longer. The longer the time period, the higher is the interest rate. At the end of the agreed upon time, the savings certificate is said to "mature" and the depositor may then withdraw his or her money, both principal and interest.

Savings certificates are sometimes called "certificates of deposit" or "time deposits." A short-term certificate for only 90 days, which is available in some banks, may pay only 2.5%, or 2.75% nominal annual rate. At the end of the 90 days, the depositor receives interest for $\frac{1}{4}$ year, since

$$90 \text{ days} = 3 \text{ months}$$

or

$$\frac{\overset{1}{\cancel{3}}}{\underset{4}{\cancel{12}}}$$

$$90 \text{ days} = 3 \text{ months} = \frac{1}{4} \text{ year}$$

Certificates for longer terms pay more, as much as a nominal annual rate of 5.25% for a certificate maturing in 6 years. The effective annual rate on such a certificate at the Money Bags Savings Bank is 5.47%.

Compound interest may be calculated *approximately* for a short period of time by using the formula studied for simple interest calculations, $I = PRT$. When using this formula, the rate of interest used is the effective annual rate. The examples on page 308 show how Mr. Ruiz and the other assistants to the vice-president of the Money Bags Savings Bank help customers to determine how much interest they can earn on different types of accounts.

EXAMPLE: Mr. Stuart has a regular savings account at the Money Bags Savings Bank that pays 2.35% nominal annual rate, equivalent to 2.38% effective annual rate. If his account had a balance of $375 on deposit during April, May, and June, how much interest did he receive when interest was credited on June 30, the end of the quarter-year?

SOLUTION: $I = PRT$, where $P = \$375$

$$R = 2.38\% = 0.0238$$

$$T = 3 \text{ months} = \frac{1}{4} \text{ year} = 0.25$$

$$I = \$375 \times 0.0238 \times 0.25$$
$$I = \$2.23125$$
$$I = \$2.23$$

$$
\begin{array}{r}
\$\quad 375 \\
\times 0.0238 \\
\hline
3000 \\
1\ 125 \\
7\ 50 \\
\hline
\$8.9250
\end{array}
\qquad
\begin{array}{r}
\$\quad 8.9250 \\
\times\ 0.25 \\
\hline
446250 \\
1\ 78500 \\
\hline
\$2.231250 = \$2.23
\end{array}
$$

EXAMPLE: Ms. Smart has a savings certificate. She deposited $1000 for 1 year at an effective annual rate of 3.92%. Calculate the amount of interest her certificate earned for the year, and determine the amount of money she will get back when the certificate matures at the end of the year.

SOLUTION: $I = PRT$, where $P = \$1000$

$$R = 3.92\% = 0.0392$$
$$T = 1 \text{ year}$$
$$I = \$1000 \times 0.0392 \times 1$$
$$I = \$39.20$$

Ms. Smart will receive $39.20 in interest, so she will receive $1000 + 39.20 = $1039.20 at the end of the year.

PRACTICE EXERCISE 5

1. Ms. Amato had $650 on deposit in a regular savings account at the **Money Bags Savings Bank**, at an effective annual rate of interest of 2.38%, for 3 months.

   a. Calculate the amount of interest earned.
   b. Find the new balance in the account after the interest is credited.

2. Mrs. Chang had $1125 on deposit in a regular savings account (2.38% effective annual rate of interest) for 1 year. How much interest did she earn?

3. Mr. Ryan had a savings certificate for $500 for $2\frac{1}{2}$ years at an effective annual rate of 4.34%.

   a. Calculate the approximate interest earned for the $2\frac{1}{2}$ years.

   b. Find the total amount of money returned to Mr. Ryan at the end of that time. (*Hint:* Write $2\frac{1}{2}$ years as 2.5 years.)

4. Mr. Stevens had a savings certificate for $2500 for 4 years at an effective annual rate of interest of 4.60%. Calculate the approximate amount of interest earned for the entire 4 years.

5. Interest is credited to savings accounts and entered in depositors' bankbooks at the Money Bags Savings Bank at the end of each quarter of the year (every 3 months). Determine the *quarterly* rates of interest for each of the following *annual* rates. To do this, change each fraction to a decimal; for example, $5\frac{1}{2}$% annual rate = 5.5% and 5.5% × $\frac{1}{4}$ = 1.375% quarterly interest rate. *Divide each example until there is no remainder.*

   a. 6%
   b. $6\frac{1}{2}$%
   c. $7\frac{1}{4}$%
   d. $7\frac{3}{4}$%
   e. 6.81%
   f. 7.90%

6. Use a calculator to check your answers to Problem 5.

CHECKING ACCOUNTS

Many customers of the Money Bags Savings Bank have checking accounts at the bank in addition to their savings accounts. They deposit cash and checks in their checking accounts, and they can then write their own checks to pay their bills or meet other expenses. Some banks charge their customers for each check they write, but others offer free checking to customers who maintain a minimum balance in their checking or savings accounts. The Money Bags Savings Bank offers its customers free checking accounts.

Some banks charge customers a penalty if their accounts are "overdrawn." An account becomes overdrawn if a check is written for more than the balance on deposit in the account. These checks are returned to the bank, unless the customer has "overdraft" privilege checking. If this is the case, the bank honors the check even if there is not enough money on deposit in the account. A fee is charged for taking a loan from the bank for the amount of the overdraft, and, of course, the loan must be repaid with interest charges.

Some people use their checking accounts to pay for many special things in addition to writing ordinary checks. They may have money subtracted automatically from their checking accounts to pay bank loan installments each month; to make monthly installment payments on credit card purchases, such as MasterCard and Visa; to make monthly, quarterly, semiannual, or annual premium payments on savings bank life insurance, sold by many savings banks; to make Christmas or Hanukkah Club payments; or to transfer money to their savings accounts in the same bank. Banks take care of all the paper work involved in such transactions.

Customers who wish to make deposits to their checking accounts must fill out deposit slips. They are given copies of these slips as receipts for their records. A sample slip is shown here:

DEPOSIT TICKET	CASH	DOLLARS	CENTS
	CHECKS LIST SINGLY	350	00
NAME _Joan Bosman_		27	35
ACCT. NO. [1][2] 05 [3][4][5][4][3][2] [1]		3	68
		172	31
January 10 19 --			
	TOTAL ITEMS TOTAL	553	34

BE SURE EACH ITEM IS PROPERLY ENDORSED

MONEY BAGS
SAVINGS BANK

Anytown, U.S.A.

⑈ 2 26071305⑈

The total deposit of $553.34 is added to Ms. Bosman's previous checking account balance to obtain the new balance. If the old balance was $257.28, the new balance is $553.34 + $257.28 = $810.62, and Ms. Bosman may now write checks up to that amount without overdrawing her account. To help avoid errors, it is important for Joe Miller and the other tellers to check the addition on deposit slips presented to them by checking account customers.

Sometimes Miller is asked by customers to check the balances in the customer's checking accounts. They wish to be certain that the records of the bank are in agreement with the balances calculated by the customers in their checkbooks.

EXAMPLE: Mr. Stein has a checking account at the Money Bags Savings Bank which had an opening balance at the beginning of the month of $654.19. He made deposits of $175, $50, $318.50, and $82 during the month. His checkbook shows that he wrote checks for $448, $165.75, $72.50, $18.75, and $10 during the month. He asks Jane to check the amount shown as the closing balance on his monthly statement from the bank. What should his balance be at the end of the month?

SOLUTION: Deposits are *credited* to his opening balance:

$$
\begin{array}{ll}
\$\ 654.19 & \text{Opening balance} \\
175.00 & \\
50.00 & \\
318.50 & \text{Deposits} \\
82.00 & \\
\hline
\$1279.69 & \text{Total credits}
\end{array}
$$

The account is *debited* or charged for checks written by Mr. Stein, the customer, and paid by the bank:

$$
\begin{array}{ll}
\$448.00 & \\
165.75 & \\
72.50 & \text{Checks (withdrawals)} \\
18.75 & \\
10.00 & \\
\hline
\$715.00 & \text{Total checks}
\end{array}
$$

Using the relationship

Closing balance = total deposits − total checks

we obtain

Closing balance = $1279.69 − $715

Closing balance = $564.69

PRACTICE EXERCISE 6

1. Mr. Sanders has a checking account at the Money Bags Savings Bank. His opening balance at the beginning of April was $963.74, and he made two deposits during the month, one for $300 and one for $175. He wrote checks for $450 to pay his income taxes on April 15, $525 to pay for new furniture, and $43.65 to pay his telephone bill for the month.

   a. Calculate his total credits.
   b. Calculate his total debits.
   c. Find the new balance in Mr. Sanders' account at the end of April.

2. Ms. Wing opened a new checking account at the beginning of January with a deposit of $300 and made no other deposits during that month. If she wrote checks totaling $172.35 and also transferred $65 to her savings account during January, what should her statement for that month show as the closing balance in her checking account?

3. Ms. Gomez had $50 on deposit in her checking account at the beginning of March. She deposited $75 during March and wrote checks for $135 in total value.

   a. Was her account overdrawn at the end of the month?
   b. If so, by how much?

4. Mr. Sweeney had $550 on deposit in his checking account at the beginning of September. He wrote checks totaling $275 and made no deposits during the month. He also transferred funds from his account as follows: $50 to his savings account, $75 to pay the monthly installment on a loan he had with the Money Bags Savings Bank, and $65 to pay the quarterly premium on the savings bank life insurance policy he bought from the bank. What should his statement show as the closing balance for September?

5. Ms. Green had $203.87 on deposit in her checking account at the beginning of February. She made no deposits during the month, but she wrote checks totaling $167.31 and transferred four weekly payments of $5 each to her Christmas Club account. Her checking account was also charged $3 for a check she wrote but did not have enough money in the account to pay for, so the check was returned to the bank. The bank made a charge for its time and trouble in handling the check. Determine Ms. Green's closing balance at the end of February.

SPECIAL SERVICES

The Money Bags Savings Bank offers its customers a variety of special services in addition to savings accounts, savings certificates, checking accounts, and bank loans. Some of these services are the following: cashing payroll checks, providing rolls of coin for change, maintaining Christmas and Hanukkah Club accounts, and providing savings bank money orders, travelers' checks, savings bank life insurance, and bank-by-mail services.

Bank tellers, like Joe Miller, and assistants to the vice-president, like Hector Ruiz, must know all these services offered to depositors and be able to provide information about them.

PRACTICE EXERCISE 7

1. Ms. White comes into the Money Bags Savings Bank every Friday afternoon to cash her paycheck of $237.69. If she does not want any bills of a denomination higher than $20, how can Joe Miller, the teller, give her the money for her check using the fewest bills ($20, $10, $5, $1) and the smallest number of coins (50¢, 25¢, 10¢, 5¢, 1¢)?

2. Customers sometimes request rolls of coins, which they need to make change in a business operation.

    a. Mr. Rubin gave Pedro Gomez, another teller, a $10 bill for a roll of quarters. How many quarters would be packed in the roll?
    b. If he gave Gomez a $5 bill for a roll of dimes, how many dimes would be packed in the roll?
    c. How many nickels are in a roll if the cost of one roll is $2?
    d. If $1 will buy two rolls of pennies, how many pennies are packed in a roll?
    e. How many coins would Mr. Rubin receive all together (quarters, dimes, nickels, and pennies) if he gave Gomez the $18 ($10, $5, $2, $1) mentioned in questions a–d above?

3. Ms. Grand usually comes into the Money Bags Savings Bank each week to make deposits to her Hanukkah Club account. Since she was on vacation for 3 weeks, she missed her payments for that time and wishes to make all of her back payments now. Her regular payment is $5 each week. She fills out her deposit slip as shown.

NAME _Helen Grand_	CLUB ACCT. NO. _54-321_
ADDRESS _327 Cardinal Street_	
Anytown, USA	DATE _August 1, 19--_
CLUB ACCOUNT DEPOSIT	
MONEY BAGS SAVINGS BANK	Previous Club Bal. $ _185.00_
Number of Payments _3_	Dollar Amount $ _____

    a. Complete the deposit slip by entering the dollar amount of Ms. Grand's deposit on the blank line.
    b. What will be the new balance in this account after the deposit is made?
    c. If Ms. Grand makes 50 deposits into her Hanukkah Club account during the year, how much money has she saved to help with her holiday shopping?

4. The Money Bags Savings Bank offers savings bank money orders to people who need them to pay bills or take care of any other expenses and who do not have checking accounts. Mr. Ryan wishes to purchase the four money orders shown on the money order application form below. The bank charges 75¢ for issuing *each* money order.

You must show the names of the persons or companies to whom you wish the money orders made out on the lines marked "payable to the order of," which means "pay to." Complete the application form by entering the fee for the four money orders and calculating the total amount of money Mr. Ryan must give José, the teller, to buy these money orders.

MONEY ORDER APPLICATION
MONEY BAGS SAVINGS BANK

PLEASE ISSUE MONEY ORDERS IN THE AMOUNTS LISTED IN RIGHT HAND COLUMN BELOW:

PAYABLE TO THE ORDER OF	AMOUNT	
1 Fancy Furniture Company	173	25
2 Edison Gas and Electric	43	68
3 Flashy Clothing Corp.	69	80
4 Best Realty Company	160	00
5		
DEPOSITORS–75¢ EACH		
NON DEPOSITORS $1.00 EACH		
TOTAL		

ACCOUNT NO. 972438

John Ryan
(PURCHASER'S SIGNATURE)

43 Dexter Street
(ADDRESS)

Anytown, USA   12345

KEEP STUB AS IT IS YOUR ONLY RECORD

5. The Money Bags Savings Bank sells travelers' checks to people who wish to use these checks while on business or pleasure trips instead of carrying large amounts of cash with them. Travelers' checks can be replaced if they are lost or stolen, and so they are safer to carry than cash. The bank makes a charge of 1% of the value of the travelers checks for issuing them and keeping records of them.

   a. Ms. Brisk wants to purchase $750 worth of travelers' checks to use while she is on vacation. How much will the bank charge for issuing these checks?

   b. How much money, all together, must Ms. Brisk give Jane, the teller, for her checks?

6. The Money Bags Savings Bank offers low-cost savings bank life insurance to its customers. Many different types of insurance policies are available to meet different needs, and the cost of these policies varies according to the type of policy and the amount of insurance purchased.
   Hector Ruiz and other assistants to the vice-president must be able to advise customers who ask questions about the best type of policy to purchase, the costs of different policies, and the savings possible when larger policies are purchased. The tables below show the cost of some term insurance policies offered by the bank. Term insurance, as the name implies, covers the purchaser for a given period or term, usually one year or five years. At the end of that time if the purchaser wishes to keep the policy in force he or she must pay a higher premium.

Low Cost Group Term Insurance				
$100,000 1ST YEAR PREMIUM*			$250,000 1ST YEAR PREMIUM*	
AGE	MALE	FEMALE	MALE	FEMALE
30	$ 89.00	$ 84.00	$ 172.50	$ 172.50
35	98.00	88.00	180.00	172.50
40	116.00	104.00	222.50	192.50
45	129.00	125.00	272.50	257.50
50	174.00	134.00	395.00	285.00
55	244.00	184.00	560.00	407.50
60	449.00	264.00	847.50	600.00

*Rates for non-smokers showing evidence of good health. Other rates also available. Rates increase annually, but once covered insured can apply for new first-year rate one time. Policy G-60, G-61.

$50,000 5 Year Term Annual Premiums*		
AGE	MALE	FEMALE
20–25	$ 64.50	$ 57.00
30	64.50	63.00
35	79.00	76.00
40	105.50	105.50
45	145.00	141.00
50	206.50	193.00

*Initial premiums. Rates increase at five year intervals. Rates shown are for non-smokers in good health. Other rates available. Policy T-8.

   a. Tom Lia is a 30 year old non-smoker who wishes to buy a $100,000 term policy. What is the annual premium on this policy?

   b. If Tom wants to buy $100,000 worth of 5-year term insurance, what would the annual premium be?

   c. Tom decides to put away some money each month so that he will have enough to pay the second year's premium. How much should he put aside each month?

   d. If he keeps the policy for 5 years and wishes to renew it, what will be his new premium?

   e. How much more is this than his original premium?

Acknowledgment: Permission granted by SBLI Fund, New York, NY 10001. The above rates are as of September 1994.

7. Note that the premiums for women are less than those for men. Sadie Lennon is a 40 year old non-smoker who wishes to purchase a one year term insurance policy.

   a. If she decides to purchase a $100,000 policy, how much less will she pay for the policy than a man of the same age?
   b. If Ms. Lennon decides to purchase a $250,000 policy, she will be getting two and one-half times as much insurance, but the premium is less than two and one-half times the cost of a $100,000 policy. How much can she save by buying the larger policy?

8. As can be seen from the table, term insurance becomes progressively more expensive as the purchaser becomes older. The Money Bags Savings Bank also offers whole life policies, which feature a premium which will never increase for the life of the policyholder. If the purchaser wishes to pay for a shorter time, policies are available which permit one to stop paying premiums after 20 years ("20 Payment Life") or at age 65 ("Life Paid Up at 65"). A table of some whole life policies is shown below.

$50,000 Whole Life Annual Premiums*			
Male Age	**Whole Life**	**20 Pay Life**	**Life Paid Up at 65**
30	$ 469.00	$ 663.00	$ 511.00
40	761.50	995.50	$ 891.00
50	1,271.00	1,510.50	$1,765.50
Female Age	**Whole Life**	**20 Pay Life**	**Life Paid Up at 65**
30	$377.00	$ 549.00	$ 420.00
40	602.00	821.50	$ 731.00
50	973.50	1,217.50	$1,441.00

*Rates shown for non-smokers in good health. Other rates also available.

   a. Richard is a 40 year old non-smoker. If he decides to buy a $50,000 whole life policy, what is his annual premium?
   b. How much more would he pay for a 20 Payment policy? for a Life Paid Up at 65 policy?
   c. Eileen is a 30 year old non-smoker who wishes to purchase $50,000 worth of insurance. How much more would she pay for a Whole Life Policy than for a 5-year Term policy?

9. The Money Bags Savings Bank has an agreement with a brokerage firm which allows depositors to purchase stocks, bonds, or mutual funds at the bank. A stock represents part ownership of a company, while a bond represents part of a loan to the company. When money is borrowed by cities or states in this way, the bonds are called municipal bonds. U.S. Government bonds are also available. A mutual fund comprises a group of investments put together by a company and managed by a professional money manager. A mutual fund permits investors to diversify their investment with one purchase. Use your calculator to do the following problems.

Acknowledgment: Permission granted by SBLI Fund, New York, NY 10001.

a. Kenneth has $2000 which he wants to invest in a mutual fund. After consulting with the bank's financial adviser, he decides to purchase a fund currently selling at $12.50 per share. How many shares of the fund can he purchase?

b. At the end of the year, the fund distributes its earnings to the shareholders. If the distribution is $1.00 per share, what percent of the original share price does this represent?

c. After a certain time, Kenneth decides to redeem his shares. At that time, the share value is $13.75 per share. How much will he receive for his shares?

d. How much profit did he make on his investment?

10. Many customers transact their business with the Money Bags Savings Bank by mail instead of coming into the bank in person. They may make deposits in and withdrawals from their savings accounts, deposits in their checking accounts, payments to Christmas or Hanukkah Club accounts, and so on, completely by mail if they wish to do so.

Bankbooks are returned to customers by mail after the completion of each transaction. Joe Miller and the other tellers must determine the correct amount of postage for each envelope before it is mailed. Postage rates for each piece of first-class mail (this is how bankbooks are mailed) are 29¢ for the first ounce and 23¢ for each additional ounce. Postcards may be mailed for 13¢ each. Banks often use postage meter machines instead of stamps for mailing to save time; also, meters are more convenient to use.

a. If 537 bankbooks must be mailed back to bank-by-mail customers on a certain day and each envelope weighs less than 1 ounce, how much postage will be required if these envelopes are stamped by a postage meter?

b. On envelopes heavier than 1 ounce postage stamps are placed by hand. What will be the total value of the postage stamps required to stamp 46 envelopes weighing 2 ounces each and 13 envelopes weighing 3 ounces each?

c. What will the cost of postage be for a mailing of 5000 postcards to the bank's customers to announce a special interest rate on bank loans for a short period of time?

d. Calculate the total cost of postage for the day for all the envelopes and postcards mentioned in questions a–c above.

BANK TELLERS' WORKING HOURS AND EARNINGS

Joe Miller and Pedro Gomez, like the other tellers at the Money Bags Savings Bank, must report to work one-half hour before the bank opens for business, and they must remain one-half hour after the bank closes for the day. This extra time allows the tellers to get ready for the day's work in the morning and permits them to complete their paper work in the afternoon.

The bank is open from 9 A.M. to 3 P.M. each weekday, Monday through Friday, and from 9 A.M. to 1 P.M. on Saturday, and from 5 P.M. to 8 P.M. on Friday.

EXAMPLE:

a. How many hours does Miller work each weekday? (He does not work on Friday evenings.)

b. How many hours does Miller work on Saturday?

c. What is the total number of hours Miller works during the week?

d. If Miller earns $280 per week, what is his average hourly rate of pay?

e. What is Joe Miller's annual salary?

f. If his weekly salary is increased by $15 during her second year of employment, how much will Miller earn during his second year at the Money Bags Savings Bank?

g. What will Miller's new hourly rate of pay be during his second year of employment?

h. If he received 2 weeks' paid vacation the first year and 3 weeks' paid vacation the second year,

   (1) determine the amount of money he was paid while on vacation the first year and then the second year.

   (2) find the total amount of his vacation pay for the 2 years.

SOLUTION:

a. The bank is open 3 hours each morning and 3 hours each afternoon. Since Miller must report and remain $\frac{1}{2}$ hour extra in the morning and in the afternoon, he works $6 + \frac{1}{2} + \frac{1}{2} = 7$ hours each weekday.

b. On Saturday the bank is open 4 hours, so Miller must work $4 + \frac{1}{2} + \frac{1}{2} = 5$ hours.

c. Seven hours each day for 5 days = 35 hours + 5 hours on Saturday = 40 hours.

d.
$$
\begin{array}{r}
7.00 \\
40\overline{)280.00} \\
\underline{280\ xx} \\
00
\end{array}
$$
or $7.00 per hour

e. $280 × 52 =
$$
\begin{array}{r}
\$\ \ 280 \\
\times\ \ 52 \\
\hline
560 \\
14\ 00 \\
\hline
\$14{,}560
\end{array}
$$

f. $280 + $15 = $295 and
$$
\begin{array}{r}
\$\ \ 295 \\
\times\ \ 52 \\
\hline
590 \\
14\ 75 \\
\hline
\$15{,}340
\end{array}
$$

g.
$$\begin{array}{r} 7.375 \\ 40\overline{)295.000} \\ \underline{280}\ \text{xxx} \\ 15\ 0 \\ \underline{12\ 0} \\ 3\ 00 \\ \underline{2\ 80} \\ 200 \end{array}$$
or $7.38 per hour

h. (1) $560 the first year, $885 the second year

(2) $1445 for the two years

PRACTICE EXERCISE 8

1. Sylvia Menendez, a student, is a part-time teller and works 4 hours each day from Monday through Friday, 3 hours on Friday evening, and 2 hours on Saturday morning. What is the total number of hours worked during the week?

2. a. Calculate the hourly rate of pay for Maria Larson, a teller, who works 35 hours each week and earns $225 per week.

   b. What is her annual salary?

3. If Pedro Gomez earns $13,000 per year, determine the amount of his weekly salary.

4. Charles Goldstein, another employee at the Money Bags Savings Bank, earned $12,000 during his first year of employment. He received a raise of $10 a week at the beginning of his second year of employment.

   a. How much was his raise per year?
   b. How much did he earn during his second year of employment?
   c. What were his total earnings for 2 years of work at the bank?

5. An employee earning $15,600 per year receives 4 weeks of paid vacation during the year. How much is she paid while on vacation?

EXAM TIME

1. Mr. Whitney has a savings account with a balance of $647.28, and he makes a deposit of $35 in cash and $83.46 in checks. What will his new balance be?

2. Ms. Quincy withdraws $57.75 from her savings account to pay a bill. If her balance before the withdrawal was $127.50, what will her balance be after the withdrawal is made?

3. Ms. Price prepares a deposit slip showing checks to be deposited in the following amounts: $23.37, $105.25, $62.18. She asks for a return of $50 in cash. How much is she actually depositing in her savings account?

4. Ms. Pacheco has $335.70 on deposit in her savings account at the beginning of September. She made a withdrawal of $75 on September 10, a deposit of $63.25 on September 15, and another withdrawal of $85 on September 30. What was the balance in her account at the end of September?

5. Mr. Schwartz withdrew $45 from his savings account, leaving a balance of $167.50. What was his balance before he made this withdrawal?

6. Change each percentage to a decimal:

   a. $6\frac{1}{2}\%$ d. 8.17%

   b. $7\frac{1}{4}\%$ e. 12.5%

   c. 8%

7. Ms. Sherry borrowed $500 for 6 months at 9% simple interest to buy a television set. She did not make any payments on this loan during the 6-month period of time the loan was outstanding (unpaid). At the end of that time, she repaid the entire loan, with interest.

   a. Calculate the amount of interest due, using the formula for finding simple interest.
   b. Find the total amount of money Ms. Sherry must repay at the end of the 6 months. (*Remember:* Time must be expressed in terms of years.)

8. Mr. Green borrowed $4000 for 3 years at 11.5% simple interest to remodel his home. He repaid the loan in equal monthly installments over the entire 3 years.

   a. Find the amount of his average unpaid balance.
   b. Calculate the approximate amount of interest he paid for his loan.
   c. How much money will Mr. Green have to pay back to the bank?
   d. How much is each monthly payment?

9. Mr. Canerossi borrowed $6000 to buy a new car and was charged 12% simple interest for this loan. How much interest did he pay at the end of the first month, when he made the first of 24 installment payments to repay the loan?

10. Ms. Sweeney has $425 in a savings account that pays an effective annual rate of interest of 2.38%. Calculate the amount of interest earned by her account for:

    a. 3 months (quarterly)   b. 1 year

11. Ms. Gray has a savings certificate for $1500 at an effective annual rate of interest of 3.92%. How much interest will she earn for 1 year?

12. Ms. Swanson has a savings certificate for $5000 for 4 years at an effective annual rate of interest of 4.60%.

    a. Calculate her approximate interest earnings for the entire 4 years.
    b. Approximately how much money will she get back at the end of that time?

13. Mr. Fermi has a checking account at the bank. If his opening balance at the beginning of March was $273.50, and he made deposits totaling $182.35 and wrote checks totaling $265.45 during March, what was his closing balance at the end of the month?

14. Ms. Moy made deposits in her checking account during the month of $50, $350, and $85. She also wrote checks against her account in the following amounts: $235.14, $94.96, $37.42, and $125.08. In addition, she transferred $25 from her checking account to her savings account automatically.

    a. Was the closing balance in her checking account at the end of the month greater or less than the opening balance?
    b. How much greater or less?

15. Mr. Snyder asks Joe Miller, the teller, to cash his paycheck for $278.40. If he wants no bills of a denomination higher than $20, how can he give him his money using the fewest bills and coins?

16. Mr. Finch is a merchant who comes into the bank regularly for change to use in his business. A roll of quarters contains 40 quarters, a roll of dimes contains 50 dimes, a roll of nickels contains 40 nickels, and a roll of pennies contains 50 pennies. How much must Mr. Finch give Pedro Gomez, the teller, in bills if he wishes to purchase 6 rolls of quarters, 5 rolls of dimes, 10 rolls of nickels, and 4 rolls of pennies?

17. Ms. Harris has two Christmas Club accounts at the bank. She pays $5 per week into one account and $2 per week into the other.

    a. How much will she have in each account when she has paid all of the 50 payments needed to complete each club?
    b. How much money will she get back, not counting interest, to help with her Christmas shopping at the end of the year?

18. Ms. Webb wishes to purchase three money orders, in amounts of $65.75, $137.50, and $8.30, to pay her bills. If the bank charges 35¢ for each money order, how much money should the teller collect from Mrs. Webb?

19. Mr. Collins wants to buy $650 worth of travelers' checks for a vacation trip. If the charge for issuing travelers' checks is 1% of their value, what is the total amount of money the teller must collect from Mr. Collins?

20. Mr. Larsen, who is 45 years old, purchases a $5000 straight life insurance policy. His annual premium for this policy is $138.30.

    a. How much is he paying for each $1000 worth of insurance protection?
    b. If a $10,000 policy has an annual premium of $271.60, how much would he have saved on each $1000 worth of protection by buying a $10,000 policy instead of a $5000 policy?

21. United States postage rates for first-class mail are 29¢ for the first ounce and 23¢ for each additional ounce. Calculate the total amount of postage used by a bank-by-mail teller to mail back to customers 365 bankbooks weighing 1 ounce or less, 8 letters weighing 2 ounces each, 4 letters weighing 3 ounces each, and 1 letter weighing 4 ounces.

22. Linda Lee works 6 hours each day from Monday through Friday and also 5 hours on Saturday. What is the total number of hours she works during the week?

23. Anne Olsen, a teller-trainee, is paid $4.25 per hour and works a 35-hour week.

   a. Calculate her weekly earnings.
   b. Calculate her annual earnings.

24. Kenneth Hayes works at the Money Bags Savings Bank and earns an annual salary of $14,560.

   a. Calculate the amount of his weekly salary.
   b. If he received 2 weeks of paid vacation during his first year of employment, how much was he paid while on vacation?

Now Check Your Answers

Now that you have completed the exam, check your answers against the correct ones, which follow the answers to the practice exercises below.

ANSWERS TO PRACTICE EXERCISES AND EXAM

PRACTICE EXERCISE 1

1. 3 hr.   2. 9 hr.   3. 4 hr.   4. 37 hr.   5. 1924 hr.

PRACTICE EXERCISE 2

1. $406.09   2. a. $61.75   b. $657.00   3. $2929.17
4. b. $206.23   c. $1079.35   5. $5071.61
6. a. Ms. Quinn   b. Mr. Chan   c. $2523.08

PRACTICE EXERCISE 3

1. $934.75   2. a. $593.65   b. $351.65   3. $232.85
4. $785.63   5. a. $437.50   b. $812.27   c. $3112.39

PRACTICE EXERCISE 4

1. $154   2. a. $3500   b. $997.50   c. $7997.50   d. 36   e. $222.15
3. a. $4000   b. $2560   c. $10,560   d. $220   4. $114.75

PRACTICE EXERCISE 5

1. a. $3.87   b. $653.87   2. $26.78
3. a. $54.25   b. $554.25   4. $460.00
5. a. 1.5%   b. 1.625%   c. 1.8125%   d. 1.9375%   e. 1.7025%   f. 1.975%

PRACTICE EXERCISE 6

1. a. $475   b. $1018.65   c. $420.09   2. $62.65
3. a. Yes   b. $10   4. $85.00   5. $13.56

PRACTICE EXERCISE 7

1. Eleven $20, one $10, one $5, two $1, one 50¢, one 10¢, one 5¢, four 1¢
2. a. 40   b. 50   c. 40   d. 50   e. 180
3. a. Fifteen and $\dfrac{xx}{100}$ b. $200   c. $250   4. $449.73
5. a. $7.50   b. $757.50
6. a. $89.00   b. $129.00   c. $10.75   d. $158.00   e. $29.00
7. a. $12.00   b. $67.50
8. a. $761.50   b. $234.00, $129.50   c. $314.00
9. a. 160 shares   b. 8%   c. $2200   d. $200.00
10. a. $155.73   b. $33.67   c. $950   d. $1139.40

PRACTICE EXERCISE 8

1. 25   2. a. $6.42   b. $11,700   3. $250
4. a. $520   b. $12,520   c. $24,520
5. $1200

EXAM TIME

1. $765.74   2. $69.75   3. $140.80   4. $238.95   5. $212.50

6. a. 0.065   b. 0.0725   c. 0.08   d. 0.0817   e. 0.125

7. a. $22.50   b. $522.50

8. a. $2000   b. $1380   c. $5380   d. $149.44

9. $60   10. a. $2.53   b. $10.12   11. $58.60

12. a. $920   b. $5920   13. $190.40   14. a. Less   b. $32.60

15. Thirteen $20, one $10, one $5, three $1, one 25¢, one 10¢, one 5¢

16. $107   17. a. $250 for $5 club, $100 for $2 club   b. $350

18. $212.60   19. $656.50   20. a. $27.66   b. $0.50

21. $113.99   22. 35   23. a. $148.75   b. $7735

24. a. $280   b. $560

How Well Did You Do?

0–25   **Poor.** Reread the unit, redo all the practice exercises, and retake the exam.

26–29   **Fair.** Reread the sections dealing with the problems you got wrong. Redo those practice exercises, and retake the exam.

30–33   **Good.** Review the problems you got wrong. Redo them correctly.

34–39   **Very good!** Continue on to the model exams. You're on your way to success!

Model Minimum Competency Examination A in Mathematics

PART A

Answer all 20 questions in this part. Write your answers on a separate sheet of paper, numbering them from 1 to 20.

1. Add: 4337
           797
          <u>2848</u>

2. From 4329 subtract 875.

3. Multiply: 562
             $\times 47$

4. Divide: $46\overline{)6164}$

5. Multiply: $\frac{1}{2} \times \frac{1}{3}$

6. Write $\frac{15}{18}$ as a fraction in its lowest terms.

7. What is $\frac{2}{5}$ of 15?

8. Add: $4.87 + $18.94 + $0.39 + $3.18

9. Subtract: 17.2
               4.8

10. Multiply: 3.73
                × 8.6

11. Divide: $34\overline{)428.4}$

12. Solve for y: $5y - 1 = 14$

13. Add: -6 and $+7$

14. Write the numeral for thirty-two thousand two hundred seventy-eight.

15. What is the average (mean) of 89, 78, and 91?

16. Without looking, Jim picked one card from part of a deck of playing cards containing only the jack, queen, king and ace of hearts. What is the probability that he picked the jack of hearts?

17. Solve for x: $\dfrac{x}{16} = \dfrac{3}{4}$

18. What is the ratio of BC to AC in the triangle below?

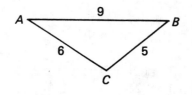

19. The circumference of a circle is found by using the formula $C = 2\pi r$. What is the number of millimeters in the circumference of a circle whose radius is 21 millimeters? $\left(\text{Use } \pi = \dfrac{22}{7}.\right)$

20. If the rate of sales tax is 8%, what is the amount of tax on a $35 purchase?

Go right on to Part B.

PART B

Answer all 40 questions in this part. Select your answers from the choices that follow each question.

21. A set of numbers is arranged as follows: 4, 4, 6, 8, 13. What is the median?

    a. 4   b. 6   c. 7   d. 8

22. In a game using dice, one die is thrown. What is the probability of getting a 2 on one throw of the die?

    a. $\dfrac{1}{6}$ b. $\dfrac{1}{3}$ c. $\dfrac{1}{2}$ d. 1

23. Which open sentence is represented by the number line below?

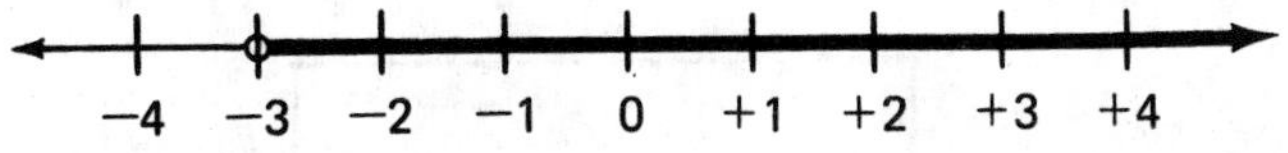

a. $x > -3$ b. $x = -3$ c. $x < -3$ d. $x = 0$

24. If two angles of a triangle measure 70° and 60°, respectively, how many degrees are there in the third angle?

a. 70° b. 60° c. 50° d. 130°

25. On the graph below, which point has the coordinates $(-4, +2)$?

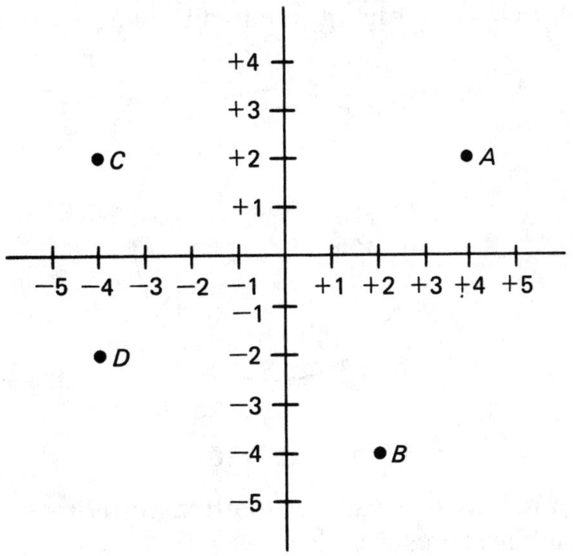

a. A b. B c. C d. D

26. Using the Pythagorean theorem, $c^2 = a^2 + b^2$, find the length of side c in the right triangle below.

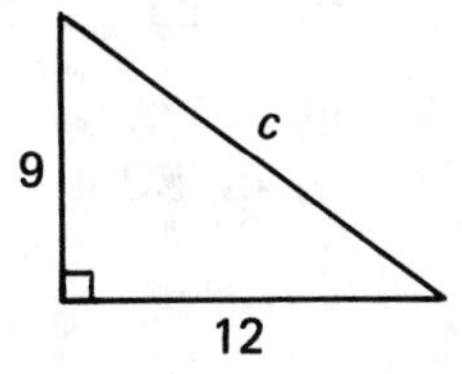

a. 21 b. 3 c. 15 d. $10\frac{1}{2}$

27. One side of a square measures 3 meters. What is the area of the square?

a. 12 m^2 b. 6 m^2 c. 9 m^2 d. 15 m^2

28. What is the perimeter of rectangle *ABCD*?

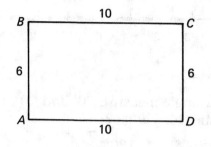

    a. 16   **b.** 32   **c.** 60   **d.** 120

29. The circle graph below shows the number of pupils enrolled by class in Northhampden High School. Which grade has the largest number of pupils?

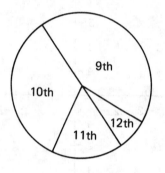

    a. 9th   **b.** 10th   **c.** 11th   **d.** 12th

30. In the diagram below, one tube of toothpaste represents 1000 tubes. What is the total number represented?

    a. 450   **b.** 4000   **c.** 4500   **d.** 5000

31. The graph below shows the ratio of polyunsaturated to saturated fat (*P/S* value) in a product. Which product has a *P/S* ratio between 5 and 6?

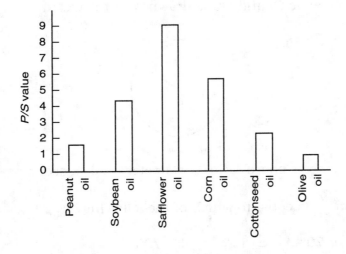

    a. Soybean oil   **b.** Safflower oil   **c.** Corn oil   **d.** Olive oil

32. The graph below shows the sales of automobiles for the Nu-Car Sales Agency during the years 1985–1994. How many more cars were sold in 1991 than in 1988?

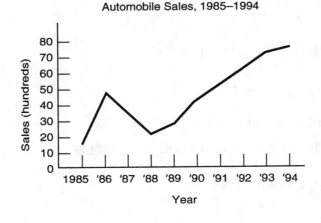

    a. 2000   **b.** 2500   **c.** 3000   **d.** 3500

33. What is the best approximation of the measure of the angle *ABC*?

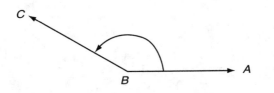

    a. 30°   **b.** 75°   **c.** 90°   **d.** 150°

34. When written as a percent, the decimal 0.9 is equal to which of the following?

    a. 0.9%   **b.** 9%   **c.** 90%   **d.** 900%

35. Point C is on circle O, and $\overline{OC}$ is drawn. What is $\overline{OC}$?

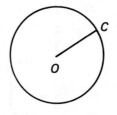

    a. a radius   **b.** a chord   **c.** a diameter   **d.** a sector

36. The fraction $\dfrac{7}{10}$ is equal to which of the following?

    a. 17%   **b.** 70%   **c.** 14%   **d.** 170%

37. What is 75% of 48?

    a. 64   **b.** 36   **c.** 12   **d.** 16

38. On a map, if $\dfrac{1}{8}$ in. represents 1 mile, how many miles does 6 in. represent?

    a. $\dfrac{3}{4}$ **b.** 48   **c.** 54   **d.** 42

39. Which fraction has the greatest value?

    a. $\dfrac{3}{4}$ **b.** $\dfrac{3}{5}$ **c.** $\dfrac{3}{7}$ **d.** $\dfrac{3}{8}$

40. The best approximation of 99×399 is which of the following?

    a. 400   **b.** 4000   **c.** 40,000   **d.** 60,000

41. Which fraction is equal to $3\dfrac{1}{2}$?

    a. $\dfrac{2}{3}$ **b.** $\dfrac{7}{3}$ **c.** $\dfrac{7}{2}$ **d.** $\dfrac{3}{7}$

42. Subtract: $\dfrac{1}{4}$
$\dfrac{1}{8}$

    a. $\dfrac{1}{4}$ **b.** $\dfrac{1}{6}$ **c.** $\dfrac{3}{8}$ **d.** $\dfrac{1}{8}$

43. Add: $\dfrac{2}{3} + \dfrac{1}{6}$

    a. $\dfrac{1}{3}$ b. $\dfrac{1}{6}$ c. $\dfrac{5}{6}$ d. $\dfrac{5}{12}$

44. Expressed as a fraction in lowest terms, $\dfrac{28}{100}$ is equal to which of the following?

    a. $\dfrac{4}{25}$ b. $\dfrac{14}{50}$ c. $\dfrac{14}{25}$ d. $\dfrac{7}{25}$

45. Which decimal has the greatest value?

    a. 0.76   b. 0.75   c. 0.748   d. 0.709

46. James earns $3.50 per hour at his part-time job. If he worked $5\dfrac{1}{2}$ hours one day, how much did he earn?

    a. $17.50   b. $19.25   c. $19.50   d. $18.75

47. Marion left a 15% tip for a restaurant check of $26. What was the amount of the tip?

    a. $390   b. $0.39   c. $0.039   d. $3.90

48. If bolts cost $0.015 each, what is the cost of 100 bolts?

    a. $15   b. $0.00015   c. $1.50   d. $150

49. Betty received a $10 bill in payment for a food bill of $6.37. How much change should Betty return?

    a. $4.37   b. $3.73   c. $3.63   d. $4.73

50. Ms. Fitzgerald has $394.05 in her checking account. If she writes a check for $56, how much is left in her account?

    a. $328.05   b. $450.05   c. $394.61   d. $338.05

51. Mary worked as a baby-sitter from 7:15 P.M. to 1:00 A.M. How many hours did she work?

    a. $6\dfrac{3}{4}$ b. $6\dfrac{1}{4}$ c. $5\dfrac{1}{2}$ d. $5\dfrac{3}{4}$

52. Mr. King bought a used car with a $4000 down payment and 12 monthly payments of $190 each. What was the total cost of the automobile?

    a. $4190   b. $2280   c. $6280   d. $4800

53. Brass drawer knobs cost $2.69 each. If the total cost was $16.14, how many knobs were purchased?

    a. 5   b. 6   c. 7   d. 8

54. The width of a rectangle is 14', and its length is 18'. What is the ratio of the width to the length?

   a. $\dfrac{7}{9}$　　b. $\dfrac{9}{7}$　　c. $\dfrac{7}{16}$　　d. $\dfrac{9}{16}$

55. The first-class rates for mailing a letter at the post office are $0.20 for the first ounce and $0.17 for each additional ounce. What will be the cost of mailing a letter weighing 4 ounces?

   a. $0.37　　b. $0.54　　c. $0.57　　d. $0.71

56. Which of the following is a prime number?

   a. 18　　b. 19　　c. 20　　d. 21

57. The population of Sherman is 2475. What is the population to the nearest thousand?

   a. 2000　　b. 2400　　c. 2480　　d. 3000

58. What is the value of 3×7^2?

   a. 42　　b. 147　　c. 441　　d. 1369

59. What is the least common denominator of $\dfrac{3}{5}$, $\dfrac{1}{2}$, and $\dfrac{3}{4}$?

   a. 10　　b. 8　　c. 20　　d. 40

60. If 7 seats out of 35 are broken in the school auditorium, what percent are broken?

   a. 20%　　b. 25%　　c. $14\dfrac{2}{7}$%　　d. 80%

ANSWERS AND EXPLANATIONS FOR MODEL EXAMINATION A

PART A

1.
```
  4337
   797
+ 2848
  7982
```

2.
```
  4329
 - 875
  3454
```

3.
```
    562
   × 47
   3934
  2248
  26414
```

4.
```
        134
  46)6164
        46xx
        156
        138
        184
        184
```

5. $\dfrac{1}{2} \times \dfrac{1}{3} = \dfrac{1}{6}$

6. $\dfrac{\overset{5}{\cancel{15}}}{\underset{6}{\cancel{18}}} \div \dfrac{\cancel{3}}{\cancel{3}} = \dfrac{5}{6}$

7. $\dfrac{2}{\underset{1}{\cancel{5}}} \times \overset{3}{\cancel{15}} = \dfrac{6}{1} = 6$

8. $\begin{array}{r} \$\ 4.87 \\ 18.94 \\ 0.39 \\ \underline{3.18} \\ \$27.38 \end{array}$

9. $\begin{array}{r} 17.2 \\ -4.8 \\ \hline 12.4 \end{array}$

10. $\begin{array}{r} 3.73 \\ \times\ 8.6 \\ \hline 2\ 238 \\ \underline{29\ 84\ \ } \\ 32.078 \end{array}$

11. $\begin{array}{r} 12.6 \\ 34\overline{)428.4} \\ \underline{34x\ x} \\ 88 \\ \underline{68} \\ 20\ 4 \\ \underline{20\ 4} \end{array}$

12. $\begin{array}{r} 5y - 1 = 14 \\ \underline{+ 1\quad +1} \\ \dfrac{5y}{5} = \dfrac{15}{5} \\ y = 3 \end{array}$

13. $(-6) + (+7) = +1$

14. $32{,}278$

15. $\dfrac{89 + 78 + 91}{3} = \dfrac{258}{3} = 86$

16. $\dfrac{1}{4}$

17. $\dfrac{x}{16} = \dfrac{3}{4}$

$\dfrac{4x}{4} = \dfrac{48}{4}$

$x = 12$

18. $\dfrac{\overline{BC}}{\overline{AC}} = \dfrac{5}{6}$

19. $C = 2\pi r$

$C = 2 \times \dfrac{22}{\underset{1}{7}} \times \overset{3}{\cancel{21}}$

$C = 132 \text{ mm}$

20. $8\% \text{ of } \$35 = \begin{array}{r} \$35 \\ \underline{\times 0.08} \\ \$2.80 \end{array}$

PART B

21. (b) Median is 6.
4, 4, ⑥, 8, 13

22. (a) $\dfrac{1}{6}$

23. (a) $x > -3$

24. (c) $\begin{array}{r} 70° \\ \underline{+60°} \\ 130° \end{array}$ $\quad \begin{array}{r} 180° \\ \underline{-130°} \\ 50° \end{array}$

25. (c) C

26. (c) $c^2 = 9^2 + 12^2$
$c^2 = 81 + 144$
$c^2 = 225$
$c\ = 15$

27. (c) $A = s^2$
$A = 3^2$
$A = 3 \times 3$
$A = 9$

28. (b) $P = 10 + 6 + 10 + 6$
$P = 32$

29. (a) 9th

30. (c) $4\frac{1}{2} \times 1000 = \frac{9}{\overset{2}{\underset{1}{\cancel{2}}}} \times \overset{500}{\cancel{1000}} = 4500$ 31. (c) Corn oil

32. (c) 1991 = 5000
         1988 = $\underline{-2000}$
                  3000

33. (d) $\angle ABC = 150°$ 34. (c) $0.9 = 90\%$ 35. (a) Radius

36. (b) $\frac{7}{10} = \frac{70}{100} = 70\%$ 37. (b) 75% of 48 = 0.75 × 48 = 36

38. (b)
$$\frac{\frac{1''}{8}}{1\text{ mile}} = \frac{6''}{x\text{ miles}}$$
$$\frac{1}{8}x = 6$$
$$(8)\frac{1}{8}x = (8)6$$
$$x = 48$$

39. (a) $\frac{3}{4}$ 40. (c) 99 × 399
                                   100 × 400 = 40,000   41. (c) $3\frac{1}{2} = \frac{3 \times 2 + 1}{2}$
$$= \frac{7}{2}$$

42. (d) $\frac{1}{4} = \frac{2}{8}$ 43. (c) $\frac{2}{3} = \frac{4}{6}$ 44. (d) $\frac{28 \div 4}{100 \div 4} = \frac{7}{25}$
         $-\frac{1}{8} = \frac{1}{8}$ $+\frac{1}{6} = \frac{1}{6}$
         $\quad\;\; = \frac{1}{8}$ $\quad\;\; = \frac{5}{6}$

45. (a) 0.76 = ⬭0.760⬭   46. (b) $\$3.50 \times 5\frac{1}{2}$
         0.75 = 0.750
         0.748 = 0.748 $\overset{1.75}{\cancel{\$3.50}} \times \frac{11}{\underset{1}{\cancel{2}}} = \$19.25$
         0.709 = 0.709

47. (d) 15% of $26 = 0.15 × $26 = $3.90

48. (c) 100 × $0.015 = 100 × 0.015 = $1.50

49. (c) $10.00   50. (d) $394.50   51. (d) 1:00 A.M. = 12:60
         $\underline{-\;\;6.37}$ $\underline{-\;\;56.00}$ $\underline{-\;\;7:15}$
         $ 3.63 $338.05 $5:45 = 5\frac{3}{4}$

52. **(c)**
$$\begin{array}{r} \$190 \\ \times\ 12 \\ \hline 380 \\ 190\ \\ \hline \$2280 \\ +\ 4000 \\ \hline \$6280 \end{array}$$

53. **(b)**
$$\$2.69 \overline{)\$16.14} \quad \begin{array}{c} 6 \\ \end{array}$$
$$16\ 14$$

54. **(a)** $\dfrac{\text{Width}}{\text{Length}} = \dfrac{14}{18} \div \dfrac{2}{2} = \dfrac{7}{9}$

55. **(d)** $\$0.20$ 1st ounce
$$\begin{array}{r} \$0.17 \times 3 = 0.51 \\ +\ 0.20 \\ \hline \$\ \ 0.71 \end{array}$$

56. **(b)** 19

57. **(a)** 2475 rounded off to the nearest thousand = 2000

58. **(b)** 3×7^2
$3 \times 7 \times 7 = 147$

59. **(c)** $5\overline{)20}$ LCD = 20
$2\overline{)20}$
$4\overline{)20}$

60. **(a)** $\dfrac{7}{35} = \dfrac{1}{5} = 5\overline{)1.00}\ \overset{.20}{} = 20\%$

Practice Exercises for Examination B

Practice the set(s) of exercises corresponding to the problems you missed in Model Examination A.

1. Add:

    a.
    $$2837 \\ 6349 \\ 3407$$

    b.
    $$2493 \\ 179 \\ 3875$$

    c.
    $$5117 \\ 684 \\ 7289$$

    d.
    $$435 \\ 2874 \\ 59$$

2. a. From 9876 subtract 3051. c. From 6271 subtract 507.
   b. From 3210 subtract 762. d. From 8963 subtract 7008.

3. Multiply:

    a.
    $$317 \\ \times\ 58$$

    b.
    $$693 \\ \times\ 27$$

    c.
    $$509 \\ \times\ 36$$

    d.
    $$825 \\ \times\ 14$$

4. Divide:

    a. $57\overline{)1824}$ b. $23\overline{)851}$ c. $63\overline{)2835}$ d. $46\overline{)17,526}$

5. Multiply:

    a. $\dfrac{1}{2} \times \dfrac{1}{4}$ b. $\dfrac{2}{3} \times \dfrac{1}{2}$ c. $\dfrac{3}{5} \times \dfrac{1}{3}$ d. $\dfrac{1}{8} \times \dfrac{1}{4}$

6. Write each fraction below in its lowest terms:

    a. $\dfrac{6}{9}$ b. $\dfrac{10}{15}$ c. $\dfrac{12}{18}$ d. $\dfrac{14}{21}$

7. What is:

    a. $\dfrac{2}{3}$ of 6?   b. $\dfrac{3}{5}$ of 15?   c. $\dfrac{5}{9}$ of 18?   d. $\dfrac{3}{7}$ of 21?

8. Add:

    a. 3.29 + 27.16 + 7.63 + 0.28   c. 0.79 + 8.75 + 13
    b. 38.79 + 69.95 + 46.17   d. 7.65 + 0.05 + 48.99 + 16.25

9. Subtract:

   a. 26.94
      18.72

   b. 93.07
      6.92

   c. 6.83
      0.9

   d. 39.1
      5.83

10. Multiply:

   a. 5.74
      × 6.3

   b. 32.8
      × 0.76

   c. 0.207
      × 0.42

   d. 9.05
      × 3.1

11. Divide:

   a. $51\overline{)91.8}$ **b.** $32\overline{)140.8}$ **c.** $27\overline{)22.41}$ **d.** $62\overline{)2.418}$

12. Solve each equation for x:

   a. $3x + 1 = 16$ **b.** $4x - 7 = 17$ **c.** $2x + 5 = 9$ **d.** $7x - 3 = 18$

13. Add:

   a. -9 and $+3$ **b.** $+6$ and $+2$ **c.** $+4$ and -7 **d.** -1 and $+7$

14. Write the numerals for:

   a. Six hundred eighty-seven
   b. Two thousand thirty-eight
   c. Seven thousand four hundred twenty-four
   d. Twelve thousand four hundred

15. For each of the following sets of values, find the average (mean):

   a. 68 and 84
   b. 45, 57, and 63
   c. 94, 73, 61, and 64
   d. 76, 76, 63, 82, and 83

16. Find the probability of drawing each of the following:

   a. the king of spades from the four kings in a deck
   b. the ace of clubs from the ace of hearts, the ace of clubs, and the ace of diamonds
   c. a red marble from a bag containing a red, a blue, and a black marble
   d. any heart from the complete deck of cards (52 cards)

17. Solve each proportion:

   a. $\dfrac{x}{15} = \dfrac{3}{5}$ **b.** $\dfrac{14}{x} = \dfrac{2}{3}$ **c.** $\dfrac{1}{2} = \dfrac{5}{x}$ **d.** $\dfrac{5}{6} = \dfrac{x}{6}$

18. Using triangle *ABC*, write the ratio of:

a. $\dfrac{AC}{BC}$

c. $\dfrac{AC}{AB}$

b. $\dfrac{BC}{AB}$

d. $\dfrac{AB}{BC}$

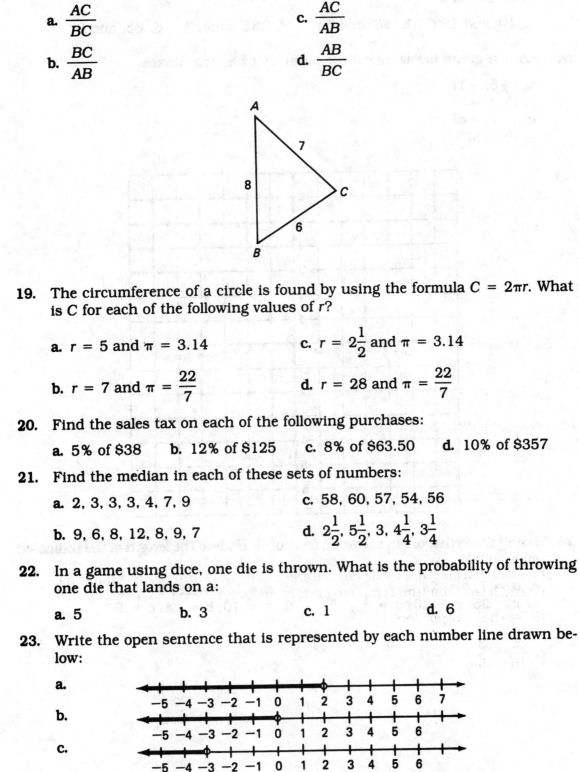

19. The circumference of a circle is found by using the formula $C = 2\pi r$. What is *C* for each of the following values of *r*?

a. $r = 5$ and $\pi = 3.14$

c. $r = 2\dfrac{1}{2}$ and $\pi = 3.14$

b. $r = 7$ and $\pi = \dfrac{22}{7}$

d. $r = 28$ and $\pi = \dfrac{22}{7}$

20. Find the sales tax on each of the following purchases:

a. 5% of $38   b. 12% of $125   c. 8% of $63.50   d. 10% of $357

21. Find the median in each of these sets of numbers:

a. 2, 3, 3, 3, 4, 7, 9

c. 58, 60, 57, 54, 56

b. 9, 6, 8, 12, 8, 9, 7

d. $2\dfrac{1}{2}$, $5\dfrac{1}{2}$, 3, $4\dfrac{1}{4}$, $3\dfrac{1}{4}$

22. In a game using dice, one die is thrown. What is the probability of throwing one die that lands on a:

a. 5   b. 3   c. 1   d. 6

23. Write the open sentence that is represented by each number line drawn below:

a.

b.

c.

d.

24. Find the measure of the third angle of a triangle if the measurements of the other two are:

    a. 10 and 130   **b.** 46 and 54   **c.** 135 and 15   **d.** 63 and 27

25. On the graph below, name the point that has coordinates of:

    a. $(+5, -1)$
    b. $(-2, -3)$
    c. $(-1, +4)$
    d. $(0, -3)$

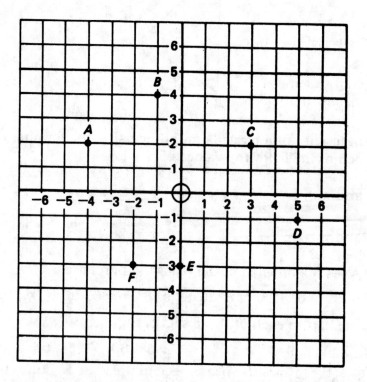

26. Using the Pythagorean theorem, $c^2 = a^2 + b^2$, find the length of the indicated side:

    a. $a = 7, b = 24, c = ?$ **c.** $a = 5, c = 13, b = ?$
    b. $c = 25, a = 20, b = ?$ **d.** $a = 10, b = 24, c = ?$

27. Find the area of each figure:

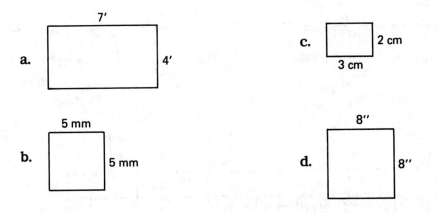

a. 7' 4'

c. 2 cm / 3 cm

b. 5 mm / 5 mm

d. 8'' / 8''

28. A room is in the shape of a rectangle. Find the number of feet in its perimeter if its dimensions are:

a. $9' \times 12'$
b. $15 \text{ m} \times 21 \text{ m}$

c. $12' \times 18'$
d. $16' \times 24'$

29. The circle graph below shows how a typical person spends his or her day.

33% Work

25% Sleep

10% Meals

16% Misc. chores

16% Recreation

a. On what activity does the person spend the least amount of time?
b. On what activity does he or she spend the most amount of time?
c. On what activity does he or she spend $\frac{1}{4}$ of his or her time?
d. On what activities does he or she spend equal amounts of time?

30. In each graph below, find the total number represented, using each legend stated:

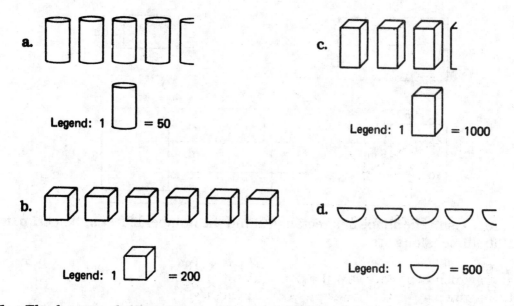

a.

Legend: 1 [] = 50

c.

Legend: 1 [] = 1000

b.

Legend: 1 [] = 200

d.

Legend: 1 ⌒ = 500

31. The bar graph below shows the average daily temperatures in Metropolis in degrees Celsius during 1 week last summer.

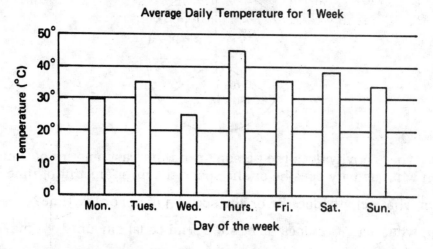

Average Daily Temperature for 1 Week

a. Which was the coolest day of the week?
b. Which was the warmest day of the week?
c. Which two days of the week had the same average temperature?
d. On which day of the week was the temperature 30°C?

32. The line graph below shows the change in population in Smallsville between 1988 and 1994.

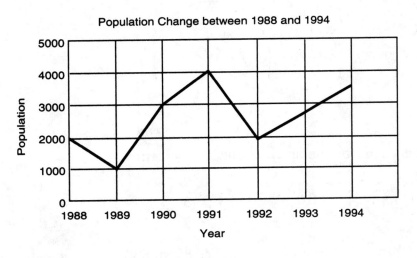

Population Change between 1988 and 1994

a. In which two years was the population the same?
b. In which year was the population the highest?
c. In what year was the population 3000?
d. In which year was the population the lowest?

33. What is the best approximation of the measurement of each angle drawn below? Also, identify each angle as acute, right, straight, obtuse, or reflex.

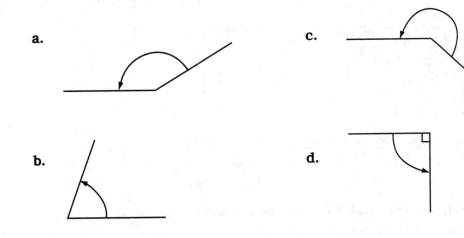

a.

c.

b.

d.

34. Write each decimal as an equivalent percent:

a. 0.27   b. 0.027   c. 2.7   d. 0.46

35. Identify and name the labeled part.

a. $\overline{BC}$
b. $\overline{BD}$
c. $\overline{OA}$
d. $\overset{\frown}{DC}$

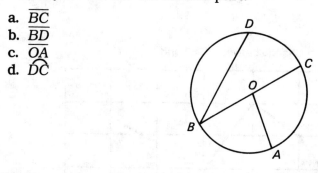

36. Express each fraction as an equivalent percent:

a. $\dfrac{23}{100}$
b. $\dfrac{7}{10}$
c. $\dfrac{2}{5}$
d. $\dfrac{1}{4}$

37. Find:

a. 5% of 45
b. 7% of 238
c. 8% of 635
d. $5\dfrac{1}{2}$% of 56

38. On a map, $\dfrac{1}{4}$ inch represents 1 mile. Find the number of miles represented by each of the following distances on the map.

a. 2 in.
b. $3\dfrac{1}{2}$ in.
c. 4 in.
d. $\dfrac{3}{4}$ in.

39. Which number in each set has the greatest value?

a. $\dfrac{1}{2}, \dfrac{1}{3}, \dfrac{1}{4}, \dfrac{1}{5}$
b. $\dfrac{2}{3}, \dfrac{3}{5}, \dfrac{7}{10}, \dfrac{5}{8}$
c. $\dfrac{1}{3}, \dfrac{3}{8}, \dfrac{3}{10}, \dfrac{2}{5}$
d. $\dfrac{7}{8}, \dfrac{4}{5}, \dfrac{5}{6}, \dfrac{9}{10}$

40. Write the best approximation for these products:

a. 19 × 49
b. 98 × 199
c. 299 × 599
d. 69 × 91

41. Write the improper fraction that has the same value as:

a. $2\dfrac{1}{2}$
b. $7\dfrac{2}{3}$
c. $4\dfrac{5}{8}$
d. $6\dfrac{3}{5}$

42. Subtract:

a. $\dfrac{1}{2}$
   $-\dfrac{1}{4}$

b. $\dfrac{3}{4}$
   $-\dfrac{5}{8}$

c. $\dfrac{2}{3}$
   $-\dfrac{1}{6}$

d. $\dfrac{3}{5}$
   $-\dfrac{1}{3}$

43. Add:

a.
$$\frac{1}{4}$$
$$+\frac{1}{2}$$

b.
$$\frac{1}{5}$$
$$+\frac{2}{3}$$

c.
$$\frac{2}{7}$$
$$+\frac{3}{14}$$

d.
$$\frac{3}{4}$$
$$+\frac{3}{8}$$

44. Express each fraction in its *lowest terms*:

a. $\frac{18}{21}$ b. $\frac{20}{45}$ c. $\frac{12}{18}$ d. $\frac{18}{32}$

45. Which decimal in each set has the greatest value?

a. 0.6, 0.61, 0.59, 0.603
b. 3.05, 3.005, 3.1, 3.01
c. 1.2, 1.19, 1.164, 1.104
d. 4.6, 4.57, 4.62, 4.499

46. Lucy works as a part-time clerk and earns $3.25 per hour. Find her wages if she works the following numbers of hours:

a. 2   b. $5\frac{1}{2}$ c. $8\frac{1}{2}$ d. 20

47. Calculate the amount of a 15% tip on each of the following restaurant checks:

a. $14   b. $18   c. $6.50   d. $22.78

48. Multiply:

a. 100 × 0.028
b. 3.5 × 100
c. 0.73 × 10
d. 10 × 0.6

49. A $20 bill was offered in payment of each of the following restaurant checks. Calculate the amount of change returned in each case:

a. $12.50   b. $16.35   c. $8.47   d. $10.75

50. Calculate the closing balance in each transaction:

a. Opening balance = $382.85; withdrawal = $32.19
b. Opening balance = $23.85; deposit = $45.60
c. Opening balance = $167.39; withdrawal = $26.39
d. Opening balance = $2689.63; deposit = $248.73

51. Find the number of hours Bill worked on each of the following days if he started and finished work at the times shown:

a. 8 A.M. to 4 P.M.
b. 9:30 A.M. to 5 P.M.
c. 10:45 A.M. to 3:45 P.M.
d. 11:15 A.M. to 4:30 P.M.

52. Calculate each of the following totals for these purchases made on the installment plan:

a. Down payment = $100
   12 payments @ $50
b. Down payment = $250
   24 payments @ $25
c. Down payment = $2000
   36 payments @ $56.50
d. Down payment = $1500
   48 payments @ $115

53. Divide:

    a. $0.7\overline{)1.47}$

    b. $0.25\overline{)12.5}$

    c. $1.2\overline{)144}$

    d. $0.03\overline{)0.216}$

54. Express the ratio of each pair indicated:

    a. If $a = 5$ and $b = 10$, find $\dfrac{a}{b}$.

    b. If $\overline{AB} = 12$ and $\overline{BC} = 28$, find $\dfrac{AB}{BC}$

    c. Find the ratio of 16" to 12".

    d. Find the ratio of the length to the width if $l = 36$ and $w = 9$.

55. First-class rates for mailing a package at the post office are $0.29 for the first ounce and $0.23 for each additional ounce. What would be the cost of mailing a package, first class, weighing each of the following?

    a. 2 oz.   b. 3 oz.   c. 4 oz.   d. 5 oz.

56. In each set of numbers, find the prime number:

    a. 13, 14, 15, 16   c. 42, 43, 44, 45

    b. 18, 19, 20, 21   d. 56, 57, 58, 59

57. Round off each of the following to the nearest thousand:

    a. 3672   b. 12,947   c. 10,455   d. 630,729

58. Find each of the following:

    a. 2×3^2 b. 4×10^3 c. $12^2 - 65$ d. 3×6^3

59. What is the least common denominator in each of these sets?

    a. $\dfrac{1}{2}, \dfrac{1}{3}, \dfrac{3}{5}$ b. $\dfrac{2}{3}, \dfrac{3}{5}, \dfrac{7}{10}$ c. $\dfrac{7}{8}, \dfrac{4}{5}, \dfrac{1}{2}$ d. $\dfrac{3}{8}, \dfrac{2}{3}, \dfrac{1}{4}$

60. Find the percent for each of the following ratios:

    a. $\dfrac{1}{2}$ b. $\dfrac{1}{5}$ c. $\dfrac{3}{4}$ d. $\dfrac{1}{3}$

ANSWERS TO PRACTICE EXERCISES FOR EXAMINATION B

1.   a. 12,593   b. 6547   c. 13,090   d. 3368

2.   a. 6825   b. 2448   c. 5764   d. 1955

3.   a. 18,386   b. 18,711   c. 18,324   d. 11,550

4.   a. 32   b. 37   c. 45   d. 381

5.   a. $\frac{1}{8}$ b. $\frac{1}{3}$ c. $\frac{1}{5}$ d. $\frac{1}{32}$

6.   a. $\frac{2}{3}$ b. $\frac{2}{3}$ c. $\frac{2}{3}$ d. $\frac{2}{3}$

7.   a. 4   b. 9   c. 10   d. 9

8.   a. 38.36   b. 154.91   c. 22.54   d. 72.94

9.   a. 8.22   b. 86.15   c. 5.93   d. 33.27

10.   a. 36.162   b. 24.928   c. 0.08694   d. 28.055

11.   a. 1.8   b. 4.4   c. 0.83   d. 0.039

12.   a. 5   b. 6   c. 2   d. 3

13.   a. -6 b. $+8$ c. -3 d. $+6$

14.   a. 687   b. 2038   c. 7424   d. 12,400

15.   a. 76   b. 55   c. 73   d. 76

16.   a. $\frac{1}{4}$ b. $\frac{1}{3}$ c. $\frac{1}{3}$ d. $\frac{1}{4}$

17.   a. 9   b. 21   c. 10   d. 5

18.   a. $\frac{7}{6}$ b. $\frac{3}{4}$ c. $\frac{7}{8}$ d. $\frac{4}{3}$

19.   a. 31.4   b. 44   c. 15.7   d. 176

20.   a. $1.90   b. $15   c. $5.08   d. $35.70

21.   a. 3   b. 8   c. 57   d. $3\frac{1}{4}$

22.   a. $\frac{1}{6}$ b. $\frac{1}{6}$ c. $\frac{1}{6}$ d. $\frac{1}{6}$

23.   a. $x < 2$ b. $x < 0$ c. $x < -3$ d. $x > 1$

24.   a. 40   b. 80   c. 30   d. 90

25.   a. D   b. F   c. B   d. E

26.   a. 25   b. 15   c. 12   d. 26

27.   a. 28 sq. ft.   b. 25 mm^2 c. 6 cm^2 d. 64 sq. in.

28.   a. 42′   b. 72 m   c. 60′   d. 80′

29.   a. Meals   b. Work   c. Sleep   d. Misc. chores and Recreation

30.   a. 225   b. 1200   c. 3250   d. 2250

31.   a. Wed.   b. Thurs.   c. Tues. and Fri.   d. Mon.

32.   a. 1988 and 1992   b. 1991   c. 1990   d. 1989

33.   a. 150°, obtuse   b. 70°, acute   c. 230°, reflex   d. 90°, right

34.   a. 27%   b. 2.7%   c. 270%   d. 46%

35.   a. diameter   b. chord   c. radius   d. arc

36.   a. 23%   b. 70%   c. 40%   d. 25%

37.   a. 2.25   b. 16.66   c. 50.8   d. 3.08

38.   a. 8   b. 14   c. 16   d. 3

39.   a. $\dfrac{1}{2}$ b. $\dfrac{7}{10}$ c. $\dfrac{2}{5}$ d. $\dfrac{9}{10}$

40.   a. 1000   b. 20,000   c. 180,000   d. 6300

41.   a. $\dfrac{5}{2}$ b. $\dfrac{23}{3}$ c. $\dfrac{37}{8}$ d. $\dfrac{33}{5}$

42.   a. $\dfrac{1}{4}$ b. $\dfrac{1}{8}$ c. $\dfrac{1}{2}$ d. $\dfrac{4}{15}$

43.   a. $\dfrac{3}{4}$ b. $\dfrac{13}{15}$ c. $\dfrac{1}{2}$ d. $1\dfrac{1}{8}$

44.   a. $\dfrac{6}{7}$ b. $\dfrac{4}{9}$ c. $\dfrac{2}{3}$ d. $\dfrac{9}{16}$

45.   a. 0.61   b. 3.1   c. 1.2   d. 4.62

46.   a. $6.50   b. $17.88   c. $27.63   d. $65

47.   a. $2.10   b. $2.70   c. $0.98   d. 3.42

48.   a. 2.8   b. 350   c. 7.3   d. 6

49.   a. $7.50   b. $3.65   c. $11.53   d. $9.25

50.   a. $350.66   b. $69.45   c. $141   d. $2938.36

51. a. 8 b. $7\frac{1}{2}$ c. 5 d. $5\frac{1}{4}$

52. a. $700 b. $850 c. $4034 d. $7020

53. a. 2.1 b. 50 c. 120 d. 7.2

54. a. $\frac{1}{2}$ b. $\frac{3}{7}$ c. $\frac{4}{3}$ d. $\frac{4}{1}$

55. a. $0.52 b. $0.75 c. $0.98 d. $1.21

56. a. 13 b. 19 c. 43 d. 59

57. a. 4000 b. 13,000 c. 10,000 d. 631,000

58. a. 18 b. 4000 c. 79 d. 648

59. a. 30 b. 30 c. 40 d. 24

60. a. 50% b. 20% c. 75% d. $33\frac{1}{3}\%$

Model Minimum Competency Examination B in Mathematics

PART A

Answer all 20 questions in this part. Write your answers on a separate sheet of paper, numbering them from 1 to 20.

1. Add: 3739
          4597
          _396_

2. From 6917 subtract 378.

3. Multiply: 389
            $\times$ 48

4. Divide: 54$\overline{)6858}$

5. Multiply: $\dfrac{1}{3} \times \dfrac{1}{4}$

6. Write $\dfrac{12}{18}$ in its lowest terms.

7. What is $\dfrac{2}{3}$ of 21?

8. Add: $3.76 + $12.07 + $0.87 + $1.27

9. Subtract: 15.1
             − 3.7

10. Multiply: 4.65
            × 8.9

11. Divide: $67\overline{)1567.8}$

12. Solve for x: $2x + 1 = 13$

13. Add: $+5$ and -3

14. Write the numeral for seventy-three thousand, three hundred sixty-five.

15. What is the average (mean) of 53, 32, and 53?

16. Without looking, Sara picked one marble from a bag containing only a red, a blue, a black, and a white marble. What is the probability that she picked a white marble?

17. Solve for x: $\dfrac{x}{12} = \dfrac{3}{4}$

18. What is the ratio of AC to AB in the triangle below?

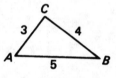

19. The circumference of a circle is found by using the formula $C = \pi d$. What is the number of centimeters in the circumference of a circle whose diameter (d) is 14 cm? $\left(\text{Use } \pi = \dfrac{22}{7}.\right)$

20. If the tax rate is 7%, what is the amount of tax on a \$35 purchase?

Go right on to Part B.

PART B

Answer all 40 questions in this part. Select your answers from the choices that follow each question.

21. A set of numbers is arranged as follows: 10, 10, 11, 12, 12, 13, 14. What is the median?

    a. 10   b. 11   c. 12   d. 13

22. In a game using dice, one die is thrown. What is the probability of getting a 3 on one throw of the die?

    a. $\dfrac{1}{6}$ b. $\dfrac{1}{4}$ c. $\dfrac{1}{3}$ d. $\dfrac{1}{2}$

23. Which open sentence is represented by the number line below?

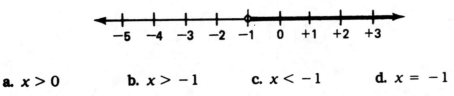

   a. $x > 0$ **b.** $x > -1$ **c.** $x < -1$ **d.** $x = -1$

24. If two angles of a triangle measure 30° and 55°, how many degrees are there in the third angle?

   a. 85°   **b.** 95°   **c.** 55°   **d.** 275°

25. On the graph below, which point has the coordinates $(+2, -3)$?

a. A
b. B
c. C
d. D

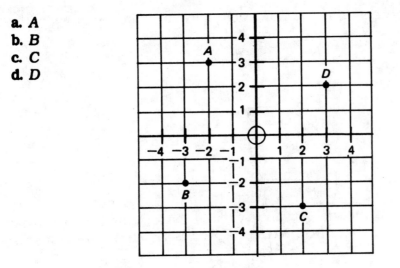

26. Using the Pythagorean theorem, $c^2 = a^2 + b^2$, find the length of side c in the right triangle below.

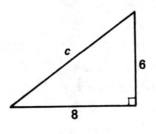

   a. 2   **b.** 10   **c.** 14   **d.** 100

27. One side of a square measures 6 meters. What is the area of the square?

   a. 24 m²   **b.** 36 m²   **c.** 12 m²   **d.** 48 m²

28.   What is the perimeter of parallelogram *ABCD*?

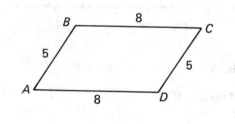

     a. 13   **b.** 26   **c.** 40   **d.** 80

29.   The circle graph below shows how each dollar of income is divided by the L & F Pizza Shop. Which is the largest item?

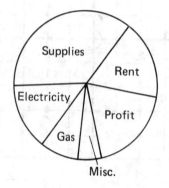

     a. Rent   **b.** Profit   **c.** Gas   **d.** Supplies

30.   In the diagram below, one bottle represents 1000 bottles. What is the total number represented?

     a. 4000   **b.** 4050   **c.** 4500   **d.** 5000

31. The graph below shows the cost per dozen of sanding belts. Which size costs between $12 and $15?

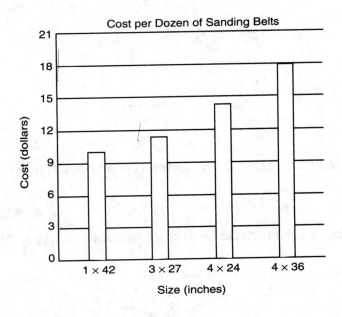

Cost per Dozen of Sanding Belts

a. $1'' \times 42''$ b. $3'' \times 21''$ c. $4'' \times 24''$ d. $4'' \times 36''$

32. The graph below shows the school population for Rose High School during the years 1988–1994. How many fewer pupils were enrolled in 1991 than in 1988?

a. 1500
b. 2000
c. 2500
d. 3000

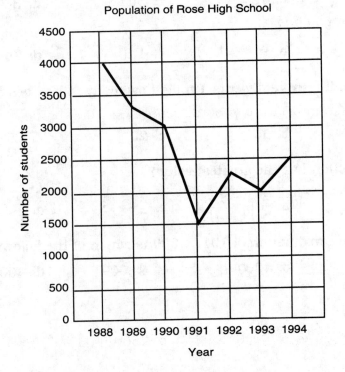

Population of Rose High School

33. What is the best approximation of the measure of angle *ABC*?

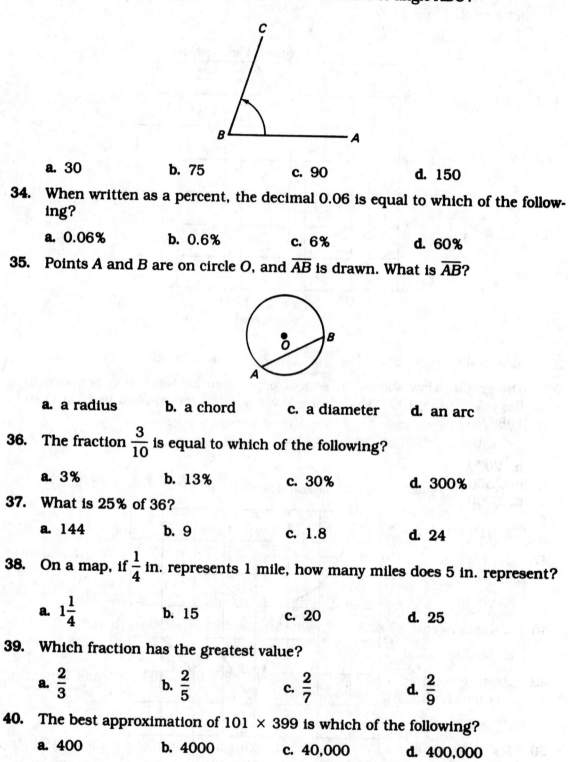

a. 30   b. 75   c. 90   d. 150

34. When written as a percent, the decimal 0.06 is equal to which of the following?

a. 0.06%   b. 0.6%   c. 6%   d. 60%

35. Points *A* and *B* are on circle *O*, and $\overline{AB}$ is drawn. What is $\overline{AB}$?

a. a radius   b. a chord   c. a diameter   d. an arc

36. The fraction $\dfrac{3}{10}$ is equal to which of the following?

a. 3%   b. 13%   c. 30%   d. 300%

37. What is 25% of 36?

a. 144   b. 9   c. 1.8   d. 24

38. On a map, if $\dfrac{1}{4}$ in. represents 1 mile, how many miles does 5 in. represent?

a. $1\dfrac{1}{4}$ b. 15   c. 20   d. 25

39. Which fraction has the greatest value?

a. $\dfrac{2}{3}$ b. $\dfrac{2}{5}$ c. $\dfrac{2}{7}$ d. $\dfrac{2}{9}$

40. The best approximation of 101 × 399 is which of the following?

a. 400   b. 4000   c. 40,000   d. 400,000

41. Which fraction is equal to $2\frac{1}{3}$?

    a. $\frac{7}{2}$ b. $\frac{3}{2}$ c. $\frac{6}{3}$ d. $\frac{7}{3}$

42. Subtract: $\begin{array}{r} \frac{3}{4} \\ -\frac{5}{8} \\ \hline \end{array}$

    a. $\frac{1}{2}$ b. $\frac{2}{3}$ c. $\frac{1}{8}$ d. $\frac{1}{4}$

43. Add: $\frac{3}{5} + \frac{1}{4}$

    a. $\frac{17}{20}$ b. $\frac{4}{9}$ c. $3\frac{2}{5}$ d. $3\frac{1}{4}$

44. Expressed as a fraction in its lowest terms, $\frac{36}{100}$ is equal to which of the following?

    a. $\frac{4}{25}$ b. $\frac{18}{50}$ c. $\frac{3}{25}$ d. $\frac{9}{25}$

45. Which decimal has the greatest value?

    a. 0.64   b. 0.638   c. 0.639   d. 0.6

46. Chuck earns $4.10 per hour at his part-time job. If he worked $4\frac{1}{2}$ hours one day, how much did he earn?

    a. $16.40   b. $16.50   c. $18.45   d. $20

47. Jane left a 15% tip for a restaurant check of $15. What was the amount of the tip?

    a. $0.25   b. $2.25   c. $2.50   d. $1

48. If washers cost $0.023 each, what is the cost of 1000 washers?

    a. $0.23   b. $2.30   c. $23   d. $230

49. Brigitte received a $20 bill in payment for a food bill of $13.65. How much change should Brigitte return?

    a. $6.35   b. $7.35   c. $7.65   d. $6.65

50. Ms. Vasquez has $297.84 in her checking account. If she makes a deposit of $94.89, how much is her new balance?

    a. $392.73   b. $202.95   c. $292.73   d. $212.95

51. Sally worked as a baby-sitter from 8:30 P.M. to 1:00 A.M. How many hours did she work?

    a. $3\dfrac{1}{2}$ b. 4   c. $4\dfrac{1}{2}$ d. 5

52. Carlos bought a new car with a down payment of $1000 and 36 monthly payments of $90 each. What was the total cost of the automobile?

    a. $3240   b. $4240   c. $1090   d. $5240

53. Air conditioner filters cost $5.25 each. If the total cost was $26.25, how many filters were purchased?

    a. 4   b. 5   c. 6   d. 7

54. There are 20 brunettes and 15 blondes in the math class. What is the ratio of blondes to brunettes in the class?

    a. $\dfrac{3}{4}$ b. $\dfrac{4}{3}$ c. $\dfrac{3}{7}$ d. $\dfrac{4}{7}$

55. The first-class rates for mailing a letter at the post office are $0.29 for the first ounce and $0.23 for each additional ounce. What will be the cost of mailing a letter weighing 3 ounces?

    a. $0.69   b. $0.75   c. $0.81   d. $0.87

56. Which of the following is a prime number?

    a. 35   b. 49   c. 53   d. 63

57. The population of New Fairfield is 12,873. What is the population to the nearest thousand?

    a. 12,800   b. 12,900   c. 12,000   d. 13,000

58. What is the value of 2×6^2?

    a. 144   b. 72   c. 24   d. 64

59. What is the least common denominator of $\dfrac{1}{2}$, $\dfrac{2}{3}$, and $\dfrac{3}{4}$?

    a. 6   b. 12   c. 24   d. 36

60. If 6 out of 24 liters of liquid in a car radiator are antifreeze, what percent of the mixture is antifreeze?

    a. 20%   b. 25%   c. 30%   d. $66\dfrac{2}{3}\%$

ANSWERS AND EXPLANATIONS FOR MODEL EXAMINATION B

PART A

1.
$$
\begin{array}{r}
3739 \\
4597 \\
+\ 396 \\
\hline
8732
\end{array}
$$

2.
$$
\begin{array}{r}
6917 \\
-\ 378 \\
\hline
6539
\end{array}
$$

3.
$$
\begin{array}{r}
389 \\
\times\ 48 \\
\hline
3112 \\
1556 \\
\hline
18672
\end{array}
$$

4.
$$
\begin{array}{r}
127 \\
54\overline{)6858} \\
\underline{54}\text{xx} \\
145 \\
\underline{108} \\
378 \\
\underline{378}
\end{array}
$$

5. $\dfrac{1}{3} \times \dfrac{1}{4} = \dfrac{1}{12}$

6. $\dfrac{12 \div 6}{18 \div 6} = \dfrac{2}{3}$

7. $\dfrac{2}{\overset{}{\underset{1}{3}}} \times \overset{7}{2\!\!\!/1} = 14$

8.
$$
\begin{array}{r}
\$\ 3.76 \\
12.07 \\
0.87 \\
+\ 1.27 \\
\hline
\$17.97
\end{array}
$$

9.
$$
\begin{array}{r}
15.1 \\
-\ 3.7 \\
\hline
11.4
\end{array}
$$

10.
$$
\begin{array}{r}
4.65 \\
\times\ 8.9 \\
\hline
4\ 185 \\
37\ 20 \\
\hline
41.385
\end{array}
$$

11.
$$
\begin{array}{r}
23.4 \\
67\overline{)1567.8} \\
\underline{134}\text{x x} \\
227 \\
\underline{201} \\
26\ 8 \\
\underline{26\ 8}
\end{array}
$$

12.
$$
\begin{array}{rcl}
2x + 1 & = & 13 \\
-\ 1 & & -\ 1 \\
\hline
\dfrac{2x}{2} & = & \dfrac{12}{2} \\
x & = & 6
\end{array}
$$

13. $(+5) + (-3) = +2$

14. $73{,}365$

15. $\dfrac{53 + 32 + 53}{3} = \dfrac{138}{3}$
$$
= 46
$$

16. $\dfrac{1}{4}$

17. $\dfrac{x}{12} = \dfrac{3}{4}$
$$
\dfrac{4x}{4} = \dfrac{36}{4}
$$
$$
x = 9
$$

18. $\dfrac{\overline{AC}}{\overline{AB}} = \dfrac{3}{5}$

19. $C = \pi d$
$$
C = \dfrac{22}{\underset{1}{7}} \times \overset{2}{\cancel{14}}
$$
$$
C = 44 \text{ cm}
$$

20. 7% of \$35
$$
0.07 \times \$35 = \$2.45
$$

PART B

21. (c) Median is 12.
10, 10, 11, (12,) 12, 13, 14

22. (a) $\dfrac{1}{6}$

23. (b) $x > -1$

24. (b)
$$\begin{array}{r} 30° \\ + 55° \\ \hline 85° \end{array} \qquad \begin{array}{r} 180° \\ - 85° \\ \hline 95° \end{array}$$

25. (c) C

26. (b)
$$c^2 = a^2 + b^2$$
$$c^2 = 8^2 + 6^2$$
$$c^2 = 64 + 36$$
$$c^2 = 100$$
$$c = 10$$

27. (b)
$$A = s^2$$
$$A = 6^2$$
$$A = 6 \times 6$$
$$A = 36 \text{ m}^2$$

28. (b) $P = 8 + 5 + 8 + 5$
$P = 26$

29. (d) Supplies

30. (c) $4\dfrac{1}{2} \times 1000$

$$\dfrac{9}{\overset{}{\underset{1}{2}}} \times \overset{500}{\cancel{1000}} = 4500$$

31. (c) $4'' \times 24''$

32. (c)
$$\begin{array}{r} 1988 = 4000 \\ 1991 = -1500 \\ \hline 2500 \end{array}$$

33. (b) $\angle ABC = 75°$

34. (c) $0.06 = \dfrac{6}{100} = 6\%$

35. (b) Chord

36. (c) $\dfrac{3}{10} = \dfrac{30}{100} = 30\%$

37. (b) 25% of $36 = 0.25 \times 36 = 9$

38. (c) $\dfrac{\frac{1}{4}''}{1 \text{ mile}} = \dfrac{5''}{x \text{ miles}}$

$$(4)\dfrac{1}{4}x = 5 \ (4)$$

$$x = 20 \text{ miles}$$

39. (a) $\dfrac{2}{3}$

40. (c) 101×399
$$100 \times 400 = 40{,}000$$

41. (d) $2\dfrac{1}{3} = \dfrac{3 \times 2 + 1}{3}$
$$= \dfrac{7}{3}$$

42. (c)
$$\begin{array}{r} \dfrac{3}{4} = \dfrac{6}{8} \\ - \dfrac{5}{8} = \dfrac{5}{8} \\ \hline \dfrac{1}{8} \end{array}$$

43. (a)
$$\begin{array}{r} \dfrac{3}{5} = \dfrac{12}{20} \\ + \dfrac{1}{4} = \dfrac{5}{20} \\ \hline \dfrac{17}{20} \end{array}$$

44. (d) $\dfrac{36 \div 4}{100 \div 4} = \dfrac{9}{25}$ **(a)** 0.64 = (0.640)
0.638 = 0.638
0.639 = 0.639
0.6 = 0.600

46. (c) $\$4.10 \times 4\dfrac{1}{2}$

$\overset{2.05}{\$4.\cancel{10}} \times \dfrac{9}{\underset{1}{\cancel{2}}} = \$18.45$

47. (b) 15% of $15 = 0.15 \times \$15 = \$2.25$

48. (c) $1000 \times \$0.023 = 1000 \times \$0.023 = \$23$

49. (a) $\begin{array}{r} \$20.00 \\ -\ 13.65 \\ \hline \$\ 6.35 \end{array}$ **50. (a)** $\begin{array}{r} \$297.84 \\ +\ 94.89 \\ \hline \$392.73 \end{array}$

51. (c) 1:00 A.M. = $\begin{array}{r} 12:60 \\ -\ 8:30 \\ \hline 4:30 \end{array} = 4\dfrac{1}{2}$ **52. (b)** $\begin{array}{r} \$\ \ 90 \\ \times\ \ \ 36 \\ \hline 540 \\ 270 \\ \hline \$3240 \\ +\ 1000 \\ \hline \$4240 \end{array}$

53. (b) $\$5.25\overline{)\$26.25}$ $\begin{array}{r} 5. \\ \hline 26\ 25 \end{array}$ **54. (a)** $\dfrac{\text{Blondes}}{\text{Brunettes}} = \dfrac{15}{20} = \dfrac{3}{4}$

55. (b) $\begin{array}{r} \$0.29 \text{ first ounce} \\ \$0.23 \times 2 = \$0.46 \\ 0.29 \\ \hline \$0.75 \end{array}$ **56. (c)** 53

57. (d) 12,873 to the nearest thousand = 13,000

58. (b) 2×6^2
$2 \times 6 \times 6 = 7$

59. (b) $2\overline{)12}$ LCD = 12
$3\overline{)12}$
$4\overline{)12}$

60. (b) $\dfrac{6}{24} = \dfrac{1}{4} = 25\%$

GOOD-BYE PUZZLE

Solve each of the problems below. Then find your answer among those given below, and write the corresponding letter in the appropriate blank or blanks on p. 363. See what the message reads.

(M)
$$3\frac{1}{2}$$
$$+ 2\frac{1}{4}$$

(A)
$$5\frac{5}{8}$$
$$- 2\frac{1}{4}$$

(X)
$$\frac{2}{3} \times \frac{5}{8}$$

(E)
$$\frac{3}{4} \div \frac{1}{2}$$

(O)
$$\frac{3}{4} \times \frac{8}{9}$$

(I)
$$4\frac{2}{3}$$
$$- 1\frac{1}{2}$$

(R)
$$7\frac{3}{8}$$
$$- 1\frac{1}{4}$$

(P)
$$4\frac{1}{2} \times \frac{2}{3}$$

(C)
$$1\frac{1}{2} \div 2$$

(Y) $5\dfrac{1}{2}$

$-\ 3\dfrac{3}{4}$

(S) $\dfrac{4}{9} \div \dfrac{1}{3}$

(G) $12\dfrac{3}{8}$

$-\ 1\dfrac{1}{2}$

(N) $2\dfrac{1}{4} \div \dfrac{1}{4}$

(T) $3\dfrac{1}{8} \times \dfrac{4}{5}$

GOOD LUCK ON THE

$\overline{}$	$\overline{}$	$\overline{}$	$\overline{}$	$\overline{}$
$6\dfrac{1}{8}$	$1\dfrac{1}{2}$	$10\dfrac{7}{8}$	$1\dfrac{1}{2}$	9

$\overline{}$	$\overline{}$	$\overline{}$	$\overline{}$	$\overline{}$	$\overline{}$	$\overline{}$	$\overline{}$
$2\dfrac{1}{2}$	$1\dfrac{1}{3}$	$\dfrac{3}{4}$	$\dfrac{2}{3}$	$5\dfrac{3}{4}$	3	$1\dfrac{1}{2}$	$2\dfrac{1}{2}$

$\overline{}$	$\overline{}$	$\overline{}$	$\overline{}$	$\overline{}$	$\overline{}$	$\overline{}$	$\overline{}$
$1\dfrac{1}{2}$	9	$\dfrac{3}{4}$	$1\dfrac{3}{4}$	$1\dfrac{1}{2}$	$\dfrac{5}{12}$	$3\dfrac{3}{8}$	$5\dfrac{3}{4}$

$\overline{}$	$\overline{}$	$\overline{}$	$\overline{}$	$\overline{}$	$\overline{}$	$\overline{}$
$3\dfrac{1}{6}$	9	$3\dfrac{3}{8}$	$2\dfrac{1}{2}$	$3\dfrac{1}{6}$	$\dfrac{2}{3}$	9

Index

MAXIMIZE YOUR MATH SKILLS!

BARRON'S EASY WAY SERIES

Specially structured to maximize learning with a minimum of time and effort, these books promote fast skill building through lively cartoons and other fun features.

ALGEBRA THE EASY WAY
Revised Third Edition
Douglas Downing, Ph.D.
In this one-of-a-kind algebra text, all the fundamentals are covered in a delightfully illustrated adventure story. Equations, exponents, polynomials, and more are explained. 320 pp. (9393-3) $12.95, Can. $16.95

CALCULUS THE EASY WAY
Revised Third Edition
Douglas Downing, Ph.D.
Here, a journey through a fantasy land leads to calculus mastery. All principles are taught in an easy-to-follow adventure tale. Included are numerous exercises, diagrams, and cartoons which aid comprehension. 228 pp. (9141-8) $12.95, Can. $16.95

GEOMETRY THE EASY WAY
Revised Second Edition
Lawrence Leff
While other geometry books simply summarize basic principles, this book focuses on the "why" of geometry: why you should approach a problem a certain way, and why the method works. Each chapter concludes with review exercises, 288 pp. (4287-5) $11.95, Can. $15.95

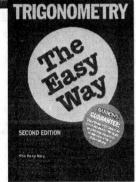

TRIGONOMETRY THE EASY WAY
Revised Second Edition
Douglas Downing, Ph.D.
In this adventure story, the inhabitants of a faraway kingdom use trigonometry to solve their problems. Covered is all the material studied in high school or first-year college classes. Practice exercises, explained answers and illustrations enhance understanding. 288 pp. (4389-8) $11.95, Can. $15.95

FM: FUNDAMENTALS OF MATHEMATICS
Cecilia Cullen and Eileen Petruzillo, editors
Volume 1 (2501-6)—
Formulas; Introduction to Algebra; Metric Measurement; Geometry; Managing Money; Probability and Statistics. 384 pp. $19.95, Can. $24.95

FM: FUNDAMENTALS OF MATHEMATICS
Cecilia Cullen and Eileen Petruzillo, editors
Volume 2 (2508-3)—
Solving Simple Equations; Tables, Graphs and Coordinate Geometry; Banking; Areas; Indirect Measurement and Scaling; Solid Geometry. The ideal text/workbooks for pre-algebra students and those preparing for state minimum competency exams. They conform with the New York State curriculum and follow units recommended by the New York City Curriculum Guide. 384 pp. $14.95, Can. $21.00

ESSENTIAL MATH
Second Edition
Edward Williams and Robert A. Atkins
Basic math principles everyone should know are made easy by being put into the context of real-life situations. Games and puzzles help make math interesting. Practice exercises appear at the end of every chapter. 384 pp., (1337-9) $11.95, Can. $15.95

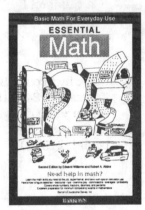

Books may be purchased at your bookstore, or by mail from Barron's. Enclose check or money order for total amount plus sales tax where applicable and 10% for postage (minimum charge $3.75, Can. $4.00). All books are paperback editions. Prices subject to change without notice.

ISBN PREFIX: 0-8120

(#44) R 7/96

Barron's Educational Series, Inc.
250 Wireless Boulevard
Hauppauge, New York 11788
In Canada: Georgetown Book Warehouse
34 Armstrong Avenue • Georgetown, Ontario L7G 4R9